Studies in Computational Intelligence

Volume 648

Series editor

Janusz Kacprzyk, Polish Academy of Sciences, Warsaw, Poland
e-mail: kacprzyk@ibspan.waw.pl

About this Series

The series "Studies in Computational Intelligence" (SCI) publishes new developments and advances in the various areas of computational intelligence—quickly and with a high quality. The intent is to cover the theory, applications, and design methods of computational intelligence, as embedded in the fields of engineering, computer science, physics and life sciences, as well as the methodologies behind them. The series contains monographs, lecture notes and edited volumes in computational intelligence spanning the areas of neural networks, connectionist systems, genetic algorithms, evolutionary computation, artificial intelligence, cellular automata, self-organizing systems, soft computing, fuzzy systems, and hybrid intelligent systems. Of particular value to both the contributors and the readership are the short publication timeframe and the worldwide distribution, which enable both wide and rapid dissemination of research output.

More information about this series at http://www.springer.com/series/7092

Svetozar Margenov · Galia Angelova
Gennady Agre
Editors

Innovative Approaches and Solutions in Advanced Intelligent Systems

 Springer

Editors
Svetozar Margenov
Institute of Information and Communication
    Technologies
Bulgarian Academy of Sciences
Sofia
Bulgaria

Gennady Agre
Institute of Information and Communication
    Technologies
Bulgarian Academy of Sciences
Sofia
Bulgaria

Galia Angelova
Institute of Information and Communication
    Technologies
Bulgarian Academy of Sciences
Sofia
Bulgaria

ISSN 1860-949X   ISSN 1860-9503   (electronic)
Studies in Computational Intelligence
ISBN 978-3-319-81223-6   ISBN 978-3-319-32207-0   (eBook)
DOI 10.1007/978-3-319-32207-0

Printed on acid-free paper

This Springer imprint is published by Springer Nature
The registered company is Springer International Publishing AG Switzerland

Preface

This volume is a selected collection of papers presented and discussed at the International Conference "Advanced Computing for Innovation (AComIn 2015)". The papers report innovative approaches and solutions in hot topics of computational intelligence—advanced computing, language and semantic technologies, signal and image processing, as well as optimisation and intelligent control.

Advanced Computing is presented by five papers. The work of I. Dimov and V. Todorov is related to *efficient methods and tools for analysis of reliability of large-scale models*. The authors present an error analysis of an almost optimal Monte Carlo algorithm based on balancing the systematic and stochastic error. This contribution to the solution of hard computational problems is relevant to environmental sciences and computational physics. Three other papers concern *high-performance computing in engineering and environmental problems*. S. Stoykov and S. Margenov propose and analyse some numerical methods for computing nonlinear frequency-response curves of plates with complex geometries. They show that parametric study on the dynamics of complex structures can be carried out deploying appropriate parallel implementation. K. Liolios et al. describe a numerical simulation of biochemical oxygen demand removal in Horizontal Subsurface Flow Constructed Wetlands (HSF CW). The simulated experimental data is obtained from five pilot-scale HSF CW units. A. Liolios et al. present an approach that can be effectively used for numerical investigation the seismic inelastic behaviour of cultural heritage industrial buildings of reinforced concrete strengthened by cable elements and subjected to multiple earthquakes. The authors propose an approach for selecting the optimal cable-bracing scheme by using computed damage indices. The paper of J. Kohler et al. is related to *application of cloud computing for processing Big Data*. The authors propose an original query rewriting approach that parallelises queries and joins in an order that has been implemented and tested for performance gains. The approach can be used for improving data security and privacy, especially in public cloud computing environments.

Five papers present innovative approaches in the areas of *Language, Semantic and Content Technologies*. The paper of V. Cantoni et al. is focused on *digital preservation of cultural heritage for research and education*. It describes an innovative use of interactive digital technologies in cultural heritage presentation practices exemplified by multimodal interaction modalities developed for the Exhibition "1525–2015. Pavia, the Battle, the Future. Nothing was the same again"—a satellite event of the Universal Exposition Expo 2015 in Milan. Two papers are devoted to *application of data mining methods for analysing Big and Educational Data*. S. Boytcheva et al. present a novel cascade data mining approach for frequent pattern mining, sequence mining and periodical events mining applied to discovery of complex temporal relationships between disorders and their treatment. An evaluation of the approach on real data is provided. I. Nikolova et al. describe application of educational data mining to a real educational web portal—UCHA.SE with the goal to improve the quality of the educational services provided by the site and its revenue generation. The extracted predictive rules are used to make recommendations to the UCHA.SE development team. Two other papers are related to *advanced methods and tools for processing of textual and semantic data*. O. Kanishcheva and G. Angelova present an original integrated approach for word sense disambiguation of image tags that can be applied for improving machine translation of tags or image similarity measurement. K. Simov et al. describe an approach for the enrichment of word sense disambiguation knowledge bases with data-driven relations from a gold standard corpus (annotated with word senses, valency information, syntactic analyses, etc.). The paper is focused on Bulgarian and English as use cases, but the approach is scalable to other languages as well.

Signal and Image Processing issues are discussed in eight papers. The work of A. Nikolov et al. is related to *advanced methods for biometric analysis*. The authors describe a novel multimodal ear database characterised by different types of ear representation, either 2D or 3D, depending on the device used for data acquisition. The database can be used as a benchmark to test different pattern recognition methods on a set of images captured in known conditions, and to highlight the strengths and the weaknesses of each approach in terms of recognition accuracy and robustness. Several problems of *smart multi-sensor signal and image processing* are treated in the following six papers. S. Ilchev and Z. Ilcheva propose a modular digital watermarking service coupled with steganalysis suitable both for commercial and non commercial users. The service acts as intermediary facilitating the payment flow for commercial uses and is able to gather and provide statistics about image distribution and popularity. S. Harizanov investigates several techniques for restoring images corrupted by a non-invertible or ill-conditioned linear transformations and Poisson noise. These techniques are based on image domain decomposition and give rise to multi-constraint optimisation problems. D. Karastoyanov et al. present a new type of graphical Braille screen containing a matrix with linear electromagnetic micro drives and non-magnetic needles, passing through the axes of the electromagnets. P. Koprinkova-Hristova et al. discuss an application of a recently developed smart approach for feature extraction from multi-dimensional data sets using Echo state networks to the focalised spectra obtained from multi-sensor measurements of an acoustic camera. The aim of the study is development of a remote diagnostic system for prediction of bearing wearing out. V. Kudriashov proposes new detection rules for multistatic reception of non-stationary random Wiener acoustic signals. The rules are suitable for such applications as monitoring aircraft engine noise at landing/take off, and testing of a car engine. I. Chirka suggests two novel techniques for the interpolation of acoustic fields generated by a single acoustic source based on sinewave model and instantaneous phase measurement, and generation of virtual microphones. The approaches allow improving the accuracy of source localisation avoiding the necessity of using expensive equipment. Some *speech analysis* problems are considered in the paper of P. Mitankin and S. Mihov, who propose an original algorithm to be integrated in the decoding stage of the speech recognition pipeline. The innovative aspect is the implicit generation of a much larger set of decoding candidates than the state-of-the-art N-best approach.

Three papers present innovative results in the areas of *Optimisation and Intelligent Control of Traffic*. S. Fidanova proposes a model of the passenger flow suitable for analysing existing public transportation systems. The task is defined as a multi-objective optimisation problem that can be solved by an ant colony optimisation algorithm. T. Stoilov et al. introduce an innovative idea for the formal description of the urban traffic control based on a bi-level model. Such an approach gives the potential for increasing the space of the control allowing simultaneously the minimisation of the waiting vehicles and maximisation of the traffic flows. A. Balabanov proposes an original algorithm based on fast finding of solutions of Riccati equations for synthesis of a steady-state Kalman estimator, which can be used for online applications such as estimations of the real traffic intensity.

We would like to thank the members of the ACOMIN 2015 Programme Committee, who reviewed the submissions thoroughly and fairly, and the AComIn project (Advanced Computing for Innovation, FP7 Capacity grant 316087) for the generous support provided to the ACOMIN 2015 conference.

February 2016

Svetozar Margenov
Galia Angelova
Gennady Agre

Organization Committee

The International Conference Advanced Computing for Innovation (ACOMIN 2015) was organised by the Institute of Information and Communication Technologies at the Bulgarian Academy of Sciences (IICT-BAS) in the frame of the FP7 project "Advanced Computing for Innovation" (AComIn) grant agreement 316087. The conference was held during 10–11 November 2015 in Sofia, Bulgaria and was aimed at providing a forum for international scientific exchange between Central/Eastern Europe and the rest of the world in several fundamental for computational intelligence topics enabling radical progress and development of novel applications.

Conference Co-chairs

Svetozar Margenov, IICT-BAS
Galia Angelova, IICT-BAS

Chair of the Organizing Committee

Gennady Agre, IICT-BAS

Program Committee

Hassane Abouaïssa, Université Lille Nord, France
Gennady Agre, IICT-BAS, Bulgaria
Kiril Alexiev, IICT-BAS, Bulgaria
Asen Asenov, University of Glasgow, UK
Christopher M. Bishop, Microsoft Research, Cambridge, UK

Contents

Part I
Advanced Computing

Error Analysis of Biased Stochastic Algorithms for the Second Kind Fredholm Integral Equation

Ivan Dimov and Venelin Todorov

Abstract In this work error analysis of biased stochastic algorithms for the second kind of Fredholm integral equation is considered. There are unbiased and biased stochastic algorithms, but the latter algorithms are more interesting, because there are two errors in there solutions—stochastic and systematic errors. An almost optimal Monte Carlo algorithm for integral equations in a combination with the idea of balancing of both systematic and stochastic errors is analysed. An optimal ratio between the number of realizations of the random variable and the number of iterations in the algorithm is studied. We considered two examples of integral equations that are widely used in computational physics and environmental sciences. We have shown that the almost optimal Monte Carlo algorithm based on balancing of the errors gives excellent results.

Keywords Almost optimal Monte Carlo algorithm · Balancing of errors · Monte Carlo methods in computational physics and environmental sciences

1 Introduction

The Monte Carlo method is a widely used tool in many fields of science. It is well known that Monte Carlo methods give statistical estimates for the functional of the solution by performing random sampling of a certain random variable whose mathematical expectation is the desired functional [3].

Monte Carlo methods are methods of approximation of the solution to problems of computational mathematics, by using random processes for each such problem, with the parameters of the process equal to the solution of the problem. The method can guarantee that the error of Monte Carlo approximation is smaller than a given

I. Dimov (✉) · V. Todorov
IICT, Bulgarian Academy of Sciences, Acad. G. Bonchev 25 A, 1113 Sofia, Bulgaria
e-mail: ivdimov@bas.bg

V. Todorov
e-mail: dvespas@mail.bg

© Springer International Publishing Switzerland 2016
S. Margenov et al. (eds.), *Innovative Approaches and Solutions in Advanced Intelligent Systems*, Studies in Computational Intelligence 648,
DOI 10.1007/978-3-319-32207-0_1

value with a certain probability [15]. Monte Carlo methods always produce an approximation to the solution of the problem or to some functional of the solution, but one can control the accuracy in terms of the probability error [8].

An important advantage of the Monte Carlo methods is that they are suitable for solving multi-dimensional problems, since the computational complexity increases polynomially, but not exponentially with the dimensionality. Another advantage of the method is that it allows to compute directly functionals of the solution with the same complexity as to determine the solution. In such a way this class of methods can be considered as a good candidate for treating innovative problems related to modern areas in quantum physics.

The simulation of innovative engineering solutions that may be treated as candidates for quantum computers need reliable fast Monte Carlo methods for solving the discrete version of the Wigner equation [10, 11, 14]. High quality algorithms are needed to treat complex, time-dependent and fully quantum problems in nano-physics and nano-engineering. These methods use estimates of functionals defined on discrete Markov chains defined on properly chosen subspaces.

In order to be able to analyze the quality of biased algorithms we need to introduce several definitions related to probability error, discrete Markov chains and algorithmic computational cost.

Definition 1 *If J is the exact solution of the problem, then the probability error is the least possible real number R_N, for which:*

$$P = Pr\{|\bar{\xi}_N - J| \leq R_N\}$$

where $0 < P < 1$. If $P = \frac{1}{2}$, then the probability error is called the probable error.

The probable error is the value r_N for which:

$$Pr\{|\bar{\xi}_N - J| \leq R_N\} = \frac{1}{2} = Pr\{|\bar{\xi}_N - J| \geq R_N\}$$

In our further considerations we will be using the above defined probable error taking into account that the chosen value of the probability P only changes the constant, but not the rate of convergence in the stochastic error estimates.

There are two main directions in the development and study of Monte Carlo algorithms [5]. The first is Monte Carlo simulation and the second is the Monte Carlo numerical algorithms. In Monte Carlo numerical algorithms we construct a Markov process and prove that the mathematical expectation of the process is equal to the unknown solution of the problem [4]. A Markov process is a stochastic process that has the Markov property. Often, the term Markov chain is used to mean a discrete-time Markov process.

Definition 2 *A finite discrete Markov chain T_k is defined as a finite set of states $\{\alpha_1, \alpha_2, \ldots, \alpha_k\}$.*

Definition 3 *A state is called absorbing if the chain terminates in this state with probability one.*

Iterative Monte Carlo algorithms can be defined as terminated Markov chains:

$$T = \alpha_{t_0} \to \alpha_{t_1} \to \alpha_{t_2} \ldots \alpha_{t_k}, \tag{1}$$

where $\alpha_{t_q}, q = 1, \ldots, i$ is one of the absorbing states.

Definition 4 *Computational cost of a randomized iterative algorithm A^R is defined by*

$$cost(A, x, w) = nE(q)t_0,$$

where $E(q)$ is the mathematical expectation of the number of transitions in the sequence (1) and t_0 is the mean time needed for value of one transition.

The computational cost is an important measure of the efficiency of Monte Carlo algorithms. If one has a set of algorithms solving a given problem with the same accuracy, the algorithm with the smallest computational cost can be considered as the most efficient algorithm. In the latter case there is a systematic error. The Monte Carlo algorithm consists in simulating the Markov process and computing the values of the random variable. It is clear, that in this case a stochastic error also appears. The error estimates are important issue in studying Monte Carlo algorithms.

In order to formalize our consideration we need to define the standard iterative process. Define an iteration of degree j as [2, 4]

$$u^{k+1} = F_k(A, b, u^k, u^{k-1}, \ldots, u^{k-j+1}),$$

where u^k is obtained from the kth iteration. It is desired that

$$u^k \to u = A^{-1}b \quad \text{as } k \to \infty.$$

The degree of j should be small because of efficiency requirement, but also in order to save computer memory storage. The iteration is called stationary if $F_k = F$ for all k, that is, F_k is independent of k. The iterative Monte Carlo process is said to be linear if F_k is a linear function of $u^k, \ldots, u^{k-j+1}$.

In this work we shall consider iterative stationary linear Monte Carlo algorithms only and will analyze both systematic and stochastic errors. We will put a special effort to set-up the parameters of the algorithm in order to have a good balance between the two errors mentioned above [1].

2 Description of the Almost Optimal Monte Carlo Algorithm for Integral Equations

In this section a description of the almost optimal Monte Carlo algorithm for integral equations will be given. Let $\mathbf{X}$ be a Banach space of real-valued functions. Let $f = f(x) \in \mathbf{X}$ and $u_k = u(x_k) \in \mathbf{X}$ be defined in $\mathbb{R}^d$ and $\mathcal{K} = \mathcal{K}(u)$ be a linear operator defined on $\mathbf{X}$.

Consider the sequence $u_1, u_2, \ldots$, defined by the recursion formula

$$u_k = \mathcal{K}(u_{k-1}) + f, \quad k = 1, 2, \ldots \tag{2}$$

The formal solution of (2) is the truncated Liouville-Neumann series

$$u_k = f + \mathcal{K}(f) + \cdots + \mathcal{K}^{k-1}(f) + \mathcal{K}^k(f), \quad k > 0, \tag{3}$$

where $\mathcal{K}^k$ means the kth iterate of $\mathcal{K}$. As an example consider the integral iterations. Let $u(x) \in \mathbf{X}, x \in \Omega \subset \mathbb{R}^d$ and $k(x, x')$ be a function defined for $x, x' \in \Omega$. The integral transformation

$$\mathcal{K}u(x) = \int_{\Omega} k(x, x') u(x') dx'$$

maps the function $u(x)$ into the function $\mathcal{K}u(x)$, and is called an iteration of $u(x)$ by the integral transformation kernel $k(x, x')$. The second integral iteration of $u(x)$ is denoted by

$$\mathcal{K}\mathcal{K}u(x) = \mathcal{K}^2 u(x).$$

Obviously

$$\mathcal{K}^2 u(x) = \int_{\Omega} \int_{\Omega} k(x, x') k(x', x'') dx' dx''.$$

In this way $\mathcal{K}^3 u(x), \ldots, \mathcal{K}^i u(x), \ldots$ can be defined. When $u^{(k)} \xrightarrow{k \to \infty} u$ then $u = \sum_{i=0}^{\infty} \mathcal{K}^i f \in \mathbf{X}$ and

$$u = \mathcal{K}(u) + f. \tag{4}$$

The truncation error of (4) is

$$u_k - u = \mathcal{K}(u_0 - u).$$

Random variables $\theta_i, i = 1, 2, \ldots, k$ are defined on spaces T_{i+1}, where

$$T_{i+1} = \underbrace{\mathbb{R}^d \times \mathbb{R}^d \times \ldots \mathbb{R}^d}_{i \; times}, \quad i = 1, 2, \ldots, k,$$

and it is fulfilled

$$E\theta_0 = J(u_0), E(\theta_1/\theta_0) = J(u_1)\ldots, E(\theta_k/\theta_0) = J(u_k).$$

An approximate value of linear functional $J(u_k)$ that is to be calculated is set up as:

$$J(u_k) \approx \frac{1}{N} \sum_{i=1}^{N} \{\theta_k\}_s, \tag{5}$$

where $\{\theta_k\}_s$ is the sth realization of θ_k.

We consider the case when $\mathcal{K}$ is an ordinary integral transform

$$\mathcal{K}(u) = \int_\Omega k(x,y)u(y)dy$$

and $u(x)$ and $f(x)$ are functions.

Equation (4) becomes

$$u(x) = \int_\Omega k(x,y)u(y)dy + f(x) \quad \text{or } u = \mathcal{K}u + f. \tag{6}$$

We want to evaluate the linear functionals of the solution of the following type

$$J(u) = \int_\Omega \varphi(x)u(x) = (\varphi(x), u(x)). \tag{7}$$

In fact (7) defines an inner product of a given function $\varphi(x) \in \mathbf{X}$ with the solution of the integral Eq. (6). The adjoint equation is

$$v = \mathcal{K}^* v + \varphi, \tag{8}$$

where $v, \varphi \in \mathbf{X}^*$, $\mathcal{K} \in [\mathbf{X}^* \to \mathbf{X}^*]$, $\mathbf{X}^*$ is the dual function space and $\mathcal{K}^*$ is an adjoint operator. We will prove that

$$J = (\varphi, u) = (f, v).$$

If we multiply (4) by v and (8) by u and integrate, then:

$$(v, u) = (v, \mathcal{K}u) + (v, f) \text{ and } (v, u) = (\mathcal{K}^*v, u) + (\varphi, u).$$

From

$$(\mathcal{K}^*v, u) = \int_\Omega \mathcal{K}^*v(x)u(x)dx = \int_\Omega \int_\Omega k^*(x, x')v(x')u(x)dx\,dx'$$
$$= \int_\Omega \int_\Omega k(x', x)u(x)v(x')dxdx' = \int_\Omega \mathcal{K}u(x')v(x')dx' = (v, \mathcal{K}u)$$

we obtain that

$$(\mathcal{K}^*v, u) = (v, \mathcal{K}u).$$

Therefore

$$(\varphi, u) = (f, v).$$

That is very important, because in practice it happens the solution of the adjoint problem to be simple than this of the original one, and they are equivalent as we have proved it above.

Usually it is assumed that $\varphi(x) \in L_2(\Omega)$, because $k(x, x') \in L_2(\Omega \times \Omega), f(x) \in L_2(\Omega)$. In a more general setting $k(x, x') \in X(\Omega \times \Omega), f(x) \in X(\Omega)$, where X is a Banach space. Then the given function $\varphi(x)$ should belong to the adjoint Banach space $\mathbf{X}^*$, and the problem under consideration may by formulated in an alternative way:

$$v = K^*v + \varphi, \qquad (9)$$

where $v, \varphi \in \mathbf{X}^*(\Omega)$, and $K^*(\Omega \times \Omega) \in [\mathbf{X}^* \to \mathbf{X}^*]$. In such a way one may compute the value $J(v) = \int f(x)v(x)dx = (f, v)$ instead of (7). An important case for practical computations is the case when $\mathbf{X} \equiv L_1$, where L_1 is defined in a standard way:

$$\|f\|_{L_1} = \int_\Omega |f(x)|dx; \quad \|\mathcal{K}\|_{L_1} \leq \sup_x \int_\Omega |k(x, x')|dx'.$$

In this case the function $\varphi(x)$ from the functional (7) belongs to L_∞, i.e. $\varphi(x) \in L_\infty$ since $L_1^* \equiv L_\infty$. It is also easy to see that $(K^*v, u) = (v, Ku)$, and also $(\varphi, u) = (f, v)$. This fact is important for practical computations since often the computational complexity for solving the adjoint problem is smaller than the complexity for solving the original one. The above consideration shows that if $\varphi(x)$ is a Dirac δ-function $\delta(x - x_0)$, then the functional $J(u) = (\varphi(x), u(x)) =$

$(\delta(x - x_0), u(x)) = u(x_0)$ is the solution of the integral equation at the point x_0, if $u \in L_\infty$.

An approximate value of the linear functional J, defined by (7) is [13]

$$J \approx \frac{1}{N} \sum_{s=1}^{N} (\theta)_s = \hat{\theta}_N.$$

The random variable whose mathematical expectation is equal to $J(u)$ is given by the following expression

$$\theta[\varphi] = \frac{\varphi(x_0)}{\pi(x_0)} \sum_{j=0}^{k} W_j f(x_j),$$

where $W_0 = 1, W_j = W_{j-1} \frac{k(x_{j-1}, x_j)}{p(x_{j-1}, x_j)}, j = 1, \ldots, k$, and $x_0, x_1, \ldots$ is a Markov chain in Ω with initial density function $\pi(x)$ and transition density function $p(x, x')$. If we consider [15]

$$p(x, x') = \frac{|\mathcal{K}(x, x')|}{\int |\mathcal{K}(x, x')| dx'}, \quad \pi(x) = \frac{|\varphi(x)|}{\int |\varphi(x)| dx},$$

then the algorithm is called an Almost Optimal Monte Carlo algorithm (MAO) and it reduces the variance [7]. We use MAO instead of optimal algorithms, because they are very expensive and give much bigger variance [8].

So, it is clear that the approximation of the unknown value (φ, u) can be obtained using a truncated Liouville-Neumann series (3) for a sufficiently large k:

$$(\varphi, u^{(k)}) = (\varphi, f) + (\varphi, Kf) + \cdots + (\varphi, K^{(k-1)}f) + (\varphi, K^{(k)}f).$$

So, we transform the problem for solving integral equations into a problem for approximate evaluation of a finite number of multidimensional integrals. We will use the following denotation $(\varphi, K^{(k)}f) = I(k)$, where $I(k)$ is a value, obtained after integration over $\Omega^{j+1} = \Omega \times \cdots \times \Omega$, $j = 0, \ldots, k$. It is obvious that the calculation of the estimate $(\varphi, u^{(k)})$ can be replaced by evaluation of a sum of linear functionals of iterative functions of the following type $(\varphi, K^{(j)}f), j = 0, \ldots, k$, which can be presented as:

$$\begin{aligned}
(\varphi, K^{(j)}f) &= \int_\Omega \varphi(t_0) K^{(j)} f(t_0) dt_0 \\
&= \int_G \varphi(t_0) k(t_0, t_1) \ldots k(t_{j-1}, t_j) f(t_j) dt_0 \ldots dt_j,
\end{aligned} \quad (10)$$

where $t = (t_0, \ldots, t_j) \in G \equiv \Omega^{j+1} \subset R^{d(j+1)}$. If we denote by $F_k(t)$ the integrand function

$$F(t) = \varphi(t_0)k(t_0, t_1)\ldots k(t_{j-1}, t_j)f(t_j), \quad t \in \Omega^{j+1},$$

then we will obtain the following expression for (10):

$$I(j) = (\varphi, K^{(j)}f) = \int_G F_k(t)dt, \quad t \in G \subset \mathbb{R}^{n(j+1)}. \tag{11}$$

Thus, we will consider the problem for approximate calculation of multiple integrals of the type (11).

The above consideration shows that there two classes of possible stochastic approaches. The first one is the so-called biased approach when one is looking for a random variable which mathematical expectation is equal to the approximation of the solution problem by a truncated Liouville-Neumann series (3) for a sufficiently large k. An unbiased approach assumes that the formulated random variable is such that its mean value approaches the true solution of the problem. Obviously, in the first class of approaches there are two errors: a systematic one (a truncation error) r_k and a stochastic (a probabilistic) one, namely r_N, which depends on the number N of values of the random variable,or the number of chains used in the estimate. In the case of unbiased stochastic methods one should analyse the only probabilistic error. In the case of biased stochastic methods more careful error analysis is needed: balancing of both systematic and stochastic error should be done in order to minimize the computational complexity of the methods (for more details, see [4]).

3   Balancing of the Errors

The error balancing is an important issue for the optimal by rate Monte Carlo algorithms.

As it was already mentioned, there are two errors in MAO algorithm-systematic r_k and stochastic error r_N. In [9] it is proven that

$$r_N \leq \frac{0.6745\|\varphi\|_{L_2}\|f\|_{L_2}}{\sqrt{N}\left(1 - \|\mathcal{K}\|_{L_2}\right)};$$

$$r_k \leq \frac{\|\varphi\|_{L_2}\|f\|_{L_2}\|\mathcal{K}\|_{L_2}^{k+1}}{1 - \|\mathcal{K}\|_{L_2}}.$$

We have shown that for an integral Eq. (6) with a preliminary given error δ, the lower bounds for N and k for the Monte Carlo algorithm with a balancing of the errors are:

$$N \geq \left(\frac{1.349 \|\varphi\|_{L_2} \|f\|_{L_2}}{\delta \left(1 - \|\mathcal{K}\|_{L_2} \right)} \right)^2, \quad k \geq \frac{\ln \frac{\delta \left(1 - \|\mathcal{K}\|_{L_2} \right)}{2\|\varphi\|_{L_2} \|f\|_{L_2} \|\mathcal{K}\|_{L_2}}}{\ln \|\mathcal{K}\|_{L_2}}. \quad (12)$$

The Eq. (12) defines the error balancing conditions. The above statement allows to find the optimal value of k for a given value of the number of realizations of the random variable N. Namely, we find the following optimal ratio between k and N:

$$k \geq \frac{\ln \frac{0.6745}{\|\mathcal{K}\|_{L_2} \sqrt{N}}}{\ln \|\mathcal{K}\|_{L_2}}.$$

We may now formulate the following corollaries:

Corollary 1 *In the next tests with a preliminary given error in Monte Carlo algorithm with a balancing of the errors we choose*

$$N = \left\lceil \left(\frac{1.349 \|\varphi\|_{L_2} \|f\|_{L_2}}{\delta \left(1 - \|\mathcal{K}\|_{L_2} \right)} \right)^2 \right\rceil, \quad k = \left\lceil \frac{\ln \frac{0.6745}{\|\mathcal{K}\|_{L_2} \sqrt{N}}}{\ln \|\mathcal{K}\|_{L_2}} \right\rceil. \quad (13)$$

Corollary 2 *If we choose k to be close to its lower bound in the error balancing conditions, then the inequality for the number of realizations of the random variable N is:*

$$N \geq \frac{0.455}{\|\mathcal{K}\|_{L_2}^{2k+2}}.$$

The two obtained lower bounds for N are equivalent because in this case one can easily see that the upper bound for the systematic error is bigger than the upper bound for the stochastic error and the above inequality directly leads.

Corollary 3 *One can also choose the following values for N and k:*

$$k = \left\lceil \frac{\ln \frac{\delta \left(1 - \|\mathcal{K}\|_{L_2} \right)}{2\|\varphi\|_{L_2} \|f\|_{L_2} \|\mathcal{K}\|_{L_2}}}{\ln \|\mathcal{K}\|_{L_2}} \right\rceil, \quad N = \left\lceil \frac{0.455}{\|\mathcal{K}\|_{L_2}^{2k+2}} \right\rceil. \quad (14)$$

The above corollaries are of great importance for the quality of the algorithm. Depending on the norm of the integral transformation kernel and the right-hand side one may chose the needed number of Monte Carlo iterations such that to reach the needed accuracy with an optimal balance between the systematic and stochastic errors.

4 Numerical Examples and Results

To illustrate the idea of the almost optimal Monte Carlo algorithm combined with the balancing of errors we give two examples from the area of environmental sciences and computational physics.

4.1 Example 1

The first example is analytically tractable model taken from biology, which describes the population growth model. This model is widely used in environmental sciences [12]:

$$u(x) = \int_{\Omega} k(x, x')u(x')dx' + f(x),$$

$\Omega \equiv [0, 1]$, $k(x, x') = \frac{1}{3}e^x$, $f(x) = \frac{2}{3}e^x$, $\varphi(x) = \delta(x)$, $u_{exact}(x) = e^x$.

We want to find the solution in the middle of Ω. In order to apply (13) we evaluate the L_2 norms: $\|\varphi\|_{L_2} = 1$, $\|\mathcal{K}\|_{L_2} = 0.3917$, $\|f\|_{L_2} = 1.1915$. The Monte Carlo Markov chain starts from $x_0 = 0.5$ so the exact solution is 1.6487 and $\pi(x) = \delta(x)$. The number of algorithmic runs is 20 and we used Intel Core i5-2410 M @ 2.3 GHz.

In Table 1 the preliminary given error δ by different values and estimates of N and k by (13) are presented. The expected relative error is obtained by divided the preliminary given error by the exact value. We compare Monte Carlo method with

Table 1 Computational time and relative error for the first example

δ	N	k	Expected rel. error	Experimental rel. error	Time (s)	Experimental rel. error	Time (s)
0.23	132	3	0.1395	0.0123	0.5	0.0121	0.2
0.037	5101	4	0.0224	0.0041	11	0.0040	7
0.025	11,172	5	0.0152	0.0014	16	0.0012	9
0.014	35,623	6	0.0085	4.5725e−04	56	4.0010e−04	34
0.0055	230,809	7	0.0033	1.5242e−04	424	9.8811e−05	346
0.0045	344,788	7	0.0027	1.5242e−04	605	1.4893e−04	592

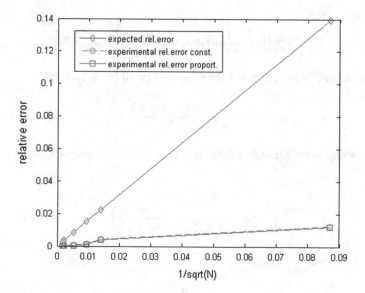

Fig. 1 Experimental and expected relative error

constant probabilities with an almost optimal Monte Carlo algorithm [2]. From the Table 1 leads that MAO gives slightly better results, but the closeness of the results is due to the fact that the initial probability is the δ function. We see that experimental relative error confirms the theoretical relative error on Fig. 1.

4.2 Example 2

It is interesting to see if the considered algorithm can be applied to non-linear integral equations with polynomial non-linearity. One may expect that if the non-linearity is not too strong instead of considering functionals on branching Markov processes (see, the analysis presented in [4]) one may use the presented balanced biased algorithm. We consider the following non-linear integral equation with a quadratic non-linearity [4, 6]:

$$u(x) = \int_\Omega \int_\Omega k(x,y,z)u(y)u(z)dydz + f(x), \tag{15}$$

where $\Omega = \mathbf{E} \equiv [0,1]$ and $x \in \mathbb{R}^1$. The kernel $k(x,y,z)$ is non-negative. In our test $k(x,y,z) = \frac{x(a_2 y - z)^2}{a_1}$ and $f(x) = c - \frac{x}{a_3}$, where $a_1 > 0, a_3 \neq 0, a_2$ and c are constants. We evaluate:

$$K_2 = \max_{x \in [0,1]} \int_0^1 \int_0^1 |k(x,y,z)| dy dz = \frac{2a_2^2 - 3a_2 + 2}{6a_1}.$$

The process converges when the following condition is fulfilled:

$$K_2 < \frac{1}{m} \Leftrightarrow \frac{2a_2^2 - 3a_2 + 2}{6a_1} < \frac{1}{2}.$$

According to the Miranda's theorem the Eq. (15) has a unique solution $u(x) = c$ when

$$c = \pm \left(\frac{6a_1}{a_3 \left(2a_2^2 - 3a_2 + 2 \right)} \right)^{\frac{1}{2}}.$$

We have that:

$$p(y,z) = \frac{6}{2a_2^2 - 3a_2 + 2} (a_2 y - z)^2.$$

We set

$$a_1 = 11, \quad a_2 = 4, \quad a_3 = 12, \quad c = 0.5,$$

and

$$\varphi(x) = \delta(x - x_0).$$

We evaluate the transition density function:

$$p(y,z) = \frac{3}{11} (4y - z)^2.$$

In order to apply the (13) we evaluate the L_2 norms:

$$\|\mathcal{K}\|_{L_2} = 0.408, \quad \|f\|_{L_2} = 0.459, \quad \|\varphi\|_{L_2} = 1.$$

	δ	N	k	Expected rel. error	Experimental rel. error
Table 2 Computational time and relative error for the second example	0.03	1487	2	0.06	0.0856
	0.01	13,381	3	0.02	0.0710
	0.005	53,522	4	0.01	0.0602
	0.002	334,512	5	0.004	0.0456

The Monte Carlo Markov chain starts from $x_0 = 0.5$, the exact solution is $u(x) = c = 0.5$ and $\pi(x) = \delta(x)$. We make 20 algorithm runs on the same CPU.

In Table 2 we give the preliminary given error δ different values and estimate N and k by (13). The initial probability is the δ-function. For problems like these scientists are happy to have error like 5 or even 10 %. Although this is multidimensional nonlinear integral equation, one can easily see that MAO based on balancing of the errors gives reliable results and experimental relative error confirms the expected theoretical error.

5 Conclusion

Error analysis of an almost optimal Monte Carlo algorithm based on balancing of the systematic and stochastic error is presented. MAO algorithm is applied to one linear and one non-linear Fredholm integral equations of the second kind. The examples that we solved are widely used in environmental sciences and computational physics. Since the algorithms are based on computing functionals on Markov chains defined in optimal Krilov subspaces one may expect that this kind of algorithms is a good candidate for treating much more complicated problems with kernels arising from nano-physics. In the latter case people are dealing with a special kind of kernels containing three components. The problem of obtaining an optimal ratio between the number of realizations of the random variable and the number of iterations in the Monte Carlo Markov chain is analyzed and solved. Error balancing conditions are obtained and studied. It is proven that the idea of combining balancing of the errors with an almost optimal Monte Carlo algorithm gives best results. It is shown that the balancing of errors reduce the computational complexity if the error is fixed.

This makes the algorithms under consideration an efficient tool for simulation advanced innovative problems arising in environmental sciences, and nano-physics as well.

Acknowledgments This work has been supported by the project Support of Young scientists from the Bulgarian Academy of Sciences under grant # 214, as well as by the Bulgarian NSF Grant DFNI I02/20.

References

1. Curtiss, J.H.: Monte Carlo methods for the iteration of linear operators. J. Math. Phys. **32**, 209–232 (1954)
2. Dimov, I.: Minimization of the probable error for some Monte Carlo methods. In: Proceedings of the International Conference on Mathematical Modeling and Scientific Computation, Albena, Bulgaria, Sofia, Publ. House of the Bulgarian Academy of Sciences, pp. 159–170 (1991)

3. Dimov, I.: Optimal Monte Carlo algorithms. In: Proceedings IEEE John Vincent Atanasoff 2006 International Symposium on Modern Computing, October 2006, Sofia, Bulgaria, IEEE, Los Alamitos, California, pp. 125–131 (2006)
4. Dimov, I.: Monte Carlo Methods for Applied Scientists, p. 291. World Scientific, New Jersey (2008)
5. Dimov, I., Atanassov, E.: What Monte Carlo models can do and cannot do efficiently? Appl. Math. Model. **32**, 1477–1500 (2007)
6. Dimov, I., Gurov, T.: Monte Carlo algorithm for solving integral equations with polynomial non-linearity. Parallel implementation, Pliska (Studia Mathematica Bulgarica), vol. 13, 2000. In: Proceedings of the 9th International Summer School on Probability Theory and Mathematical Statistics, Sozopol, pp. 117–132 (1997)
7. Dimov, I., Karaivanova, A.: Error analysis of an adaptive Monte Carlo method for numerical integration. Math. Comput. Simul. **47**, 201–213 (1998)
8. Dimov, I., Karaivanova, A., Philippe, B., Weihrauch, C.: Robustness and applicability of Markov Chain Monte Carlo algorithms for eigenvalue problem. J. Appl. Math. Model. **32**, 1511–1529 (2008)
9. Dimov, I., Georgieva, R., Todorov, V.: Balancing of systematic and stochastic errors in Monte Carlo algorithms for integral equations, LNCS89622. In: Proceeding of 8th International Conference Numerical Methods and Applications, NMA 2014, Borovets, Bulgaria, Aug 20–24 2014, pp. 44–51 (2014)
10. Dimov, I., Nedjalkov, M., Selberherr, S., Sellier, J.M.: Boundary conditions and the Wigner equation solution. J. Comput. Electr. **14**(4), 859–863 (2015)
11. Dimov, I., Nedjalkov, M., Sellier, J.M.: An introduction to applied quantum mechanics in the Wigner Monte Carlo formalism. Phys. Rep. **577**, 1–34 (2015)
12. Doucet, A., Johansen, A.M., Tadic, V.B.: On solving integral equations using Markov chain Monte Carlo methods. Appl. Math. Comput. **216**, 2869–2880 (2010)
13. Georgieva, R.: Computational complexity of Monte Carlo algorithms for multidimensional integrals and integral equations. Ph.D. Thesis, IICT-BAS, Sofia (2003)
14. Nedjalkov, M., Ferry, D.K., Vasileska, D., Dollfus, P., Querlioz, D., Dimov, I., Schwaha, P., Selberherr, S.: Physical scales in the Wigner-Boltzmann equation. Ann. Phys. **328**, 220–237 (2013)
15. Sobol, I.: Numerical Methods Monte Carlo. Nauka, Moscow (1973)

Finite Element Method for Nonlinear Vibration Analysis of Plates

Stanislav Stoykov and Svetozar Margenov

Abstract Plates are structures which have wide applications among engineering constructions. The knowledge of the dynamical behavior of the plates is important for their design and maintenance. The dynamical response of the plate can change significantly due to the nonlinear terms at the equation of motion which become essential in the presence of large displacements. The current work presents numerical methods for investigating the dynamical behavior of plates with complex geometry. The equation of motion of the plate is derived by the classical plate theory and geometrical nonlinear terms are included. It is discretized by the finite element method and periodic responses are obtained by shooting method. Next point from the frequency-response curve is obtained by the sequential continuation method. The potential of the methods is demonstrated on rectangular plate with hole. The main branch along the fundamental mode is presented and the corresponding time responses and shapes of vibration are shown.

1 Introduction

Plates are thin structures and due to strong external loads their displacements can become large. Linear theories are not appropriate for modeling large displacements thus one should include geometrical nonlinear terms at the equation of motion for obtaining more accurate and reliable results. The nonlinearities can change drastically

S. Stoykov (✉) · S. Margenov
Institute of Information and Communication Technologies,
Bulgarian Academy of Sciences, Acad. G. Bonchev Str., bl. 25A,
1113 Sofia, Bulgaria
e-mail: stoykov@parallel.bas.bg

S. Margenov
e-mail: margenov@parallel.bas.bg

© Springer International Publishing Switzerland 2016   17
S. Margenov et al. (eds.), *Innovative Approaches and Solutions in Advanced
Intelligent Systems*, Studies in Computational Intelligence 648,
DOI 10.1007/978-3-319-32207-0_2

the behavior of the system, thus additional tools for analyzing such systems need to be used. The aim of the work is to present efficient numerical methods for analyzing nonlinear dynamical systems which arise from space discretization of elastic plates with complex geometry.

The equation of motion of the plate is derived assuming classical plate theory and including geometrically nonlinear terms. It is discretized into a system of ordinary differential equations (ODE) by the finite element method. The variation of the periodic steady-state responses with the excitation frequency are of interest for the dynamical analysis, thus periodic responses due to external harmonic excitations are obtained by shooting method. Newmark's time integration scheme is used for solving the ODE in time domain. The resulting nonlinear algebraic system is solved by Newton's method. Prediction for the next point from the frequency-response curve is defined by the sequential continuation method. The process involves simultaneously solvers for sparse and dense matrices. The complete process of computing the frequency-response curve becomes computationally slow and burdensome when the resulting ODE system is large. A suitable parallel implementation used for beams and three-dimensional structures in [6] is also applied here.

2 Equation of Motion of Plates

The nonlinear equation of motion of plate is derived in Cartesian coordinate system assuming classical plate theory, also known as Kirchoff's hypotheses. Only transverse displacements are considered on the middle plane. Kirchoff's hypotheses states that stresses in the direction normal to the plate middle surface are negligible and strains vary linearly within the plate thickness [5].

Assuming Kirchoff's hypotheses, the in-plane displacements $u(x,y,z,t)$ and $v(x,y,z,t)$ and the out-of-plane displacement $w(x,y,z,t)$ are expressed by the out-of-plane displacement on the middle plane $w_0(x,y,t)$:

$$u(x,y,z,t) = -z\frac{\partial w_0(x,y,t)}{\partial x},$$
$$v(x,y,z,t) = -z\frac{\partial w_0(x,y,t)}{\partial y}, \tag{1}$$
$$w(x,y,z,t) = w_0(x,y,t).$$

The middle plane is defined for $z=0$. Using the nonlinear strain-displacement relations from Green's strain tensor and assuming that ε_z, γ_{xz} and γ_{yz} are negligible, i.e. $\varepsilon_z = \gamma_{xz} = \gamma_{yz} = 0$, the following expressions are obtained:

$$\varepsilon_x = \frac{\partial u}{\partial x} + \frac{1}{2}\left(\frac{\partial w}{\partial x}\right)^2 = -z\frac{\partial^2 w_0}{\partial x^2} + \frac{1}{2}\left(\frac{\partial w_0}{\partial x}\right)^2,$$

$$\varepsilon_y = \frac{\partial v}{\partial y} + \frac{1}{2}\left(\frac{\partial w}{\partial y}\right)^2 = -z\frac{\partial^2 w_0}{\partial y^2} + \frac{1}{2}\left(\frac{\partial w_0}{\partial y}\right)^2, \tag{2}$$

$$\gamma_{xy} = \frac{\partial u}{\partial y} + \frac{\partial v}{\partial x} + \frac{\partial w}{\partial x}\frac{\partial w}{\partial y} = -2z\frac{\partial^2 w_0}{\partial x \partial y} + \frac{\partial w_0}{\partial x}\frac{\partial w_0}{\partial y}.$$

The stresses are related to the strains by the constitutive relations written in reduced form. For isotropic materials this relation is given by:

$$\left\{\begin{array}{c} \sigma_x \\ \sigma_y \\ \tau_{xy} \end{array}\right\} = \frac{E}{1-v^2}\begin{bmatrix} 1 & v & 0 \\ v & 1 & 0 \\ 0 & 0 & \frac{1-v}{2} \end{bmatrix}\left\{\begin{array}{c} \varepsilon_x \\ \varepsilon_y \\ \gamma_{xy} \end{array}\right\}, \tag{3}$$

where E is the Young's modulus and v is the Poisson's ratio. The equation of motion is derived by the Hamilton's principle:

$$\int_{t_1}^{t_2} (\delta T - \delta \Pi + \delta W)dt = 0, \tag{4}$$

where T is the kinetic energy, Π is the potential energy and W is the work done by external loads:

$$T = \rho \int_V (u\ddot{u} + v\ddot{v} + w\ddot{w})dV, \tag{5}$$

$$\Pi = \int_V \left(\varepsilon_x\sigma_x + \varepsilon_y\sigma_y + \gamma_{xy}\tau_{xy}\right)dV, \tag{6}$$

$$W = \int_V q(x,y,t)w\,dV, \tag{7}$$

where by double dot is denoted the second derivative with respect to time, V is the volume of the plate, ρ is the density and $q(x,y,t)$ is the applied external load in transverse direction. The equation of motion is obtained in the following form:

$$\rho h\frac{\partial^2 w_0}{\partial t^2} + \frac{Eh^3}{12(1-v^2)}\left(\frac{\partial^4 w_0}{\partial x^4} + 2\frac{\partial^4 w_0}{\partial x^2 \partial y^2} + \frac{\partial^4 w_0}{\partial y^4}\right)$$
$$= q + \frac{\rho h^3}{12}\frac{\partial^2}{\partial t^2}\left(\frac{\partial^2 w_0}{\partial x^2} + \frac{\partial^2 w_0}{\partial y^2}\right) + N_x\frac{\partial^2 w_0}{\partial x^2} + 2N_{xy}\frac{\partial^2 w_0}{\partial x \partial y} + N_y\frac{\partial^2 w_0}{\partial y^2}, \tag{8}$$

where h is the thickness and N_x, N_y and N_{xy} are the stress resultants given by:

$$N_x = \frac{Eh}{1-v^2}\left(\frac{1}{2}\left(\frac{\partial w_0}{\partial x}\right)^2 + \frac{v}{2}\left(\frac{\partial w_0}{\partial y}\right)^2\right),$$

$$N_y = \frac{Eh}{1-v^2}\left(\frac{1}{2}\left(\frac{0}{\partial y}\right)^2 + \frac{v}{2}\left(\frac{\partial w_0}{\partial x}\right)^2\right), \tag{9}$$

$$N_{xy} = \frac{Eh}{2(1+v)}\left(\frac{\partial w_0}{\partial x}\frac{\partial w_0}{\partial y}\right).$$

In the current work are considered simply-supported boundary conditions. The essential boundary condition, which is imposed in the space of admissible functions, is given by:

$$w_0(x, y, t) = 0. \tag{10}$$

The natural boundary condition is related with the second derivatives of the transverse displacement w_0. It is not imposed in the finite element model, but it is automatically satisfied. Details about the derivation of the equation of motion can be found, for example in [4].

3   Finite Element Method and Computation of Periodic Responses

The partial differential Eq. (8) is discretized by the finite element method [7]. A mesh of rectangular elements is used. The reference finite element is square element with local coordinates denoted by ξ and $\eta \in [-1, 1]$. The element has 4 nodes, each node is considered to have 4 degrees of freedom (DOF), i.e. w_0, $\frac{\partial w_0}{\partial \xi}$, $\frac{\partial w_0}{\partial \eta}$ and $\frac{\partial^2 w_0}{\partial \xi \partial \eta}$. Hence, each element has 16 DOF. The transverse displacement w_0 is approximated by the shape functions within each element by:

$$w^e(\xi, \eta, t) = \sum_{i=1}^{16} q_i(t)\psi_i(\xi, \eta), \tag{11}$$

where $\psi_i(\xi, \eta), i = 1, \ldots, 16$ are the shape functions and $q_i(t)$ are the local unknowns (DOF). The finite element is conforming, the FEM space is $\mathcal{V}_h \subset \mathcal{H}_2(\Omega)$.

The equation of motion is written in variational form and application by parts is applied. Assuming the approximation (11) for the transverse displacement w_0, and using the same expression for the test functions, a system of nonlinear ordinary differential equations of the following type is obtained:

$$\mathbf{M}\ddot{\mathbf{q}}(t) + \mathbf{C}\dot{\mathbf{q}}(t) + \mathbf{K_L}\mathbf{q}(t) + \mathbf{K_{NL}}(\mathbf{q}(t))\mathbf{q}(t) = \mathbf{F}(t), \qquad (12)$$

where $\mathbf{M}$ represents the mass matrix, $\mathbf{K_L}$ represents the stiffness matrix of constant terms, $\mathbf{K_{NL}}(\mathbf{q}(t))$ represents the stiffness matrix that depends on the global vector of unknowns $\mathbf{q}(t)$, $\mathbf{F}(t)$ is the global vector of external forces and $\mathbf{C}$ is the damping matrix. Stiffness proportional damping is considered in the model, hence the damping matrix is expressed as $\mathbf{C} = \alpha\mathbf{K_L}$, where α is the factor of proportionality.

The external force $\mathbf{F}(t)$ is assumed to be harmonic:

$$\mathbf{F}(t) = \mathbf{A}cos(\omega t), \qquad (13)$$

where $\mathbf{A}$ is the force vector and ω is the excitation frequency.

Equation (12) is solved in frequency domain where each point from the frequency-response curve presents a periodic solution. Shooting method is used to compute the periodic responses. The method consists of iterative correction of the initial conditions that lead to periodic solution.

Of interest are the initial conditions $\mathbf{q}(0) = \mathbf{q}_0$ and $\dot{\mathbf{q}}(0) = \dot{\mathbf{q}}_0$ which satisfy the equations:

$$\mathbf{q}(T, \mathbf{q}_0, \dot{\mathbf{q}}_0) = \mathbf{q}_0, \qquad (14)$$

$$\dot{\mathbf{q}}(T, \mathbf{q}_0, \dot{\mathbf{q}}_0) = \dot{\mathbf{q}}_0, \qquad (15)$$

where T is the minimal period of vibration which is determined from the excitation frequency, $\mathbf{q}(T, \mathbf{q}_0, \dot{\mathbf{q}}_0)$ and $\dot{\mathbf{q}}(T, \mathbf{q}_0, \dot{\mathbf{q}}_0)$ present the response and velocity of the equation of motion (12) at time T due to initial conditions $\mathbf{q}_0$ and $\dot{\mathbf{q}}_0$. The initial conditions are also written in the response and velocity, in order to outline the dependence of the solution on the initial conditions.

The method finds corrections of the initial conditions $\mathbf{q}_0$ and $\dot{\mathbf{q}}_0$ in such a way that the response of the system with the corrected initial conditions performs periodic steady-state vibration. I.e. of interest is to find $\delta\mathbf{q}_0$ and $\delta\dot{\mathbf{q}}_0$ such that

$$\begin{aligned} \|\mathbf{q}(T, \mathbf{q}_0 + \delta\mathbf{q}_0, \dot{\mathbf{q}}_0 + \delta\dot{\mathbf{q}}_0) - \mathbf{q}_0 - \delta\mathbf{q}_0\| < \epsilon, \\ \|\dot{\mathbf{q}}(T, \mathbf{q}_0 + \delta\mathbf{q}_0, \dot{\mathbf{q}}_0 + \delta\dot{\mathbf{q}}_0) - \dot{\mathbf{q}}_0 - \delta\dot{\mathbf{q}}_0\| < \epsilon, \end{aligned} \qquad (16)$$

for some small ϵ.

The corrections $\delta\mathbf{q}_0$ and $\delta\dot{\mathbf{q}}_0$ are obtained by solving the following linear system:

$$\begin{bmatrix} \mathbf{Q_d}(T) - \mathbf{I} & \mathbf{Q_v} \\ \dot{\mathbf{Q}}_\mathbf{d}(T) & \dot{\mathbf{Q}}_\mathbf{v}(T) - \mathbf{I} \end{bmatrix} \begin{Bmatrix} \delta\mathbf{q}_0 \\ \delta\dot{\mathbf{q}}_0 \end{Bmatrix} = \begin{Bmatrix} \mathbf{q}_0 - \mathbf{q}(T, \mathbf{q}_0, \dot{\mathbf{q}}_0) \\ \dot{\mathbf{q}}_0 - \dot{\mathbf{q}}(T, \mathbf{q}_0, \dot{\mathbf{q}}_0) \end{Bmatrix}. \qquad (17)$$

The matrix in Eq. (17) is obtained by solving another initial value problem, i.e. $\mathbf{Q_d}(T)$ and $\dot{\mathbf{Q}}_\mathbf{d}(T)$ are solutions at time T of the system:

$$\mathbf{M}\ddot{\mathbf{Q}}(t) + \mathbf{C}\dot{\mathbf{Q}}(t) + \mathbf{K_L}\mathbf{Q}(t) + \mathbf{J}(\mathbf{q}(t))\mathbf{Q}(t) = \mathbf{0} \tag{18}$$

due to initial conditions $\mathbf{Q_d}(0) = \mathbf{I}$ and $\dot{\mathbf{Q}}_\mathbf{d}(0) = \mathbf{0}$. $\mathbf{Q_v}(T)$ and $\dot{\mathbf{Q}}_\mathbf{v}(T)$ are solutions at time T of the system (18) due to initial conditions $\mathbf{Q_v}(0) = \mathbf{0}$ and $\dot{\mathbf{Q}}_\mathbf{v}(0) = \mathbf{I}$. $\mathbf{J}(\mathbf{q}(t))$ is the Jacobian of the nonlinear terms of the equation of motion (12), i.e.

$$\mathbf{J}(\mathbf{q}(t)) = \frac{\partial \mathbf{K_{NL}}(\mathbf{q}(t))\mathbf{q}(t)}{\partial \mathbf{q}(t)}. \tag{19}$$

The shooting method requires time integration scheme for obtaining the responses of Eqs. (12) and (18) from time 0 to T. Newmark's method is used for the time integration. Newton's method is used for solving the resulting nonlinear algebraic system of Eq. (12) after application of Newmark's method. Equation (18) presents linear system of ODE, thus Newton's method is applied only to Eq. (12). Next point from the frequency-response curve is computed by increasing the excitation frequency and repeating the shooting procedure. This parametric analysis is known as sequential continuation method.

In the cases of plates with complex geometries (Fig. 1), the necessity of obtaining sufficiently accurate results requires the usage of enough elements. The resulting system of ODE (12) becomes large and the process of computing periodic responses and frequency-response curves becomes computationally burdensome. The parallel implementation of the shooting method, presented in [6] is also applied here. It consists of efficient parallel algorithms for solving large systems of sparse and dense matrices. UMFPACK library [2], which is a direct solver, is used for solving the sparse systems which result from the application of Newmark's method to systems (12) and (18). ScaLAPACK [1] library, which is library of high-performance linear algebra routines for parallel distributed memory machines, is used for solving the dense system (17).

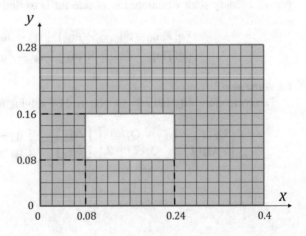

Fig. 1 Plate with complex geometry and mesh of quadratic finite elements

4 Numerical Tests

4.1 Linear Free Vibration

The developed finite element model is validated by comparing the natural frequencies of rectangular plate with analytical solutions. Rectangular plate with dimensions: $a = 0.3$ m, $b = 0.6$ m and thickness $h = 0.001$ m is considered. The material is assumed to be isotropic with material properties (Aluminum): $E = 70$ GPa (Young modulus), $\rho = 2778$ kg/m^3 (density) and $v = 0.3$ (Poisson's ratio). The results are presented in Table 1. Three different meshes are used in order to investigate the convergence of the FEM. Meshes with rectangular elements of sizes $\Delta h = 0.05$ m, $\Delta h = 0.025$ m and $\Delta h = 0.0125$ m are considered. Figure 2 shows the convergence rates of the natural frequencies. The results correspond to the theoretical estimates and confirm that the FEM model is implemented correctly. The theoretical evaluation of the error of the natural frequencies is $O(\Delta h^4)$ [3], which is also obtained by the implemented FEM (Fig. 2).

Further to investigate the dynamical behavior of plates with complex geometry, a model of rectangular plate with whole is generated. The plate and its dimensions are presented in Fig. 1, the thickness is $h = 0.001$ m and the same material properties are assumed (Aluminum). The natural frequencies of the plate are presented in Table 2 and first four mode shapes are presented in Fig. 3.

Table 1 Comparison of the natural frequencies (rad/s) of FEM model and analytical solutions—$\omega_{mn} = \pi^2[(\frac{m}{a})^2 + (\frac{n}{b})^2]\sqrt{\frac{D}{\rho h}}, D = \frac{Eh^3}{12(1-v^2)}$

Mode	FEM, 364 DOF	FEM, 1300 DOF	FEM, 4900 DOF	Analytical
1	208.2338	208.2283	208.2279	208.2279
2	333.1668	333.1650	333.1646	333.1646
3	541.4399	541.3965	541.3927	541.3924
4	708.4704	708.0070	707.9768	707.9747
5	833.2874	832.9382	832.9132	832.9114

Fig. 2 Rates of convergence of first (– – – –), third (· – · – ·) and fourth (· · ·) natural frequencies, ●—computed values

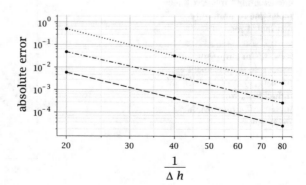

Table 2 Natural frequencies (rad/s) of plate from Fig. 1

Mode	FEM
1	848.69
2	1214.39
3	1382.06
4	1831.07
5	2249.98
6	2489.70

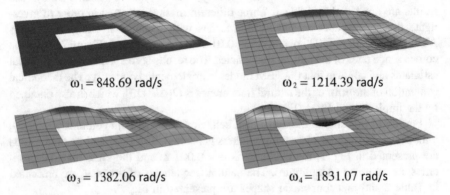

$\omega_1 = 848.69$ rad/s $\omega_2 = 1214.39$ rad/s

$\omega_3 = 1382.06$ rad/s $\omega_4 = 1831.07$ rad/s

Fig. 3 First four natural modes of vibration of plate from Fig. 1

4.2 Nonlinear Forced Vibration

Uniformly distributed harmonic loads, with amplitudes of 4, 5 and 6 kN/m^2, are applied on the plate (Fig. 1) in order to study steady-state forced vibrations. Shooting method is started with small excitation frequency, which is continuously increased. The frequency-response curve is shown in Fig. 4. The amplitude is

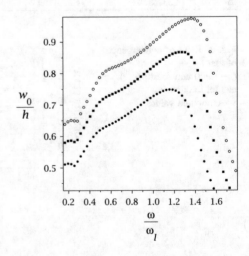

Fig. 4 Frequency-response curves around the first linear mode due to uniformly distributed load of ○—6 kN/m^2, ■—5 kN/m^2, ●—4 kN/m^2, w_0—maximum transverse displacement at point $(x, y) = (0.32, 0.16$ m), h thickness, ω/ω_l fundamental dimensionless frequency

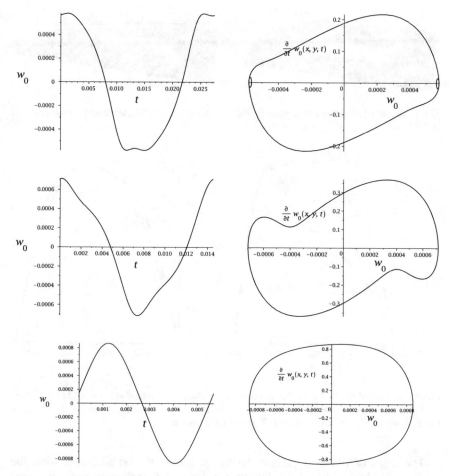

Fig. 5 Time and phase plots for different excitation frequencies of point $(x, y) = (0.32 \text{ m}, 0.16 \text{ m})$, uniformly distributed load of 5 kN/m², first row $\omega/\omega_l = 0.27$, second row $\omega/\omega_l = 0.5$, third row $\omega/\omega_l = 1.3$

measured at point $(x, y) = (0.32$ m, 0.16 m$)$ which is the point with maximum displacement on the first linear mode. The figure shows that there is decrease of the amplitude for small values of the excitation frequency and then the amplitude increases for values of the excitation frequency close to the linear mode.

Figure 5 presents time responses and phase plots of the plate for different excitation frequencies. It can be seen from the time response and the phase plot that the third harmonic appears in the response for $\omega/\omega_l = 0.27$, which is due to a super-harmonic resonance. When the excitation frequency is close to the linear frequency, the first harmonic is dominant and the response is similar to harmonic (Fig. 5, $\omega/\omega_l = 1.3$). Figure 5 confirms that the plate vibrates in nonlinear regime.

Fig. 6 Shapes of vibration of the plate for different times, $\omega = 1.3\omega_l$, uniformly distributed load of 5 kN/m², (*right*) $t = 0$, (*left*) $t = T/4$

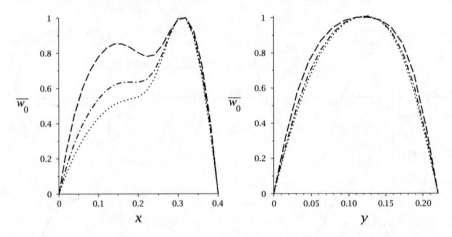

Fig. 7 Sections of the plate for different times, $\omega = 1.3\omega_l$, uniformly distributed load of 5 kN/m², (*right*) xz plane, $y = 0.2$ m, (*left*) yz plane, $x = 0.32$ m, $----t = 0$, $\ldots t = T/12$, $-\cdot-\cdot-$ $t = T/4$, $\overline{w_0}$ normalized displacement, T period of vibration

The shape of vibration changes during the period of vibration, even when the response is close to harmonic. Shapes for different times within one period are shown in Fig. 6, where the amplitude is normalized. The change of the shape can be better seen in Fig. 7, where different sections are presented.

5   Conclusion

Numerical methods for computing nonlinear frequency-response curves of plates with complex geometries are presented. The equation of motion is derived by the Hamilton's principle and the classical plate theory and discretized by the finite element method. Geometrical nonlinear terms are included in the model. Periodic responses are computed by the shooting method and next points from the frequency-response curve are defined by the continuation method.

A plate with complex geometry is analyzed in order to investigate the potential of the proposed methods. The main branch of solutions is obtained for loads with different amplitudes. The effects of the nonlinear terms on the response of the plate and the influence of the excitation frequency are outlined. The proposed numerical methods and the results show that parametric study on the dynamics of complex structures which lead to large-scale dynamical systems can be carried out with appropriate parallel implementation.

Acknowledgments This work was supported by the project AComIn "Advanced Computing for Innovations", grant 316087, funded by the FP7 Capacity Programme.

References

1. Blackford, L.S., Choi, J., Cleary, A., D'Azevedo, E., Demmel, J., Dhillon, I., Dongarra, J., Hammarling, S., Henry, G., Petitet, A., Stanley, K., Walker, D. Whaley, R.C.: ScaLAPACK users' guide. Soc. Ind. Appl. Math. (1997)
2. Davis, T.: Algorithm 832: Umfpack, an unsymmetric-pattern multifrontal method. ACM Trans. Math. Softw. **30**(2), 196–199 (2004)
3. Fix, G., Strang, G.: An Analysis of the Finite Element Method. Wellesley-Cambridge Press (2008)
4. Nayfeh, A., Pai, P.: Linear and Nonlinear Structural Mechanics. Wiley (2004)
5. Reddy, J.N.: Mechanics of Laminated Composite Plates and Shells: Theory and Analysis. CRC Press (2004)
6. Stoykov, S., Margenov, S.: Scalable parallel implementation of shooting method for large-scale dynamical systems. application to bridge components. J. Comput. Appl. Math. **293**, 223–231 (2016)
7. Zienkiewicz, O., Taylor, R. Zhu, J.: The Finite Element Method: Its Basis and Fundamentals. Elsevier (2005)

A Computational Investigation of the Optimal Reaction Type Concerning BOD Removal in Horizontal Subsurface Flow Constructed Wetlands

Konstantinos Liolios, Vassilios Tsihrintzis, Krassimir Georgiev
and Ivan Georgiev

Abstract A numerical simulation of Biochemical Oxygen Demand (BOD) removal in Horizontal Subsurface Flow Constructed Wetlands (HSF CW) is presented. Emphasis is given to select the optimal type of the reaction concerning the BOD removal. For this purpose, a computational investigation is realized by comparing the most usual reaction type, the first-order one, and the recently proposed Monod type, with simulated experimental data obtained from five pilot-scale HSF CW units. These units were operated for 2 years in the facilities of the Laboratory of Ecological Engineering and Technology (LEET), Democritus University of Thrace (DUTh), Xanthi, Greece. For the numerical simulation the Visual MODFLOW family computer code is used, and especially the RT3D code.

Keywords Computational environmental engineering · Groundwater flow · Constructed wetlands · Biochemical Oxygen Demand removal · MODFLOW computer code · Optimal reaction type

K. Liolios (✉) · K. Georgiev · I. Georgiev
Institute of Information and Communication Technologies,
Bulgarian Academy of Sciences, Sofia, Bulgaria
e-mail: kliolios@parallel.bas.bg

K. Georgiev
e-mail: georgiev@parallel.bas.bg

V. Tsihrintzis
Department of Infrastructure and Surveying Engineering,
School of Rural and Surveying Engineering,
National Technical University of Athens, Athens, Greece

I. Georgiev
Institute of Mathematics and Informatics, Bulgarian Academy of Sciences,
Sofia, Bulgaria

© Springer International Publishing Switzerland 2016
S. Margenov et al. (eds.), *Innovative Approaches and Solutions in Advanced Intelligent Systems*, Studies in Computational Intelligence 648,
DOI 10.1007/978-3-319-32207-0_3

29

1 Introduction

Constructed Wetlands (CW) are recently used in the Environmental Engineering as a good alternative solution for small settlements, in order to treat municipal wastewater and to remove groundwater pollutants in contaminated soils [1]. The numerical simulation of CW operation is based on concepts of both, groundwater flow and contaminant transport and removal through porous media, and requires the choice of the relevant optimal reaction type. Thus, it seems necessary to investigate computationally the optimal design characteristics of CW, in order to maximize their removal efficiency and keep their area and construction cost to a minimum.

The present research treats with a numerical simulation of BOD removal in HSF CW. Emphasis is given to select the optimal type of the relevant reaction by using inverse problem procedures [2]. For this purpose, a computational investigation is realized by comparing the most usual reaction type, the first-order one [1, 3] and the recently proposed Monod type [4], with available experimental data. These data have been obtained from five pilot-scale HSF CW units, which were operated for 2 years in the facilities of LEET, DUTh, Xanthi, Greece. For more details concerning the description of the above units and the experimental procedures, see [3, 5]. Concerning the numerical simulation, the Visual MODFLOW family computer code [6] is used, and especially the RT3D code [7] for the Monod type reaction. The comparison between first-order and Monod reaction types is based on the experimental and computational values of BOD concentration at selected points along the length of the pilot-scale DUTh units.

2 The Mathematical Formulation of the Problem

The partial differential equation which describes the fate and transport of contaminants of species k, without adsorption, in 3-dimensional transient groundwater flow systems, can be written by using tensorial notation ($i, j = 1, 2, 3$), as follows [8, 9]:

$$\frac{\partial(\varepsilon C^k)}{\partial t} = \frac{\partial}{\partial x_i}\left(\varepsilon D_{ij}\frac{\partial C^k}{\partial x_j}\right) - \frac{\partial}{\partial x_i}\left(\varepsilon v_i C^k\right) + q_v C_s^k + \sum R_n \qquad (1)$$

where ε is the porosity of the subsurface medium [dimensionless]; D_{ij} is the hydrodynamic dispersion coefficient tensor, in $[L^2T^{-1}]$; C^k is the dissolved concentration of species k, in $[ML^{-3}]$; v_i is the seepage or linear pore water velocity, in $[LT^{-1}]$, which is related to the specific discharge or Darcy flux through the relationship: $q_i = v_i\varepsilon$; q_V is the volumetric flow rate per unit volume of aquifer representing fluid sources (positive) and sinks (negative), when precipitation and evapotraspiration effects are taken into account respectively, in $[T^{-1}]$; C_s^k is the concentration of the source or sink flux for species k, in $[ML^{-3}]$; and ΣR_n is the chemical reaction term, in $[ML^{-3}T^{-1}]$.

The above Eq. (1) is the governing equation underlying in the transport model and contaminant removal. The required velocity field v_i is computed through the Darcy relationship:

$$v_i = -\frac{K_{ij}}{\varepsilon}\frac{\partial h}{\partial x_j} \tag{2}$$

where K_{ij} is a component of the hydraulic conductivity tensor, in $[LT^{-1}]$; and h is the hydraulic head, in $[L]$. The hydraulic head $h = h(x_i;t)$ is obtained from the solution of the 3-dimensional groundwater flow partial differential equation:

$$\frac{\partial}{\partial x_i}(K_{ij}\frac{\partial h}{\partial x_j}) + q_v = S_y\frac{\partial h}{\partial t} \tag{3}$$

where S_y is the specific yield of the porous materials.

As concerns the reaction term ΣR_n, the optimal type according to available experimental data must be chosen. For the usual linear reaction case, ΣR_n depends linearly on the concentration C via the first-order removal coefficient λ, in $[T^{-1}]$, and is given by the formula:

$$\sum R_n = -\lambda \varepsilon C \tag{4}$$

For non-linear reaction cases, the Monod reaction type has been alternatively proposed recently [4]. Then, the following formula is used for the reaction term, which is now not depending linearly on the concentration C:

$$\sum R_n = -K_{max}\frac{C}{C+K_s} \tag{5}$$

Here, K_{max} is the zero-order removal capacity, in $[ML^{-3}T^{-1}]$; and K_s is the half-saturation constant for the considered contaminant, in $[ML^{-3}]$. Depending on the relation between C and K_s, the Monod reaction type degradation involves as limited cases the first-order reactions, when $K_s \gg C$, and the zero-order reaction, when $K_s \ll C$. Also, the two unknown Monod parameters are the K_s and the K_{max}.

The above Eqs. (1)–(3), combined with either, Eq. (4) or Eq. (5), and appropriate initial and boundary conditions, describe the 3-dimensional flow of groundwater and the transport and removal of contaminants in a heterogeneous and anisotropic medium. So, for the case of one only pollutant species ($k = 1$), the unknowns of the problem are the following five space-time functions: The hydraulic head: $h = h(x_i;t)$, the three velocity components: $v_i = v_i(x_i;t)$ and the concentration: $C = C(x_i;t)$.

The problem is linear when the reaction Eq. (4) for the first-order type is used, whereas becomes a highly non-linear one when the reaction Eq. (5) for the Monod type is used.

3 The Numerical Treatment of the Problem

For the numerical treatment of the problem described in the previous paragraph, the Finite Difference Method (FDM) is chosen among the other available numerical methods. The reason is that on the one hand CW usually have a rectangular scheme, something that happens also in our case for the DUTh pilot-scale units, and on the other hand this method is the basis for the computer code family MODFLOW. This code is widely used for the simulation of groundwater flow and mass transport, see e.g. [3, 8, 9]. In the present study the MODFLOW code, accompanied by the effective computer packages MT3DMS and RT3D modules, is used. Concerning RT3D, it is a general purpose, multispecies, reactive transport code [7]. The double Monod model will be applied herein, where BOD is the electron donor and oxygen is the electron acceptor.

4 Short Description of Experimental Data

As mentioned, the experimental data were collected after the operations of five similar pilot-scale HSF CW. A detailed description of these units, and their experimental procedures, are presented in [5]. Briefly, each tank has a rectangular scheme, with dimensions 3 m of length, 0.75 m of width and 1 m of depth. Three units contained medium gravel (MG), one contained fine gravel (FG) and one cobbles (CO). Common reeds (R) were planted in three tanks (MG-R, FG-R and CO-R), while one unit was planted with cattail (MG-C) and one was kept unplanted (MG-Z) for comparison reasons. The above five pilot-scale units operated continuously for 2 years (2004–2006). The space discretization of the computation domain in each unit was performed by using 210 columns and 40 layers [3, 9], resulted to 8400 cells. This grid was also used for the computation of both, the hydraulic head and the concentration fields. For more details about the numerical simulation, see [3].

5 The Problem of Transport and Removal
for Interacting Pollutants

For the general case of biological treatment of a pollutant, the biodegradation contains the influence of bacteria, with are dissolved either in aqueous phase or attached to the porous material. Assuming interacting pollutants [10, 11], the fate and transport equation for an electron donor D (in our case, BOD) in a 3-dimensional saturated porous media, can be written as [12]:

$$R_d \frac{\partial[D]}{\partial t} = \frac{\partial}{\partial x_i}\left(D_{ij}\frac{\partial[D]}{\partial x_j}\right) - \frac{\partial}{\partial x_i}(v_i[D]) + \frac{q_s}{\varepsilon}[D_s] - M_1 \tag{6a}$$

where:

$$M_1 = \mu_m\left(X_a + \frac{\rho_b X_s}{\varepsilon}\right)\left(\frac{[D]}{K_D + [D]}\right)\left(\frac{[A]}{K_A + [A]}\right) \tag{6b}$$

In the above equations, R_d is the retardation factor; $[D]$ is the electron donor concentration in the aqueous phase, in $[ML^{-3}]$; $[D_s]$ is the donor concentration in the sources/sinks, in $[ML^{-3}]$; μ_m is the contaminant utilization rate, in $[T^{-1}]$; X_a is the concentration of the bacteria in the aqueous phase, in $[M_X L^{-3}]$; ρ_b is the bulk density, in $[ML^{-3}]$; X_s is the concentration of the bacteria in the solid phase, in $[M_X/M_s]$; K_D is the half saturation coefficient for the electron donor, in $[ML^{-3}]$; $[A]$ is the electron acceptor concentration in the aqueous phase, in $[ML^{-3}]$; and K_A is the half saturation coefficient for the electron acceptor, in $[ML^{-3}]$.

The differential equation with partial derivatives, which describes the fate and transport of the electron acceptor A (in our case, oxygen), is:

$$R_A \frac{\partial[A]}{\partial t} = \frac{\partial}{\partial x_i}\left(D_{ij}\frac{\partial[A]}{\partial x_j}\right) - \frac{\partial}{\partial x_i}(v_i[A]) + \frac{q_s}{\varepsilon}[A_s] - M_2 \tag{7a}$$

where:

$$M_2 = Y_{A/D}\mu_m\left(X_a + \frac{\rho_b X_s}{\varepsilon}\right)\left(\frac{[D]}{K_D + [D]}\right)\left(\frac{[A]}{K_A + [A]}\right) \tag{7b}$$

In these equations, R_A is the retardation coefficient of the electron acceptor; $[A_s]$ is the electron acceptor source concentration in the aqueous phase, in $[ML^{-3}]$; and $Y_{A/D}$ is the stoichiometric yield coefficient that relates the donor and acceptor concentrations.

The fate and transport of bacteria concentration in the aqueous phase, X_a, is described by the following differential equation with partial derivatives:

$$\frac{\partial X_a}{\partial t} = \frac{\partial}{\partial x_i}\left(D_{ij}\frac{\partial X_a}{\partial t}\right) - \frac{\partial}{\partial x_i}(v_i X_a) + \frac{q_s}{\varepsilon}[X_a] - K_{att}X_a + \frac{K_{det}\rho_b X_s}{\varepsilon} + M_3 \tag{8a}$$

where:

$$M_3 = Y_{X/D}\mu_m X_a\left(\frac{[D]}{K_D + [D]}\right)\left(\frac{[A]}{K_A + [A]}\right) - K_e X_a \tag{8b}$$

In these Equations, K_{att} is the bacterial attachment coefficient, in $[T^{-1}]$; K_{det} is the bacterial detachment coefficient, in $[T^{-1}]$; $Y_{X/D}$ is the stoichiometric yield coefficient that relates the bacteria and donor concentrations; and K_e is the endogenous decay coefficient, in $[T^{-1}]$.

The growth of attached-phase bacteria concentration, X_s, can be described by using the following differential equation:

$$\frac{\partial X_s}{\partial t} = \frac{K_{att}\varepsilon X_\alpha}{\rho_b} - K_{det}X_s + Y_{X/D}\mu_m X_s \left(\frac{[D]}{K_D + [D]}\right)\left(\frac{[A]}{K_A + [A]}\right) - K_e X_s \qquad (9)$$

Thus, in the system of the above partial differential Eqs. (6)–(9), the unknown space-time functions are the four concentrations: $[D]$, $[A]$, X_a and X_s.

The numerical solution of this system is realized by considering appropriate initial and boundary conditions. For this solution, the reaction-operator splitting strategy is used, see [8, 13, 14]. Thus, the original problem first requires the separate solution of the following differential equations system concerning the terms reaction M_1, M_2 and M_3:

$$\frac{d[D]}{dt} = -\frac{\mu_m}{R_d}\left(X_\alpha + \frac{\rho_b X_s}{\varepsilon}\right)\left(\frac{[D]}{K_D + [D]}\right)\left(\frac{[A]}{K_A + [A]}\right) \qquad (10a)$$

$$\frac{d[A]}{dt} = -\frac{Y_{A/D}\mu_m}{R_A}\left(X_\alpha + \frac{\rho_b X_s}{\varepsilon}\right)\left(\frac{[D]}{K_D + [D]}\right)\left(\frac{[A]}{K_A + [A]}\right) \qquad (10b)$$

$$\frac{dX_a}{dt} = -Y_{X/D}\mu_m\left(X_\alpha + \frac{\rho_b X_s}{\varepsilon}\right)\left(\frac{[D]}{K_D + [D]}\right)\left(\frac{[A]}{K_A + [A]}\right) + \frac{K_{det}\rho_b X_s}{\varepsilon} \qquad (10c)$$
$$- K_{att}X_a - K_e X_a$$

$$\frac{dX_s}{dt} = \frac{K_{att}\varepsilon X_\alpha}{\rho_b} - K_{det}X_s + Y_{X/D}\mu_m X_s\left(\frac{[D]}{K_D + [D]}\right)\left(\frac{[A]}{K_A + [A]}\right) - K_e X_s. \qquad (10d)$$

The solution of the above equations is possible by using the Visual MODFLOW family computer code. In more details, these equations are coded into the RT3D code as described in [11, 12]. First, the problem of flow is solved and then, by using the Darcy law, the velocity field is calculated. Finally, for the interacting pollutants the unknowns, as described above, are the four space-time functions of the concentrations $[D]$, $[A]$, X_a and X_s. On the contrary, for individual pollutants the unknown was only one concentration.

6 Simulation Results and Discussion

6.1 Input Parameters

According to previous paragraph 5, for each one cell of the total number 8400 in a DUTh pilot-scale unit, eight unknown parameters should be determined: μ_m, K_D, K_A, $Y_{X/D}$, $Y_{A/D}$, K_e, K_{att} and K_{det}. In order to estimate the plausible range of these parameters, first a literature review was performed concerning the BOD removal. In the international literature, some typical parameter values are given for the degradation of pollutants by using Monod reactions. Especially for BOD, the following values were suggested by [15]: $K_D = 10$–40 mg BOD/L and $Y_{X/D} = 0.30$–0.65. Similarly, the following values, concerning degradation of BOD by using Monod reactions, were suggested by [16]: $K_D = 25$–100 mg BOD/L and $Y_{X/D} = 0.4$–0.8.

About the determination of other parameters, an inverse problem procedure [2], based on the least square method, was used. Finally, the optimal combination for the input parameters values was the following: $\mu_m = 5$ d^{-1}, $K_D = 70$ mg/L, $K_A = 0.001$ mg/L, $Y_{X/D} = 0.005$, $Y_{A/D} = 0.011$, $K_e = 0.001$ d^{-1}, $K_{att} = 0.010$ d^{-1} and $K_{det} = 0.001$ d^{-1}.

Especially for the parameter K_D, an average value was selected ($K_D = 70$ mg/L), which is in agreement with the range ($K_D = 25$–100 mg/L) recommended by [16]. For the boundary conditions, the experimental data [5] were used for the incoming concentration C_{in}, in [mg BOD/L], and for the incoming oxygen concentration an average measured value of $C_{in,O2} = 4$ mg O$_2$/L was adopted. About the dissolved oxygen (DO), the value of 9 mg DO/L was referred by [15], as maximum concentration.

6.2 Representative Results

The first question was to determine the optimal value of the maximum reaction rate K_{max}, when BOD is considered as individual pollutant. In order to achieve this, the inverse problem procedure was adopted [2], when $K_s = 70$ mg/L. In the following Table 1, the estimated values of K_{max} are presented for each one of the five HSF CW and for Hydraulic Residence Time (*HRT*) of 6, 8, 14 and 20 days. The experimental values of inlet and outlet concentrations, C_{in} and C_{out} respectively, in [mg BOD/L], are also presented.

In Fig. 1, the distribution of BOD concentration along a representative HSF CW tank is presented, after the simulation by using Monod kinetics. The diagram is for the MG-R tank and for *HRT* = 8 days.

In Figs. 2 and 3 are presented the distributions of BOD and oxygen concentrations, respectively, in the MG-C tank and for *HRT* = 14 days. The green curve

Table 1 Input parameters in MODFLOW for the simulation of BOD removal by using Monod kinetics, when $K_D = 70$ mg/L

Tank	*HRT* (days)	C_{in} (mg/L)	C_{out} (mg/L)	K_{max} (mg/L/day)
MG-R	6	332.6	83.7	57.4
	8	364.4	57.3	54.6
	14	389.4	47.3	35.0
	20	357.6	35.0	22.4
MG-C	6	332.6	76.2	59.5
	8	364.4	29.7	64.4
	14	389.4	36.3	37.1
	20	357.6	29.5	25.2
MG-Z	6	332.6	67.2	63.0
	8	364.4	58.2	54.6
	14	389.4	40.0	36.4
	20	357.6	33.3	24.5
FG-R	6	332.6	65.4	63.7
	8	364.4	28.7	64.4
	14	389.4	36.8	37.1
	20	357.6	27.7	25.9
CO-R	6	332.6	84.2	58.7
	8	364.4	45.1	60.9
	14	389.4	42.9	36.4
	20	357.6	24.2	26.6

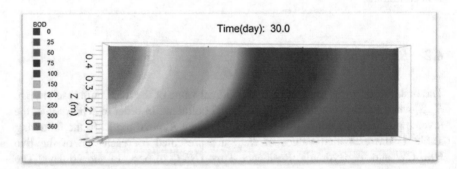

Fig. 1 BOD concentration distribution, in (mg/L), along the MG-R tank (length: $L = 3$ m) after the simulation by using Monod kinetics, for *HRT* = 8 days

represents the temporal variation of the concentration at distance $x_1 = 1$ m from the inlet of the pollutant, the purple curve at position $x_2 = 2$ m and the blue curve at the outlet of the tank ($x = L = 3$ m). In Fig. 3, the declining form of curves show the consumption of oxygen, which is necessary for the degradation of BOD.

Fig. 2 BOD concentration distribution, in (mg/L), at selected points along the MG-C tank, after the simulation by using Monod kinetics, for *HRT* = 14 days

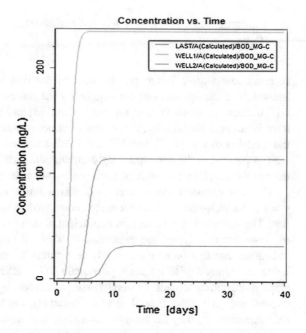

Fig. 3 Oxygen concentration distribution, in (mg/L), at selected points along the MG-C tank, after the simulation by using Monod kinetics, for *HRT* = 14 days

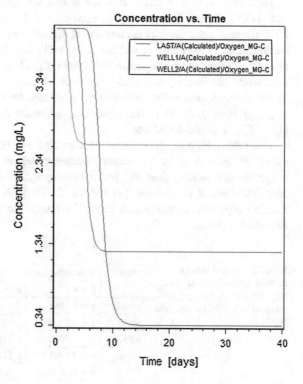

7 Comparison Between First-Order and Monod Reaction Types

In order to compare which one of the two reaction types, the first-order or the Monod, was the optimal one for simulating the removal of BOD in the HSF CW DUTh units, a sensitivity analysis was realized [3] and inverse problem procedures have been used [2]. Briefly, the values of concentration which were computed by the model at distances 1/3 and 2/3 of the unit length, by using both reaction types, were compared to the corresponding experimental data which were collected at the respective sampling locations of each pilot-scale HSF CW.

The first criterion which was used was a linear regression line of equation: $y = ax$. As well-known, the best match occurs when all data fall on the 1:1 slope line. The values of slope a and for coefficient of determination (R^2) for all tanks and for both reaction types, are presented in Table 2. These results give us a first indication that first-order were clearly better than Monod type reactions for simulating the removal of BOD in the pilot-scale DUTh HSF CW. Indeed, the values of slope a are closer to unit. Generally, the simulation by Monod kinetics seems to overestimate the experimental results. Regarding the values of the coefficient of determination R^2, they do not give us a clear conclusion.

Next Figs. 4, 5, 6, 7 and 8 show the spatial distribution of concentrations along each HSF CW unit. In more details, each pilot tank and for each standard value of *HRT*, the concentrations at the entrance, at distances $x_1 = 1$ m and $x_2 = 2$ m from the entrance and at the outlet ($x = 3$ m) of the tanks are shown. More specifically, the blue lines relate to the experimental values C_{exp} [5], the red curves to the concentrations obtained by simulating first-order reactions (C_{first}) and the green curves to the concentration values which were simulated by using Monod type reactions (C_{Monod}). Finally, the black line represents the concentrations if zero-order reaction types (C_{zero}) would be adopted.

From these Figures, the optimal reaction type can be easily concluded.: The red curves (representing the first-order reaction type) are closer to the blue curves (experimental values) than the green curves, which represents the Monod type reaction. Also, it is obvious that for the DUTh HSF CW facilities, the optimal reaction type which approaches in the best way the experimental results, is the first-order reaction.

Table 2 Comparison of reaction types (first order or Monod) for simulating the removal of BOD in pilot-scale HSF CW

Tank	First-order reaction type		Monod reaction type	
	$y = ax$	R^2	$y = ax$	R^2
MG-R	y = 1.1840x	0.5716	y = 1.6417x	0.5218
MG-C	y = 1.2155x	0.5250	y = 1.8470x	0.7038
MG-Z	y = 1.4572x	0.2186	y = 2.1123x	0.4347
FG-R	y = 1.0150x	0.5123	y = 1.6111x	0.4636
CO-R	y = 1.0995x	0.6218	y = 1.5489x	0.5171

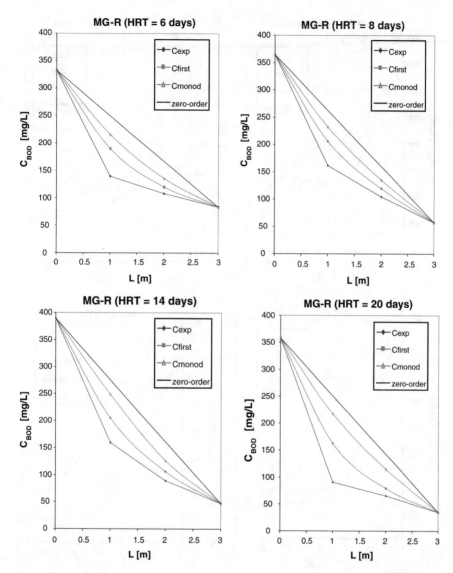

Fig. 4 Comparison between first-order and Monod reaction types for simulating the removal of BOD, in MG-R tank

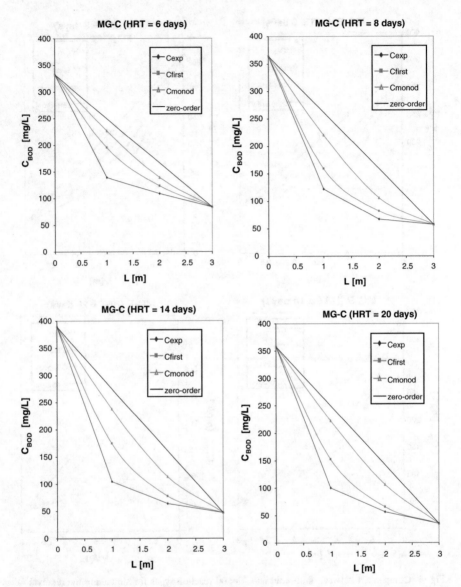

Fig. 5 Comparison between first-order and Monod reaction types for simulating the removal of BOD, in MG-C tank

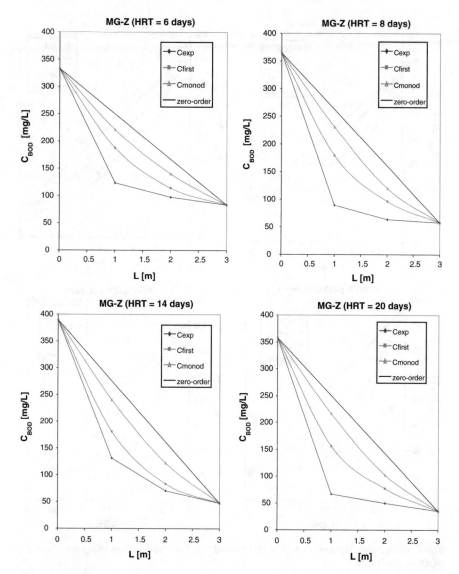

Fig. 6 Comparison between first-order and Monod reaction types for simulating the removal of BOD, in MG-Z tank

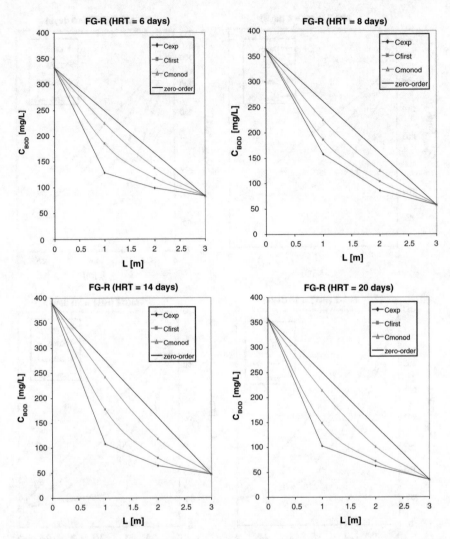

Fig. 7 Comparison between first-order and Monod reaction types for simulating the removal of BOD. in FG-R tank

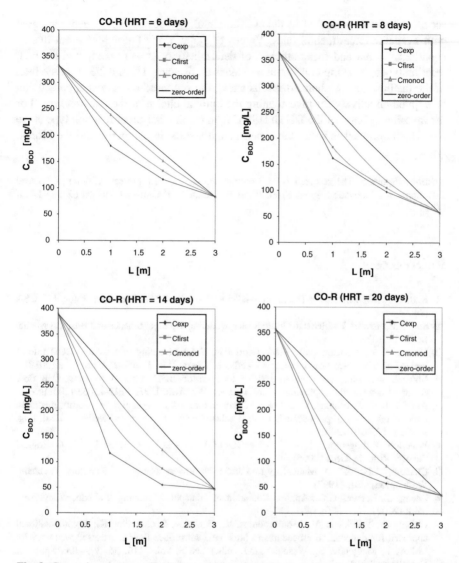

Fig. 8 Comparison between first-order and Monod reaction types for simulating the removal of BOD, in CO-R tank

8 Concluding Remarks

A computational investigation to select the optimal reaction type, concerning the BOD removal in HSF CW, has been presented. For this purpose, a simulation by using on the one hand the first-order and on the other hand the Monod reaction type has been realized. The proper modeling procedure using the MODFLOW—RT3D set of codes was presented in detail. The simulation results have been compared with available experimental ones. Values of slope a, for linear regression line of equation: $y = ax$, and for coefficient of determination R^2 were computed for both reaction types, by using concentration values at distances 1/3 and 2/3 from the inlet of the facility. As the obtained results show, the proposed numerical approach can be applied effectively for investigating the optimal choice of the reaction type. For the investigated case of the DUTh HSF CW units, the first-order reaction type is the one which approaches better the experimental results, in comparison to the Monod type.

Acknowledgments The research is partly supported by the FP7 project AComIn: Advanced Computing for Innovation, grant 316087 and Bulgarian NSF Grants, DMU 03-62 and DFNI I-01/5.

References

1. Kadlec, R.H., Wallace, S.: Treatment Wetlands, 2nd edn. CRC Press, Boca Raton, FL, USA (2009)
2. Sun, N.Z.: Inverse Problems in Groundwater Modeling. Springer Science and Business Media, Berlin (2013)
3. Liolios, K.A., Moutsopoulos, K.N., Tsihrintzis, V.A.: Modeling of flow and BOD fate in horizontal subsurface flow constructed wetlands. Chem. Env. J. **200–202**, 681–693 (2012)
4. Mitchell, C., McNeviv, D.: Alternative analysis of BOD removal in subsurface flow constructed wetlands employing monod kinetics. Wat. Res. J. **35**, 1295–13033 (2001)
5. Akratos, C.S., Tsihrintzis, V.A.: Effect of temperature, HRT, vegetation and porous media on removal efficiency of pilot-scale horizontal subsurface flow constructed wetlands. Ecol. Eng. J. **29**(2), 173–191 (2007)
6. Waterloo Hydrogeologic Inc.: Visual MODFLOW v. 4.2. Users Manual. U.S. Geological Survey, Virginia, USA (2006)
7. Clement, T.P.: RT3D Manual, v.1. Pacific Northwest National Laboratory Richland, Washington, USA (1997)
8. Zheng, C., Bennett, G.D.: Applied Contaminant Transport Modelling, 2nd edn. Wiley, New York (2002)
9. Liolios, K., Tsihrintzis, V., Moutsopoulos, K., Georgiev, I., Georgiev, K.: A computational approach for remediation procedures in horizontal subsurface flow constructed wetlands. In: Lirkov, I., Margenov, S., Wasniewski, J. (eds.) LNCS, vol. 7116, pp. 299–306. Springer, Heidelberg (2012)
10. Clement, T.P., Sun, Y., Hooker, B.S., Petersen, J.N.: Modeling multispecies reactive transport in ground water. Groundwater Mon. Rem. J. **18**(2), 79–92 (1998)
11. Sun, Y., Clement, T.P., Petersen, J.N., Hooker, B.S.: Effect of reaction kinetics on predicted concentration profiles during subsurface bioremediation. J. Cont. Hyd. **31**(3), 359–372 (1998)
12. Clement, T.P., Jones, N.L.: RT3D Tutorial for GMS Users. Pacific Northwest National Laboratory Richland, Washington, USA (1998)
13. Kaluarachchi, J.J., Morshed, J.: Critical assessment of the operator-splitting technique in solving the advection-dispersion-reaction equation: 1. First-order reaction. Adv. Water Res. J. **18**(2), 89–100 (1995)

14. Morshed, J., Kaluarachchi, J.J.: Critical assessment of the operator-splitting technique in solving the advection-dispersion-reaction equation: 2. Monod kinetics and coupled transport. Adv. Wat. Res. J. **18**(2), 101–110 (1995)
15. Tchobanoglous, G., Schroeder, E.: Water Quality. Addison-Wesley Pub. Comp, Amsterdam (1987)
16. Tchobanoglous, G., Burton, F.L., Stensel, H.D.: Wastewater Engineering. Treatment and Use, 4th edn. McGraw-Hill, New York (1991)

A Computational Approach for the Seismic Sequences Induced Response of Cultural Heritage Structures Upgraded by Ties

Angelos Liolios, Antonia Moropoulou, Asterios Liolios,
Krassimir Georgiev and Ivan Georgiev

Abstract The seismic upgrading of Cultural Heritage structures under multiple earthquakes excitation, using materials and methods in the context of Sustainable Construction, is computationally investigated from the Civil Engineering praxis point of view. A numerical approach is presented for the seismic response of Cultural Heritage industrial buildings of reinforced concrete (RC), which are seismically strengthened by using cable elements (tension-ties). A double dis-cretization, in space by the Finite Element Method and in time by an incremental approach, is used for the system of the governing partial differential equations (PDE). The unilateral behaviour of the cable-elements, as well as the other non-linearities of the RC frame-elements, are strictly taken into account and result to inequality problem conditions. A non-convex linear complementarity problem is solved in each time-step by using optimization methods. The seismic assessment of the RC structure and the decision for the optimal cable-strengthening scheme are obtained on the basis of computed damage indices.

A. Liolios · A. Liolios
Democritus-University of Thrace, Department of Civil Engineering,
Institute of Structural Mechanics and Earthquake Engineering,
GR67100 Xanthi, Greece
e-mail: aliolios@civil.duth.gr

A. Liolios
e-mail: liolios@civil.duth.gr

A. Moropoulou
National Technical University of Athens, School of Chemical Engineering,
9 Iroon Polytechniou Street, GR15780 Zografou Campus, Athens, Greece
e-mail: amoropul@central.ntua.gr

K. Georgiev · I. Georgiev
Bulgarian Academy of Sciences, Institute of Information
and Communication Technologies, Acad. G. Bonchev St. bl. 25A, 1113 Sofia, Bulgaria
e-mail: georgiev@parallel.bas.bg

I. Georgiev (✉)
Bulgarian Academy of Sciences, Institute of Mathematics and Informatics,
Sofia, Bulgaria
e-mail: ivan.georgiev@parallel.bas.bg

© Springer International Publishing Switzerland 2016
S. Margenov et al. (eds.), *Innovative Approaches and Solutions in Advanced Intelligent Systems*, Studies in Computational Intelligence 648,
DOI 10.1007/978-3-319-32207-0_4

47

Keywords Computational structural mechanics · Finite element method · Cultural heritage structures · Seismic upgrading by ties · Multiple earthquakes · Optimization methods

1 Introduction

In Civil Engineering praxis, recent Cultural Heritage includes, besides the usual historic monumental structures (churches, monasteries, old masonry buildings etc.), also existing industrial buildings of reinforced concrete (RC), e.g. old factory premises [1]. As concerns their global seismic behaviour of such RC structures, it often arises the need for seismic upgrading. For the recent built Cultural Heritage, this upgrading must be realized by using materials and methods in the context of the Sustainable Construction [1, 26–28, 35].

The use of cable-like members (tension-ties) can be considered as an alternative strengthening method in comparison to other traditional methods (e.g. RC mantles) [2, 8, 22, 39]. As well-known, ties have been used effectively in monastery buildings and churches arches [1]. The ties-strengthening approach can be considered as an alternative one in comparison to other well-known traditional seismic-strengthening methods, e.g. the RC mantles [5, 8, 33], and has the advantages of "cleaner" and "more lenient" operation, avoiding as much as possible the unmaking, the digging, the extensive concreting and "nuisance" functionality of the existing building.

These cable-members (ties) can undertake tension but buckle and become slack and structurally ineffective when subjected to a sufficiently large compressive force. Thus the governing conditions take equality as well as an inequality form and the problem becomes a highly nonlinear one [13–20, 29].

As concerns the seismic upgrading of existing RC structures, modern seismic design codes adopt exclusively the use of the isolated and rare 'design earthquake', whereas the influence of repeated earthquake phenomena is ignored. But as the results of recent research have shown [11], multiple earthquakes generally require increased ductility design demands in comparison with single isolated seismic events. Especially for the seismic damage due to multiple earthquakes, this is accumulated and so it is higher than that for single ground motions.

The present research treats with a computational approach for the seismic analysis of Cultural Heritage existing industrial RC frame-buildings that have been strengthened by cable elements and are subjected to seismic sequences. Damage indices [6, 25, 30] are computed for the seismic assessment of historic and industrial structures and in order the optimum cable-bracing strengthening version to be chosen. Finally, an application it is presented for a simple typical example of a three-bay three-story industrial RC frame strengthened by bracing ties under multiple earthquakes.

2 The Computational Approach

2.1 Mathematical Formulation of the Problem

Due to unilateral behavior of the tie-elements, the dynamic hemivariational inequalities approach of Panagiotopoulos [29] can be used for the mathematical formulation of the seismic problem concerning Cultural Heritage RC structures strengthened by ties. For the so-resulted system of the governing partial differential equations (PDE), a double discretization, in space and time, is used as usually in structural dynamics [3, 4, 42]. Details of the developed numerical approaches given in [20] are briefly summarized herein.

First, the structural system is discretized in space by using frame finite elements [3, 42]. Pin-jointed bar elements are used for the cable-elements. The behavior of these elements includes loosening, elastoplastic or/and elastoplastic-softening-fracturing and unloading—reloading effects. Non-linear behavior is considered as lumped at the two ends of the RC frame elements, where plastic hinges can be developed. All these non-linear characteristics, concerning the ends of frame elements and the cable constitutive law, can be expressed mathematically by the subdifferential relation [29]:

$$s_i(d_i) \in \hat{\partial} S_i(d_i). \tag{1}$$

Here s_i and d_i are the (tensile) force (in [kN]) and the deformation (elongation) (in [m]), respectively, of the ith cable element, $\hat{\partial}$ is the generalized gradient and S_i is the superpotential function, see Panagiotopoulos [29].

Next, dynamic equilibrium for the assembled structural system with cables is expressed by the usual matrix relation:

$$\mathbf{M}\ddot{\mathbf{u}} + \mathbf{C}(\dot{\mathbf{u}}) + \mathbf{K}(\mathbf{u}) = \mathbf{p} + \mathbf{As}. \tag{2}$$

Here $\mathbf{u}$ and $\mathbf{p}$ are the displacement and the load time dependent vectors, respectively, and $\mathbf{s}$ is the cable stress vector. $\mathbf{M}$ is the mass matrix and $\mathbf{A}$ is a transformation matrix. The damping and stiffness terms, $\mathbf{C}(\dot{\mathbf{u}})$ and $\mathbf{K}(\mathbf{u})$, respectively, concern the general non-linear case. Dots over symbols denote derivatives with respect to time. For the case of ground seismic excitation $\mathbf{x}_g$, the loading history term $\mathbf{p}$ becomes $\mathbf{p} = -\mathbf{M}\mathbf{r}\ddot{\mathbf{x}}_g$, where $\mathbf{r}$ is the vector of stereostatic displacements [3, 4].

The above relations (1) and (2), combined with the initial conditions, consist the problem formulation, where, for given p and/or $\ddot{\mathbf{x}}_g$, the vectors $\mathbf{u}$ and $\mathbf{s}$ have to be computed. From the strict mathematical point of view, using (1) and (2), we can formulate the problem as a dynamic hemivariational inequality one by following [24, 29, 36, 37] and investigate it.

2.2 Numerical Treatment of the Problem

In Civil Engineering practical cases, a numerical treatment of the problem is realized by applying to the above constitutive relations (1) a piecewise linearization based on experimental investigations [9, 12]. So, simplified stress-deformation constitutive diagrammes are used [9, 12, 40, 41]. Such a force-displacement constitutive diagramme is shown in Fig. 1, where F is the force in [kN], Δ is the displacement in [m], and K_e is the elastic stiffness in [kN/m]. By lower indices y, c, r and u the yield, cracking, remaining and ultimate quantities, respectively, are denoted. Hardening and softening correspond to positive α (≥ 0) and negative γ (≤ 0), respectively.

Further, for the above piecewise linearization of the constitutive relations, use is made of the plastic multipliers approach as in elastoplasticity [21, 29]. Applying a direct time-integration scheme, in each time-step a relevant non-convex linear complementarity problem [18–20] of the following matrix form is solved:

$$\mathbf{v} \geq \mathbf{0}, \qquad \mathbf{D}\mathbf{v} + \mathbf{a} \leq \mathbf{0}, \qquad \mathbf{v}^{\mathrm{T}} \cdot (\mathbf{D}\mathbf{v} + \mathbf{a}) = 0. \tag{3}$$

So, the nonlinear Response Time-History (RTH) can be computed for a given seismic ground excitation.

An alternative approach for treating numerically the problem is the incremental one. So, the matrix incremental dynamic equilibrium is expressed by relation:

$$M\Delta\ddot{u} + C\Delta\dot{u} + K_T\Delta u = -M\Delta\ddot{u}_g + A\Delta s \tag{4}$$

where C and $K_T(u)$ are the damping and the tangent stiffness matrix, respectively, and $\ddot{u}_g$ is the ground seismic acceleration. On such incremental approaches is based the structural analysis software Ruaumoko [3], which is applied hereafter.

Ruaumoko software [3] uses the finite element method and permits an extensive parametric study on the inelastic response of structures. For practical applications, an efficient library of hysteretic behaviour models is available. Concerning the

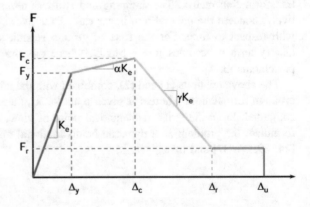

Fig. 1 Representative simplified force-displacement constitutive diagramme (backbone)

time-discretization, implicit or explicit approaches can be used. Here the Newmark implicit scheme is chosen and Ruaumoko is used to provide results which are related to the following critical parameters: local or global structural damage, maximum displacements, interstorey drift ratios, development of plastic hinges and response using the incremental dynamic analysis (IDA) method [40, 41].

Details of the approach are described in [19, 20], where the seismic response of cable-braced RC systems to multiple earthquakes is investigated. As concerns multiple earthquakes, it is reminded [11] that current seismic codes suggest the exclusive adoption of the isolated and rare "design earthquake", while the influence of repeated earthquake phenomena is ignored. This is a significant drawback for the realistic design of building structures in seismically active regions, because, as it is shown in [11], real seismic sequences have accumulating effects on various damage indices.

2.3 Comparative Investigations for the Cable-Strengthening Versions by Damage Indices

The decision about a possible strengthening for an existing structural, damaged by a seismic event, can be taken after a relevant assessment of the existing Cultural Heritage structure [8]. This is obtained by using in situ structural identifications and evaluating suitable damage indices [6, 8, 30, 38]. Further, a comparative investigation of structural responses due to various seismic excitations can be used. So, the system is considered for various cases, with or without strengthening by cable-bracings.

Among the several response parameters, the focus is on the overall structural damage index (OSDI) [6, 25, 30]. This is due to the fact, that this parameter summarises all the existing damages on columns and beams of reinforced concrete frames in a single value, which is useful for comparison reasons.

In the OSDI model after Park and Ang [30], the global damage is obtained as a weighted average of the local damage at the section ends of each structural element or at each cable element. First the *local* damage index DI_L is computed by the following relation:

$$DI_L = \frac{\mu_m}{\mu_u} + \frac{\beta}{F_y d_u} E_T \qquad (5)$$

where μ_m is the maximum ductility attained during the load history, μ_u the ultimate ductility capacity of the section or element, β a strength degrading parameter, F_y the yield force of the section or element, E_T the dissipated hysteretic energy, and d_u the ultimate deformation.

Next, the dissipated energy E_T is chosen as the weighting function and the *global* damage index DI_G is computed by using the following relation:

$$DI_G = \frac{\sum_{i=1}^{n} DI_{Li} E_i}{\sum_{i=1}^{n} E_i} \qquad (6)$$

where DI_{Li} is the local damage index after Park and Ang at location i, E_i is the energy dissipated at location i and n is the number of locations at which the local damage is computed.

The proposed numerical approach has been successfully calibrated in [15, 16] by using available experimental results from literature [23].

3 Numerical Example

3.1 Description of the Considered Cultural Heritage RC Structural System

The old industrial reinforced concrete frame F0 of Fig. 2 is considered to be subjected to a multiple ground seismic excitation. The frame, of concrete class C 16/20, was initially constructed without cable-bracings.

Due to various extreme actions (environmental etc.), corrosion and cracking has been taken place, which has caused a strength and stiffness degradation. The stiffness reduction due to cracking results [7–10, 31, 33, 34] to effective stiffness of 0.60 I_g for the two external columns, 0.80 I_g for the internal columns and 0.40 I_g for the beams, where I_g is the gross inertia moment of their cross-section. Using Ruaumoko software [3], the columns and the beams are modeled using prismatic frame elements [4]. Nonlinearity at the two ends of RC members is idealized using one-component plastic hinge models, following the Takeda hysteresis rule. Interaction curves (M-N) for the critical cross-sections of the examined RC frame have been computed.

After the seismic assessment [8], it was decided the frame F0 to be strengthened by ties. The X-cable-bracings system, shown in Fig. 3, has been proposed as the optimal one in order the frame F0 to be seismically upgraded. The system of the frame with the X-bracing diagonal cable-elements shown in Fig. 3 is denoted as system F6 (the dimensions are: L = 5 m and H = 3 m).

The cable constitutive law, concerning the unilateral (slackness), hysteretic, fracturing, unloading-reloading etc. behavior, is depicted in Fig. 4. The cable elements have a cross-sectional area $F_c = 18$ cm^2 and they are of steel class S220 with elasticity modulus $E_c = 200$ GPa.

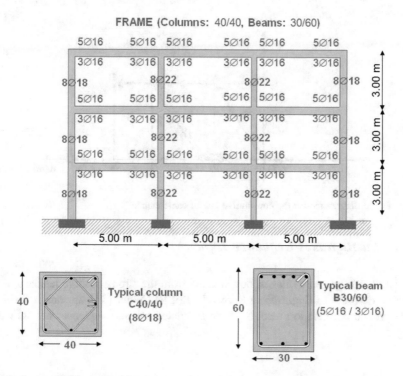

Fig. 2 System F0: the industrial RC frame without cable-strengthening

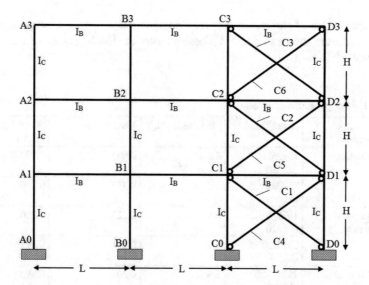

Fig. 3 System F6: the industrial RC frame cable-strengthened with X-bracings

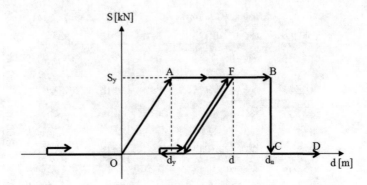

Fig. 4 The diagramme for the constitutive law of cable-elements

3.2 Earthquakes Sequence Input

A list of multiple earthquakes, downloaded from the strong motion database of the
Pacific Earthquake Engineering Research (PEER) Center [32], appears in Table 1.
The ground acceleration records of the simulated seismic sequences are shown in
Fig. 5.

3.3 Representative Results

After application of the herein proposed computational approach, some representative results are shown in Table 2. These concern the Coalinga case of the seismic
sequence of Table 1.

Table 1 Multiple earthquakes data

No	Seismic sequence	Date (Time)	Magnitude (M_L)	Recorded PGA(g)	Normalized PGA(g)
1	Coalinga	1983/07/22 (02:39)	6.0	0.605	0.165
		1983/07/25 (22:31)	5.3	0.733	0.200
2	Imperial valley	1979/10/15 (23:16)	6.6	0.221	0.200
		1979/10/15 (23:19)	5.2	0.211	0.191
3	Whittier narrows	1987/10/01 (14:42)	5.9	0.204	0.192
		1987/10/04 (10:59)	5.3	0.212	0.200

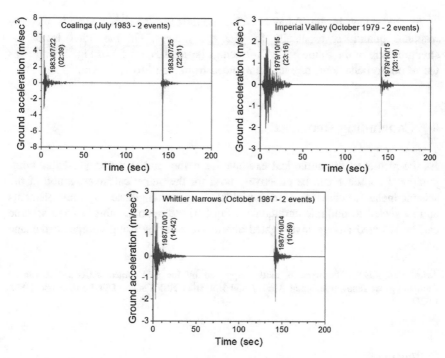

Fig. 5 Ground acceleration records of the simulated seismic sequences

System	Events	DI_G	DI_L	u_{top} (cm)
(0)	(1)	(2)	(3)	(4)
F0	Event E_1	0.2534	0.179	2.087
	Event E_2	0.508	0.522	2.984
	Event $(E_1 + E_2)$	0.5474	0.588	3.517
F6	Event E_1	0.108	0.037	1.126
	Event E_2	0.202	0.114	1.278
	Event $(E_1 + E_2)$	0.228	0.125	1.422

Table 2 Representative response quantities for the systems F0 and F6

In column (1) of the Table 2, the Event E_1 corresponds to Coalinga seismic event of 0.605 g PGA, and Event E_2 to 0.733 g PGA, (g = 9.81 m/s^2). The sequence of events E_1 and E_2 is denoted as Event $(E_1 + E_2)$.

In table column (2) the Global Damage Indices DI_G and in column (3) the Local Damage Index DI_L for the bending moment at the left fixed support A0 of the frame are given. Finally, in the column (4), the absolutely maximum horizontal top roof displacement u_{top} is given.

As the values in the Table 2 show, multiple earthquakes generally increase response quantities, especially the damage indices. On the other hand, the strengthening of the frame F0 by X-bracings (system Frame F6 of Fig. 3) improves the response behaviour, e.g. u_{top} is reduced from 3.517 to 1.422 cm.

4 Concluding Remarks

As the results of the numerical example have shown, the herein presented computational approach can be effectively used for the numerical investigation of the seismic inelastic behaviour of industrial RC frames strengthened by cable elements and subjected to multiple earthquakes. Further, the optimal cable-bracing scheme can be selected among investigated alternative ones by using computed damage indices.

Acknowledgments The work is partly supported by the FP7 project AComIn: Advanced Computing for Innovation, grant 316087 and Bulgarian NSF Grants, DFNI I-01/5 and DFNI I-02/9.

References

1. Asteris, P.G., Plevris, V. (eds.): Handbook of Research on Seismic Assessment and Rehabilitation of Historic Structures. IGI Global (2015)
2. Bertero, V.V., Whittaker, A.S.: Seismic upgrading of existing buildings. 5as Jornadas Chilenas de Sismología e Ingeniería Antisísmica, **1**, 27–46 (1989)
3. Carr, A.J.: RUAUMOKO—Inelastic Dynamic Analysis Program. Dep. Civil Engineering, University of Canterbury, Christchurch, New Zealand (2008). 2008
4. Chopra, A.K.: Dynamics of Structures: Theory and Applications to Earthquake Engineering. Pearson Prentice Hall, New York (2007)
5. Dritsos, S.E.: Repair and Strengthening of Reinforced Concrete Structures (in greek). University of Patras, Greece (2001)
6. Elenas, A., Liolios, Ast., Vasiliadis, L., Favvata, M., Liolios, A.: Seismic intensity parameters as damage potential descriptors for life-cycle analysis of buildings. In: IALCCE 2012: Proceedings Third International Symposium Life-Cycle Civil Engineering, Vienna, Austria, Oct 2012, CRC Press, Taylor & Francis, London (2012)
7. Eurocode 8.: Design of Structures for Earthquake Resistance, Part 3: Assessment and Retrofitting of Buildings, (EC8-part3), EN 1998–3, Brussels (CEN 2004)
8. Fardis, M.N.: Seismic Design, Assessment and Retrofitting of Concrete Buildings: Based on EN-Eurocode 8. Springer, Berlin (2009)
9. FEMA P440A.: Effects of Strength and Stiffness Degradation on the Seismic Response of Structural Systems. U. S. Department of Homeland Security, Federal Emergency Management Agency (2009)
10. Greek Retrofitting Code-(KANEPE).: Greek Organization for Seismic Planning and Protection (OASP), Greek Ministry for Environmental Planning and Public Works. Athens, Greece, 2013. (In Greek): www.oasp.gr (2013)

11. Hatzigeorgiou, G., Liolios, A.: Nonlinear behaviour of RC frames under repeated strong ground motions. Soil. Dyn. Earthq. Eng. **30**, 1010–1025 (2010)
12. Ibarra, L.F., Medina, R.A., Krawinkler, H.: Hysteretic models that incorporate strength and stiffness deterioration. Earthq. Eng. Struct. Dynam. **34**(12), 1489–1511 (2005)
13. Liolios A.: A numerical approach for the earthquake response of historic monuments under unilateral and non-smooth behavior of structural elements. In: Biscontin, G., Moropoulou, A., Erdik, M., Delgado Rodrigues, J. (eds.) Proceedings INCOMARECH-RAPHAEL 2nd Workshop, Istanbul, Bogazici University, 2–3 Oct 1998, PACT 56, vol. 2, pp. 69–77, Techn. Chamber of Greece (TEE), Athens (1998)
14. Liolios, A., Yeroyianni, M., Stavrakakis, E., Karavezyroglou, M.: On the analysis of historical stone arch bridges-the inequality numerical approach. In: Moropoulou, A., Erdik, M. et al. (Sci. Eds.), PACT 59, Journal of European Study Group on Physical, Chemical and Mathematical Techniques Applied to Archaeology, vol. 59, pp. 123–131 (2000)
15. Liolios, A.: A computational investigation for the seismic response of RC structures strengthened by cable elements. In: Papadrakakis, M., Papadopoulos, V., Plevris, V. (eds.) Proceedings of COMPDYN 2015: Computational Methods in Structural Dynamics and Earthquake Engineering, 5th ECCOMAS Thematic Conference, Crete Island, Greece, 25–27 May 2015, vol. II, pp. 3997–4010 (2015)
16. Liolios A., Chalioris, C.: Reinforced concrete frames strengthened by cable elements under multiple earthquakes: A computational approach simulating experimental results. In: Proceedings of 8th GRACM International Congress on Computational Mechanics, Volos, 12–15 July 2015. (http://8gracm.mie.uth.gr/Papers/Session%20D3-B1/A.%20Liolios.pdf) (2015)
17. Liolios, A., Chalioris, C., Liolios, A., Radev, S., Liolios, K.: A computational approach for the earthquake response of cable-braced reinforced concrete structures under environmental actions. In: Lecture Notes in Computer Science, LNCS, vol. 7116, pp. 590–597. Springer, Berlin Heidelberg (2012)
18. Liolios, A.: An inequality numerical modeling for the seismic analysis of monumental complexes and masonry structures: The Hagia Sophia case. In: Moropoulou, A., Erdik, M. et al (Sci. Eds.), PACT 59, Journal of European Study Group on Physics Chemistry Biology and Mathematical Techniques Applied to Archaeology, vol. 59, pp. 65–74 (2000)
19. Liolios, Ast, Liolios, Ang, Hatzigeorgiou, G.: A numerical approach for estimating the effects of multiple earthquakes to seismic response of structures strengthened by cable-elements. J. Theor. Appl. Mech. **43**(3), 21–32 (2013)
20. Liolios, A., Elenas, A., Liolios, A., Radev, S., Georgiev, K., Georgiev, I.: Tall RC buildings environmentally degradated and strengthened by cables under multiple earthquakes: a numerical approach. In: Dimov, I. et al (eds.) Numerical Methods and Applications, NMA2014, (pp. 187–195). vol. LNCS 8962. Springer, Berlin (2015)
21. Maier, G.: Incremental elastoplastic analysis in the presence of large displacements and physical instabilizing effects. Int. J. Solids Struct. **7**, 345–372 (1971)
22. Markogiannaki, O., Tegos, I.: Strengthening of a multistory R/C building under lateral loading by utilizing ties. Appl. Mech. Mater. **82**, 559–564 (2011)
23. Massumi, A., Absalan, M.: Interaction between bracing system and moment resisting frame in braced RC frames. Arch. Civil Mech. Eng. **13**(2), 260–268 (2013)
24. Mistakidis, E.S., Stavroulakis, G.E.: Nonconvex optimization in mechanics. In: Smooth and nonsmooth algorithmes, heuristic and engineering applications. Kluwer, London (1998)
25. Mitropoulou, C.C., Lagaros, N.D., Papadrakakis, M.: Numerical calibration of damage indices. Adv. Eng. Softw. **70**, 36–50 (2014)
26. Moropoulou, A.: Current trends towards the sustainable construction. The relation between environment and concrete. In: Technical chamber of Greece -TEE (ed.) Proceedings of 16th Hellenic Concrete Conference, Cyprus (2009)
27. Moropoulou, A., Labropoulos, K.C., Delegou, E.T., Karoglou, M., Bakolas, A.: Non-destructive techniques as a tool for the protection of built cultural heritage. Constr. Build. Mater. **48**, 1222–1239 (2013)

28. Moropoulou, A., Bakolas, A., Spyrakos, C., Mouzakis, H., Karoglou, A., Labropoulos, K., Delegou, E.T., Diamandidou, D., Katsiotis, N.K.: NDT investigation of holy sepulchre complex structures. In: Radonjanin, V., Crews, K. (eds) Proceedings of Structural Faults and Repair 2012, Proceedings in CD-ROM (2012)
29. Panagiotopoulos, P.D.: Hemivariational inequalities. In: Applications in Mechanics and Engineering. Springer, Berlin, New York (1993)
30. Park, Y.J., Ang, A.H.S.: Mechanistic seismic damage model for reinforced concrete. J. Struct. Division ASCE 111(4), 722–739 (1985)
31. Paulay, T., Priestley, M.J.N.: Seismic design of reinforced concrete and masonry buildings. Wiley, New York (1992)
32. PEER (2011) Pacific Earthquake Engineering Research Center. PEER Strong Motion Database. http://peer.berkeley.edu/smcat
33. Penelis, G., Penelis, G.: Concrete Buildings in Seismic Regions. CRC Press (2014)
34. Priestley, M.J.N., Seible, F.C., Calvi, G.M.: Seismic Design and Retrofit of Bridges. Wiley (1996)
35. Spyrakos, C.C., Maniatakis Ch. A.: Retrofitting of a historic masonry building. In: 10th national and 4th international scientific conference on planning, design, construction and renewal in the construction industry (iNDiS 2006), Novi Sad, 22–24 Nov 2006, 535–544 (2006)
36. Stavroulaki, M.E., Stavroulakis, G.E., Leftheris, B.: Modelling prestress restoration of buildings by general purpose structural analysis optimization software. Comput. Struct. 62(1), 81–92 (1997)
37. Stavroulakis, G.E.: Computational nonsmooth mechanics: variational and hemivariational inequalities. Nonlinear Anal. Theory Methods Appl. 47(8), 5113–5124 (2001)
38. Strauss, A., Frangopol, D.M., Bergmeister, K.: Assessment of existing structures based on identification. J. Struct. Eng. ASCE 136(1), 86–97 (2010)
39. Tegos, I., Liakos, G., Tegou, S., Roupakias, G., Stilianidis, K.: An alternative proposal for the seismic strengthening of existing R/C buildings through tension-ties. In: Proceedings, 16th Pan-Hellenic Concrete Conference, Cyprus (2009)
40. Vamvatsikos, D., Cornell, C.A.: Incremental dynamic analysis. Earthq. Eng. Struct. Dyn. 31, 491–514 (2002)
41. Vamvatsikos, D., Cornell, C.A.: Direct estimation of the seismic demand and capacity of oscillators with multi-linear static pushovers through IDA. Earthq. Eng. Struct. Dyn. 35(9), 1097–1117 (2006)
42. Zienkiewicz, O.C., Taylor, R.L.: The finite element method for solid and structural mechanics. Butterworth-heinemann (2005)

On the Performance of Query Rewriting in Vertically Distributed Cloud Databases

Jens Kohler, Kiril Simov, Adrian Fiech and Thomas Specht

Abstract Cloud Computing with its dynamic pay as you go model and scalability characteristics promises computing on demand with associated cost savings compared to traditional computing architectures. This is a promising computing model especially in the context of *Big Data*. However, renting computing capabilities from a cloud provider means the integration of external resources into the own infrastructure and this requires a great amount of trust and raises new data security and privacy challenges. With respect to these still unsolved problems, this work presents a *fixed vertical partitioning and distribution approach* that uses traditional relational data models and distributes the corresponding partitions vertically across different cloud providers. So, every cloud provider only gets a small, but defined (and therefore *fixed*) logically independent chunk of the entire data, which is useless without the other parts. However, the distribution and the subsequent join of the data suffer from great performance losses, which are unbearable in practical usage scenarios. The novelty of our approach combines the well-known vertical database partitioning technique with a distribution logic that stores the vertical partitions at different cloud computing environments. Traditionally, vertical as well as horizontal partitioning approaches are used to improve the access to database data, but these approaches use dynamic and automated partitioning algorithms and schemes based on query workloads, data volumes, network bandwidth, etc. In contrast to

J. Kohler (✉) · K. Simov
Linguistic Modelling Department, IICT-BAS, Sofia, Bulgaria
e-mail: jkohler@hs-mannheim.de

K. Simov
e-mail: kivs@bultreebank.org

A. Fiech
Computer Science Department, Memorial University St. John's,
Newfoundland, Canada
e-mail: afiech@mun.ca

J. Kohler · T. Specht
Institute for Enterprise Computing, University of Applied Sciences Mannheim,
Mannheim, Germany
e-mail: t.specht@hs-mannheim.de

© Springer International Publishing Switzerland 2016
S. Margenov et al. (eds.), *Innovative Approaches and Solutions in Advanced Intelligent Systems*, Studies in Computational Intelligence 648,
DOI 10.1007/978-3-319-32207-0_5

this, our approach uses a *fixed user-defined vertical partitioning* approach, where no two critical attributes of a relation should be stored in a single partition. Thus, our approach aims at improving data security and privacy especially in public Cloud Computing environments, but raises the challenging research question of how to improve the data access to such fixed *user-partitioned and distributed database environments*. In this paper, we outline a query rewriting approach that parallelizes queries and joins in order to improve the query performance. We implemented our *fixed partitioning and distribution approach* based on the TPC-W benchmark and we finally present the performance results in this work.

Keywords Vertically distributed cloud databases · Query performance · Query rewriting

1   Introduction

Cloud Computing with its dynamic pay as you go model and scalability characteristics promises computing on demand with associated cost savings compared to traditional computing architectures. This is a promising computing model especially in the context of *Big Data*. Here, large computing capabilities can be rented on demand for analyzing, processing or storing *Big Data* volumes. Besides the increasing network bandwidth nowadays, new ways of storing such vast amounts of data (i.e. In-Memory) and new methods of representing information and knowledge (i.e. real-time analytics) have to be investigated. With respect to the latter topic, this work presents an approach that uses traditional relational data models and distributes the corresponding data vertically across different cloud providers. It has to be noted that renting computing or storage capabilities from external cloud providers requires the integration of a 3rd party into the entire architecture. It demands a great amount of trust when sensitive data should be stored at an external location somewhere in the cloud. This data security and protection challenges are still unsolved despite the fact that there are various approaches and research works that try to find adequate solutions. Basically, there are two directions that address such data security and privacy issues. Firstly, there is encryption of data and secondly, there is data distribution. Our work is based on security by data distribution, namely a vertical data distribution. In general, we vertically partition and distribute database data across various cloud providers in a way such that every provider only receives a small chunk of the data that is useless without the other chunks. This distribution suffers actually from a great performance loss, which is unbearable in practical usage scenarios.

The novelty of our approach combines vertically database partitioning with a logic that distributes the vertical partitions to different cloud computing environments. Encryption of data is intentionally not used in this approach at the moment because of the additional encryption effort. However, the future development with different data access strategies will show, if the overall performance can be improved to a level that is in the same order of magnitude as a non-partitioned approach. Then, encryption becomes a viable approach to further enhance the level of security. Traditionally, vertical as well as horizontal partitioning approaches are used to improve the access to database data (i.e. queries and data manipulations), but these approaches e.g. [1–4] use dynamic and automated partitioning algorithms and schemes based on query workloads, data volumes, network bandwidth, etc. In contrast to these works, our approach uses a *fixed user-defined vertical partitioning* approach, where no two critical attributes of a relation should be stored in a single partition. Thus, our approach aims at improving data security and privacy especially in public Cloud Computing environments, but raises the challenging research question of how to improve the data access to such a *fixed user-partitioned and distributed database environment*. Finally, this fixed data partitioning and distribution approach makes it difficult to compare the performance with the above-mentioned dynamic horizontal and vertical partitioning approaches.

In this paper we present a distributed data access strategy that fetches data in parallel to increase the overall access performance. This query rewriting approach has the promising possibility to run the distributed queries in parallel in order to increase the overall query performance. We used a similar approach in one of our caching works [5] to build a client-based cache without adapting the queries but with transferring entire database tables into the cache.

Motivating Example

To illustrate our approach in more detail we consider the following motivating example, based on the TPC-W CUSTOMER database table [6]:

CUSTOMER (C_ID, C_UNAME, C_PASSWD, C_FNAME C_LNAME, C_ADDR_ID, C_PHONE, C_EMAIL, C_SINCE, C_LAST_LOGIN, C_LOGIN, C_EXPIRATION, C_DISCOUNT, C_BALANCE, C_YTD_PMT, C_BIRTHDATE, C_DATA)

Basically, the table represents customer data that is necessary in a common web shop with a customer id (C_ID) as primary key. In our vertically distributed cloud setup, we store this table (containing i.e. sensitive data as year-to-day payment (C_YTD_PMT) or the current account balance (C_BALANCE)) logically distributed in two different public clouds. An exemplified distribution with 2 partitions could be stated as follows (whereas in a real world application scenario the distribution should consider that no sensitive columns are stored in the same partition):

CUSTOMER_p1 (C_ID, C_UNAME, C_PASSWD, C_FNAME C_LNAME, C_ADDR_ID, C_PHONE, C_EMAIL, C_SINCE)

CUSTOMER_p2 (C_ID, C_LAST_LOGIN, C_LOGIN, C_EXPIRATION, C_DISCOUNT, C_BALANCE, C_YTD_PMT, C_BIRTHDATE, C_DATA)

In this example the replicated primary key (C_ID) is used to join the customer data together. A user query is now formulated in *Hibernate Query Language* (HQL) to take advantage of the database abstraction of Hibernate. Thus, we are able to support different database systems with their different SQL implementations. It is even possible to use different database systems for the respective partitions simultaneously.

A sample query in Java could be formulated as follows:

List <Customer> customers = session.createCriteria(Customer.class).list();

This query clearly shows the main advantage of HQL, as it deals with objects rather than tables and relations. Above that, we use *Hibernate Interceptors* to adapt the HQL queries to our vertical distribution setup. The foundation for our approach is a relational database table, which is vertically partitioned in two relations and these relations are then distributed across (ideally) different clouds. Due to the usage of Hibernate as an *object-relational mapper* (ORM) and its so-called *domain classes*, which bridge the gap between the object oriented and the relational paradigm, both the partitioning and the distribution are transparent for client applications. Furthermore, as the partitioned and distributed relations are only accessed via their corresponding domain class, the entire SQL query and manipulation standard can be used. Thus, integrating the framework into existing applications requires to define a partitioning and distribution scheme, which is done via an XML file, then the partitioning and distribution scheme must be applied to the relation(s) and finally, the corresponding domain classes must be enhanced with a framework specific annotation. Currently, the framework supports MySQL and Oracle as database systems for the partitioning and Amazon EC2 and Eucalyptus as *Infrastructure as a Service* (*IaaS*) cloud vendors. Moreover, it is limited to database tables that have a defined primary key column and further constraints, such as foreign keys, unique constraints, etc. are currently not maintained during the partitioning and distribution phase.

This current implementation resulted in the SeDiCo framework (A Secure and Distributed Cloud Data store). With this we demonstrated the technological feasibility of the presented vertical distribution. However, in practical real-world usage scenarios our approach suffers from great performance losses [7]. At the moment we follow two different directions in order to accelerate the data query and manipulation. The first approach uses caching [8] and its variations, i.e. server-based, client-based, distributed caching etc. Our second approach follows the data distribution principle and distributes the queries against the databases as well [5]. This work follows the query distribution principle and aims at finding a generic way to rewrite queries such that they fit into our vertical distribution environment.

The main contributions of this paper can be summarized as follows:

- We analyze and measure the performance of distributed queries against vertically distributed cloud databases. Therefore, we implement the introductory example based on the TPC-W benchmark and its CUSTOMER table. We partition this table vertically and distribute the partitions across two different clouds.
- We also evaluate the 3 well-known basic join algorithms (i.e. *nested loops, hash and sorted-merge join*) and compare their performance with our *fixed vertical partitioning and distribution setup*.
- This work outlines an approach how HQL and traditional SQL can be transferred and rewritten such that queries can be adapted to vertically partitioned data models.

- Finally, this work is a foundation for our future work, in which we aim at finding a generic method that is able to rewrite arbitrary SQL queries and adapt them to our vertical distribution approach. Based on the performance results of this work, we are able to decide whether the query rewriting approach is worth pursuing any further.

The remainder of this work is organized as follows: after this introduction we present other work that uses different approaches in order to address similar challenges. We then define and formalize our problem in Sect. 3, before we present our approach and its implementation to tackle the problem of the slow data access in vertically distributed databases in Sect. 4. In Sect. 5, we evaluate our Java-based implementation and present our achieved performance gain. We also contrast the results of this work to our previous work in Sect. 6 and we interpret the results and give an outlook about our future work plans.

2 Related Work

Other work in the context of manipulating or adapting user queries can be classified into 3 main fields: *query manipulation* based on user profiles [9] or due to different vendor-specific SQL implementations [10], *query adaption* in order to adapt the query to a generic database scheme [11] use a so-called *Universal Table* as a generic database mapping scheme and query *decompositioning* to answer user queries with an optimal subset of sub-queries or with the creation of views such as e.g. *Bucket* or *MiniCon* [12–15]. These are all interesting approaches and the algorithms to analyze and adapt the queries are of particular interest for our query rewriting approach. However, the generation of a generic mapping scheme that merges different database partitions in one unique scheme or the generation of different views that contain the complete requested tuples would contradict our distribution approach because of the one single location where all data are joined together.

Concerning the concrete query rewriting algorithms, [9] augments the queries with data from previously collected user data based on a fuzzy logic. Other approaches like [11] are inspired by database optimizers that generate a query tree in order to gain some speed-up. Intercepting a database optimizer and modifying the query tree is also a viable approach but transferred to our distributed approach with different database vendors and thus, different and distributed database optimizers, it is not very promising.

Above that, on a more technological level, there are approaches like [16] that analyze the query in compiled Java byte code and generate optimized SQL queries. This is insofar an interesting approach as it also deals with an object-relational mapper to abstract from the relational database model. However, in order to recognize a SQL query in the compiled byte code, Queryll introduces a new SQL-like query language. We consider this as a major drawback as we aim to rely completely on standard SQL. Another technologically focused approach are vendor-specific SQL rewrite plugins such as MySQL Query Rewrite Plugins [17] or Oracle Query Rewrite [18]. However, there are two major drawbacks of these approaches, firstly, the restricted access and modifiability of these tools and secondly, the fact that they are vendor-specific and not compatible with various different database systems. With respect to this approach, there is another work originating from the field of data provenance that aims to trace meta-information of database data (creation, updates and deletion, etc.) [19]. Here, the authors augment the SQL standard with just one keyword in order to identify queries that should be rewritten. Finally, the rewritten queries are answered with views like the approaches discussed above, but we consider the extension of the SQL queries with just a single keyword as a viable approach for our work as well.

Besides these query rewriting approaches, this paper uses the 3 basic well-known and exhaustively examined join algorithms, namely, *nested loops, hash and sorted merge joins* for the evaluation of the approach. Discussing and outlining these approaches would go beyond the scope of this work hence the following literature is recommended for the interested reader [1–4].

3   Problem Formulation

The basic problem that we address in this work is the transformation of a database query into vertically distributed queries that run against vertically partitioned data sets in order to increase the overall query performance. We first have to mention that we focus our approach on conjunctive queries (i.e. queries that have their predicates connected with an AND) as they are the most common forms of queries used in practical scenarios for both, OLTP and OLAP workloads [13].

The formalization of the problem according to relational algebra [20] is as follows:

Our starting point is an HQL query:

$$\text{List} < \text{Classname} > \text{objects} = \text{Hibernate.session.createQuery(}$$
$$\text{``SELECT columnX, columnY FROM Classname} \qquad (1)$$
$$\text{WHERE columnX} = \;'\text{x}' \text{ AND columnY} = \;'\text{y}')$$

which can be formalized as:

$$\text{Objects(objectID, columnX, columnY)} < \text{-}$$
$$\pi_{(\text{objectID, columnX, columnY})} \left(\sigma_{(\text{columnX=x AND columnY=y})} \right) (\text{TABLE\_NAME}) \qquad (2)$$

The Hibernate framework then transfers this query into native standard SQL:

$$
\begin{aligned}
&\text{SELECT id, columnX, columnY} \\
&\text{FROM TABLE\_NAME} \\
&\text{WHERE columnX} = {}'X'\text{ AND columnY} = {}'Y'.
\end{aligned}
\tag{3}
$$

In relational calculus this query is written as:

$$
\pi_{(\text{objectID, columnX, columnY})}\left(\sigma_{(\text{columnX}=x\text{ AND columnY}=y)}\right)^{(\text{TABLE\_NAME})}
\tag{4}
$$

Based on this query we aim at finding a generic rewriting rule, such that the query can be run against our vertically distributed database tables. Finally, the resulting queries should have the following forms:

$$
\text{Q1: ResultSet\_partition1} <\!\!-\pi_{(\text{id, columnX})}\left(\sigma_{(\text{columnX}=x)}\right)^{(\text{TABLE\_NAME\_partition1})}
\tag{5}
$$

$$
\text{Q2: ResultSet\_partition2} <\!\!-\pi_{(\text{id, columnY})}\left(\sigma_{(\text{columnY}=y)}\right)^{(\text{TABLE\_NAME\_partition2})}
\tag{6}
$$

And the final result set (RS_{final}) of the query is the intersection of the distributed queries based on their unique identifies (id):

$$
RS_{final} = \text{ResultSet\_partition1}\ \Pi\ \text{ResultSet\_partition2}
\tag{7}
$$

The intersection described in (7) is a great advantage of conjunctive queries as non-conjunctive queries (e.g. query parameters that are connected with OR, or *average* or *sum* operations) use the union of the partitions. This shows that non-conjunctive queries are slower than conjunctive ones as optimization possibilities for the creation of a union are limited [13].

The procedure from step 4 to steps 5 and 6 is done via query rewriting. Here, the original query is analyzed and based on the previously defined partitioning and distribution scheme, which is done by the framework user in an XML file, we are able to distinguish which attribute belongs to which partition. Moreover, this XML file determines the location of the partitions in the clouds and thus both, the query rewriting and the tuple reconstruction are done.

Another issue that we have to consider is the fact that Hibernate deals with objects instead of tables and relations. This is commonly referred to as *impedance mismatch* which is well-described and analyzed in [21]. Hence, it is not enough to just merge the result sets into an overall result set that meets all conjunctive criteria; we also have to create domain objects from the overall result set. Recall that query (1) uses the class name in the FROM clause instead of the table name and it also uses class properties instead of table columns as query parameters. Thus, we retrieve concrete domain objects instead of a generic result set. Nonetheless, the *Query-Parser* (Fig. 1)

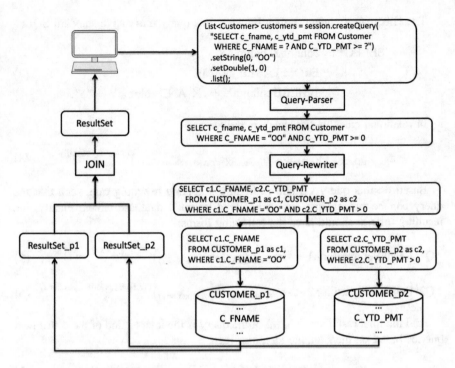

Fig. 1 Query-rewriting approach

analyzes and distributes the native SQL statement derived from the original HQL query (1) and here (in (3) and also in (5) and (6)) we receive generic result sets. So basically the last step is the transformation of the result set (RS_{final}) into a list of domain objects and this can be formalized as follows:

$$List < Classname > objects =$$
$$RS_{final} <-$$
$$\pi_{(id, columnX, columnY)} \left(\sigma_{(columnX=x \: AND \: columnY=y)} \right) \quad (8)$$
$$(TABLE\_NAME\_partition1, TABLE\_NAME\_partition2)$$

Finally, this step verifies that the merged result set RS_{final} is equal to the list of domain objects (List<Classname> objects), i.e.

$$List < Classname > objects = RS_{final} \quad (9)$$

which can be used to prove that the distributed queries return the same result as the non-distributed original query.

In the following section we now outline our approach to solve this formalized problem.

4 Approach

Figure 1 depicts our syntactic query rewriting approach with a concrete scenario derived from our motivating example above.

The starting point is a query formulated in either HQL or native SQL. In order to remain independent from a specific database vendor, HQL queries should be preferred. The *Query-Parser* then analyzes the query syntactically and incorporates the actual query parameters. After that the *Query-Rewriter* parses the query and augments it with the partitioning information provided by the SeDiCo framework in form of an XML file. Based on this XML file the Query-Rewriter is able to decide which query predicate belongs to which partition and thus we are able to split the original query into sub-queries that then run against their corresponding partition. The entire approach is based on the assumption that this distribution has two substantial performance improvements. Firstly, the results are restricted faster and more easily as the potential overall result set is bounded to the number of tuples that both partitions return. As we deal with conjunctive queries, only the intersection (ResultSet_p1 ∏ ResultSet_p2, see Fig. 1) of the two result sets have to be checked for the AND condition and not the entire data set. Secondly, the distributed queries and the join of the result sets are performed in parallel in different threads. Thus, the partitions are queried in two parallel threads and the result sets are merged (based on the AND condition) in parallel threads based on the number of available CPU cores of the querying client.

5 Evaluation

The implementation of our approach is based on an adaption of the TPC-W Benchmark [6]. As we deal with vertically distributed data, we just used the *Customer* table of TPC-W and partitioned it into 2 partitions that are equal in the number of the columns (see also the motivating example in Sect. 1). The query we used for this evaluation is as follows (see also Fig. 1):

> SELECT C_FNAME, C_YTD_PMT
>
> FROM CUSTOMER
>
> WHERE C_FNAME = "OO" AND C_YTD_PMT > 0

where we made sure that one query parameter (C_FNAME) is from one partition and the other parameter (C_YTD_PMT) from the other partition. Above that, we restricted the query to just retrieve the parameters in the result set (RS_{final}) and not the entire domain object (*customer*). The possibility to restrict columns for a query can further improve the query performance as it might be the case that only indexed columns are queried, which results in a great performance gain. However, in any other case, restricting the retrieved columns only has a little effect on the overall query performance. Nevertheless, in order to achieve results that are comparable with our previous works [8, 7], we renounced the usage of any indexes or other database tuning opportunities.

Finally, we used our own MySQL implementation of the TPC-W data model based on [22], which randomly generates 288 K tuples (i.e. *customers*) in its standard configuration and inserts them into the *Customer* table. We then partitioned the table (see above) and run our distributed queries against these partitions. Before we actually performed the queries, we adapted the randomly generated table data such that always all tuples matched the query, i.e. we updated all customer tuples and set C_FNAME to "OO" and C_YTD_PMT to 999. Thus, we were able to measure a defined number of tuples and verify that the retrieved result set is correct and complete. Above that, we performed each query 3 times and built the average of the query times. Thus, all tables present the query times of our approach in milliseconds (ms). We performed the entire evaluation on one physical machine with 2×2.9 GHz processors (4 cores available with hyper threading), 8 GB RAM and 500 GB hard disk storage. The approach was implemented in form of a Java client-server application based on CentOS 6 and the logic was compiled with Java 1.7_79 (×64). We used this local setup in order to measure the pure query performance and to omit side effects like the unknown host utilization in a virtualized cloud environment or network overheads (Tables 1 and 2).

Table 1 Nested loops query performance

# Tuples	Fetch time (ms)	Join time (ms)	Total time (ms)
288.000	10.168	14.213	24.381
88.000	3.559	1.322	4.881
50.000	2.345	538	2.883
25.000	1.685	195	1.879
20.000	1.486	186	1.672
15.000	1.326	144	1.470
10.000	1.325	136	1.462
5.000	1.089	101	1.190
4.000	1.015	98	1.113
3.000	900	55	955
2.000	697	84	781
1.000	525	51	576
500	313	38	351
250	191	13	204
125	162	5	167
100	146	2	148
50	83	4	87
25	59	1	60
15	57	2	59
10	51	1	52
5	48	1	49
1	45	2	46

Table 2 Hash join query performance

# Tuples	Build time (ms)	Probe time (ms)	Total time (ms)
288.000	12.812	560	13.372
88.000	6.488	381	6.869
50.000	4.221	427	4.648
25.000	2.226	241	2.467
20.000	2.285	209	2.494
15.000	2.271	153	2.425
10.000	1.676	98	1.775
5.000	1.559	127	1.686
4.000	1.530	97	1.627
3.000	926	52	978
2.000	797	48	845
1.000	620	18	638
500	561	64	626
250	390	18	408
125	320	6	326
100	265	7	272
50	227	6	233
25	206	2	208
15	201	2	203
10	164	2	166
5	153	2	154
1	121	1	122

6   Conclusion and Outlook

Our results show that the *sorted merge* outperforms the *hash* and the *nested loops join*. Considering our fixed vertically partitioned and distributed dataset and the average runtime complexity of the 3 compared algorithms, this is not surprising. As the *nested loops join* is $O(n^2)$ with its 2 nested loops to collect and to join matching tuples, it is clear that it is outperformed by the hash and the sorted merge join, which are $O(n + m)$ and $O(n*\log(n))$. This also reflects our fixed vertically partitioning approach, where every matching tuple has to be reconstructed via joins to the other partitions. More specific: if a tuple matches the query parameters of the 1st partition, there has to be at least one corresponding part in the 2nd partition and, due to the fact that we join on the primary key, there has to be at most on corresponding part in the 2nd partition. Thus, because of the join on the primary key, it is guaranteed that there is exactly one corresponding part of a matching tuple in each of the vertically distributed partitions. Otherwise there would be inconsistencies in the dataset. We used this in the implementation of the *nested loops join*, as we can advance with the next matching tuple as soon as a corresponding part (of another partition) is found and the tuple is entirely reconstructed. Therefore, we do not have to search entirely through all partitions. Nevertheless, the *nested loops join* is the slowest algorithms in our setup.

Another interesting result is related to the *fetch times* of the compared algorithms, as they are all very similar. Thus, a parallelization of this step would promise even further performance improvements.

Above that, the *hash* and the *sorted merge join* showed almost similar query times, but the *sorted merge join* was a little faster in the end. This proved that the probe resp. the merge phases were also almost similar and so the performance depends on the build phase (*hash join*), in which the hash table is build resp. the sorting phase of the result sets (*sorted merge join*). Here it has to be noted that sorting the result sets is faster than building a hash table. This is not surprising, as we join on the primary keys of the result sets and for sorting primary keys the underlying index can be used.

Another issue concerns *data skewness*, in which one result set contains only few and the other one contains a lot of tuples. Recent experiments with the 3 join algorithms with an extreme case of *skewed data* where one result set contains 1 and the other 288 K tuples show that the performance with heavily *skewed data* can be calculated by the *fetch, build or sort time* of the bigger *result set* and the *join, probe* or *merge time* of the smaller *result set*. Thus, the *sorted merge join* performance of two result sets with 1 and 288 K tuples respectively, is sort time for 288 K tuples (9.831 ms) and the merge time for 1 tuple (1 ms) (see Table 3).

Table 3 Sorted merge join query performance	# Tuples	Sort time (ms)	Merge time (ms)	Total time (ms)
	288.000	9.831	1.147	10.977
	88.000	3.660	242	3.902
	50.000	2.346	206	2.552
	25.000	1.594	147	1.741
	20.000	1.530	139	1.669
	15.000	1.410	111	1.522
	10.000	1.229	101	1.330
	5.000	1.088	90	1.178
	4.000	945	85	1.030
	3.000	882	98	980
	2.000	702	81	783
	1.000	515	60	575
	500	304	34	338
	250	200	9	210
	125	134	4	138
	100	127	3	130
	50	94	2	96
	25	72	1	73
	15	63	1	64
	10	57	4	61
	5	49	4	53
	1	51	1	52

In summary we can say that compared to the original framework setup [7], we received a remarkable performance gain with our presented approach (~92 % compared to sorted-merge join). Although this approach is slower than our caching approach [8] (on the average ~20 % compared to sorted-merge join), it avoids further synchronization, invalidation and replacement strategies, which are complex to implement and which also slow down the performance of the cache to a certain extent. We were surprised that our query rewriting almost achieved the performance of our caching approach. However, it has to be noted that the evaluation of our caching approach [8] was based on a transaction-aware cache implementation that guaranteed all ACID (atomicity, consistency, isolation and durability) criteria. This also reflects our current and future research work in which we investigate efficient synchronization and other caching strategies that we possibly could integrate into our *SeDiCo* framework. Recent experiments with a read-only cache which can be useful in online analytical processing (OLAP) scenarios show that the query performance of a read-only cache increases the performance by factor 10 compared to our results in [8] and in Table 4. However, this is not directly comparable to this approach, as there is no need to use cache synchronization, replacement and

Table 4 Non-distributed query performance	# Tuples	Total time pure framework (ms) [7]	Total time with caching (ms) [8]
	288.000	3155.194	12.346
	88.000	954.316	2.832
	50.000	545.730	1.503
	25.000	280.577	787
	20.000	222.842	642
	15.000	170.582	590
	10.000	117.718	445
	5.000	62.323	145
	4.000	51.241	114
	3.000	39.040	84
	2.000	27.586	37
	1.000	14.769	40
	500	7.872	15
	250	4.369	5
	125	2.410	3
	100	1.900	2
	50	1.112	2
	25	750	2
	15	549	1
	10	429	1
	5	221	1
	1	92	1

invalidation mechanisms. All in all, the results are promising and encourage us to follow our vision of creating a distributed cloud data storage, which has performance characteristics similar to non-partitioned cloud data stores.

References

1. Balkesen, C., Alonso, G., Teubner, J., Ozsu, M.T.: Multi-core, main-memory joins : sort vs. hash revisited. Proc. Endowment. **7**(1), 85–96 (2014)
2. Jindal, A., Palatinus, E., Pavlov, V., Dittrich, J.: A comparison of knives for bread slicing. Proc. VLDB **6**(6), 361–372 (2013)
3. Ngo, H., Re, C., Rudra, A.: Skew Strikes Back: New Developments in the Theory of Join Algorithms. arXiv preprint arXiv:1310.3314 (2013)
4. Doshi, P., Raisinghani, V.: Review of dynamic query optimization strategies in distributed database. In: ICECT 2011–2011 3rd International Conference on Electronics Computer Technology, vol. 6, pp. 145–149 (2011)
5. Kohler, J., Specht, T.: A performance comparison between parallel and lazy fetching in vertically distributed cloud databases. In: International Conference on Cloud Computing Technologies and Applications—CloudTech 2015, Marrakesh, Morocco (2015)
6. TPC.: TPC Benchmark W (Web Commerce) Specification Version 2.0r. http://www.tpc.org/tpcw/default.asp (2003). Accessed 15 July 2015
7. Kohler, J., Specht, T.: Vertical query-join benchmark in a cloud database environment. In: Proceedings of 2nd World Conference on Complex Systems 2014, Agadir, Marocco, pp. 143–150 (2014)
8. Kohler, J., Specht, T.: Performance analysis of vertically partitioned data in clouds through a client-based in-memory key-value store cache. In: Proceedings of The 8th International Conference on Computational Intelligence in Security for Information Systems, Burgos, Spain (2015)
9. Hachani, N., Ounelli, H.: A knowledge-based approach for database flexible querying. In: 17th International Conference on Database and Expert Systems Applications (DEXA'06) (2006)
10. Hossain, M.I., Ali, M.M.: SQL query based data synchronization in heterogeneous database environment. In: 2012 International Conference on Computer Communication and Informatics, ICCCI 2012, Coimbatore, India, pp. 1–5 (2012)
11. Liao, C.-F., Chen, K., Tan, D.H., Chen, J.-J.: Automatic query rewriting schemes for multitenant SaaS applications. Autom. Softw. Eng. **1**(2015), 1–34 (2015)
12. Halevy, Y.: Answering queries using views: a survey. VLDB J. **10**(4), 270–294 (2001)
13. Nash, A., Segoufin, L., Vianu, V.: Views and queries. ACM Trans. Database Syst. **35**(3), 1–41 (2010)
14. Konstantinidis, G., Ambite, J.L.: Scalable query rewriting: a graph-based approach. In: Proceedings of the 2011 ACM SIGMOD International Conference on Management of Data, New York, USA, pp. 97–108 (2011)
15. Chirkova, R., Li, C., Li, J.: Answering queries using materialized views with minimum size. VLDB J. **15**(3), 191–210 (2005)
16. Ming-Yee Lu, W.Z.: Queryll: java database queries through bytecode rewriting. In: ACM/IFIP/USENIX 7th International Middleware Conference, Melbourne, Australia, pp. 201–218 (2006)
17. MySQL.: MySQL 5.6 reference manual including MySQL cluster NDB 7.3 reference guide. http://dev.mysql.com/doc/ (2014). Accessed 15 July 2015
18. Oracle.: Oracle® database VLDB and partitioning guide 11 g release 1 (11.1). http://docs.oracle.com/cd/B28359_01/server.111/b32024/partition.htm#i460833 (2007). Accessed 15 July 2015

19. Glavic, B., Alonso, G.: Perm: processing provenance and data on the same data model through query rewriting. In: Proceedings—International Conference on Data Engineering, Shanghai, China, pp. 174–185 (2009)
20. Codd. E.F.: A relational model of data for large shared data banks. Commun. ACM **13**(6), 377–387 (1970)
21. Ireland, C., Bowers, D., Newton, M., Waugh. K.: A classification of object-relational impedance mismatch. In: Proceedings—2009 1st International Conference on Advances in Databases, Knowledge and Data Applications, DBKDA 2009, Gosier, Guadeloup, pp. 36–43 (2009)
22. Bezenek, T., Cain, T., Dickson, R., Heil, T., Martin, M.: Characterizing a Java implementation of TPC-W. In: Proceedings of 3rd Workshop On Computer Architecture Evaluation Using Commercial Workloads (CAECW). HPCA Conference http://pharm.ece.wisc.edu/tpcw.shtml (2000). Accessed 15 July 2015

Part II
Language, Semantic and Content Technologies

Interactive, Tangible and Multi-sensory Technology for a Cultural Heritage Exhibition: The Battle of Pavia

Virginio Cantoni, Luca Lombardi, Marco Porta and Alessandra Setti

Abstract New generation multimedia may have a great impact on exhibition visit experience. This contribution focuses on the innovative use of interactive digital technologies in cultural heritage practices. "Live" displays steered by visitors support the creation of various content formats, smartly adapt the content delivered to the visitor, stimulate self-motivated learning, and lead to a memorable and effective experience. Multimodal interaction modalities have been developed for the exhibition "1525–2015. Pavia, the Battle, the Future. Nothing was the same again", a satellite event of the Universal Exhibition in Milan (Expo 2015). The Computer Vision & Multimedia Lab of the University of Pavia, in cooperation with the Bulgarian Academy of Sciences, in the framework of the European project "Advanced Computing for Innovation", has contributed to set up the exhibition, enriching an educational and experiential room with products and targeted applications. Visitors can observe and analyze seven ancient tapestries, illustrating different phases of the battle, through 3D reconstructions, virtual simulations, eye interaction and gesture navigation, along with transpositions of the tapestries into tactile images that enable the exploration by partially sighted and blind people. In the near future, we may assess the impact of this interactive experience. Due to the novelty of the approach, new insights can be potentially derived about the effectiveness and

Some parts of this paper have been previously published in "1525–2015. Pavia, the Battle, the Future. Nothing was the same again. The Computer Vision & Multimedia Lab at the exhibition", Cantoni V. and Setti A. eds. (by kind permission of the copyright holder).

V. Cantoni · L. Lombardi · M. Porta · A. Setti (✉)
Department of Electrical, Computer and Biomedical Engineering,
University of Pavia, Via A. Ferrata 5, 27100 Pavia, Italy
e-mail: alessandra.setti@unipv.it

V. Cantoni
e-mail: virginio.cantoni@unipv.it

L. Lombardi
e-mail: luca.lombardi@unipv.it

M. Porta
e-mail: marco.porta@unipv.it

© Springer International Publishing Switzerland 2016
S. Margenov et al. (eds.), *Innovative Approaches and Solutions in Advanced Intelligent Systems*, Studies in Computational Intelligence 648,
DOI 10.1007/978-3-319-32207-0_6

77

manageability of each specific system component. Under this scope, not only the exhibition success is important, but also the augmented learning experience in cultural heritage contexts.

Keywords IT for cultural heritage · Human-machine interaction · Natural user interfaces · Gesture and gaze interaction · 3D modeling and printing · Tactile images

1 Introduction

During the last years, museums have welcomed millions of visitors to galleries and exhibitions. However, museums have made little progress towards a concrete understanding of what a visitor actually sees when he/she looks at a work of art. While some visitors clearly have an experience of great impact, the average visitor stops at less than half the exhibition objects, spending less than 30 s at single components, and in most cases spending even less time [1, 2].

New generation multimedia is a powerful, emerging technology that has the potential to play a central role on the visitor-artwork interactions. New opportunities are opened for all exhibitions, engaging with a public which is increasingly familiar with the new modes of multimedia technology. Nearly two decades of research on the cognitive and emotional impact of art and creative work are providing solutions for configuring custom and technology-enhanced spaces to improve and personalize the visitor experience. Often in the past, digital technology for cultural heritage created relations between visitors and art in which technology overshadowed art. Instead of attracting visitors, often this has given rise to a gap between visitors and heritage assets. Adverse effects could be produced by a dominance of information over semantic content, attentional issues and cognitive overload, visitors' seclusion, etc. This unfavorable behavior mainly arises from a mismatch between the objectives of the exhibition curators and visitors' expectations [3]. Nowadays, the continuous growth in importance of the digital age has changed the habits of exhibitions and even the concept of visitor [4]. Multimedia applications have gained their place in Cultural Heritage contexts leveraging emerging tools to foster more interactive experiences [5]. Natural User Interfaces (NUIs) make technology transparent, transforming the visitor's experience into an interactive hands-on way to experiment the content. In this connection, modern digital technologies for cultural and historical materials are an emerging research field that bridges people, culture, and technologies [6–8].

The New Media Consortium (NMC) is an international not-for-profit consortium of learning-focused organizations that can be regarded as the world's longest-running exploration on new media and their emerging technology trends. Almost every year a panel of about 50 experts discuss on the five-year horizon for museums [9, 10]. According to the "NMC Horizon Report: 2015 Museum Edition" [11], over the next 5 years (2015–2019) topics very likely impacting on the core missions of museums

Fig. 1 Key trends, challenges, and technological developments that will impact on museums, galleries and exhibitions over the next 5 years

are the 18 represented in Fig. 1. The discussions have been organized into three time-related categories: short-term, related to the first year; mid-term, approximately related to the second and third years; and far-term, related to the fourth and fifth years. Each term is characterized by two main trends, challenges and developments in technology. The short-term trends consist in visitors' more participatory experiences, which call for a new extended notion of "visitor". Mid-term trends identify as a major topic of interest the analysis of statistical data about digitization and digital preservation, with new strategic cross institutional collaborations. Long-term trends, lastly, foresee a future in which emerging tools will be leveraged by creativity and experimentation, as well as by an increased involvement of private companies. In particular, three challenging pairs of goals can be identified: the already going on development of digital strategies, combined with the expanded competence and advanced skills of museum staff, to better understand the use of a variety of digital tools; the development of standard protocols for measuring the success of technologies, in parallel with technology at the center of many daily activities (museum strategies must meet visitors' needs, through a proper balance between online and offline worlds); finally, when tracking visitor transactions and behavior, the adoption of technical infrastructures to implement digital learning, asset digitization and content production, combined with privacy protection. As regards technology developments, the current highly connected community is likely to lead to Natural User Interfaces, and even to the Internet of Things.

In line with these objectives, the European Commission has addressed the value of museums as creative educational forces with a report entitled "Towards an Integrated Approach to Cultural Heritage for Europe" [12], which highlights the importance of digitization and e-learning tools as a way to support new models of cultural heritage governance.

The impact of this interactive experience can be qualitative and often also quantifiable. New insights can be potentially derived, for example by reporting the visual attention of visitors in front of an exhibition-specific content [13]. The transition from an insight to the individual's thoughts about works of art, to find relations between visitor's viewing and thinking, is still challenging.

2 Natural User Interfaces

The development of research and practice on Human-Computer Interaction (HCI) started between the 1970s and 80s, initially involving experts in cognitive sciences, human factors and ergonomics [14]. The exploratory works that led to user interfaces were developed at first in university research laboratories. In the 90s, HCI experienced an explosive growth, and fundamentally changed computing interaction modalities at the end of the millennium. Figure 2 shows the approximate

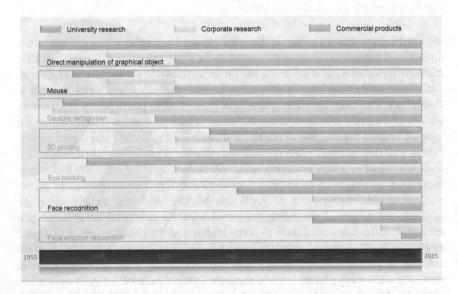

Fig. 2 Approximate time lines on the major interactive technologies. The basic ones exploited in the exhibition are highlighted in *orange* color

timelines related to the major interactive technologies. The figure is an updated version from Myers [14] and summarizes the modalities closely related to the new media technologies applied in our exhibition at the Visconti Castle of Pavia. This interaction among universities, corporate research and commercial activities is constantly increasing, and this trend will be likely confirmed by future interfaces.

The idea to provide users with natural interactions, with movements similar to those performed in the real world, was not new. However, only recent devices allow sufficiently robust and intuitive interaction experiences. Exploiting the convergence of gesture-sensing technologies, humans can interact with computers by 'communicating their intentions' with gestures, facial expressions, eye input and voice. The full potential of these NUIs has not been achieved yet, but their fast and constant improvements are leading to inexpensive and robust interaction devices for content manipulation.

NUIs have profound effects in educational contexts. Children use NUIs tools and mechanisms quite naturally, without thinking about them, without help or instruction. Together with learning, leisure activities have been the major focus of NUIs. Perhaps, even more than for leisure and learning environments, as NUIs become increasingly common, museums and exhibitions have now the opportunity to use these facilities to create new forms of interpretation and presentation. NUIs, by providing the visitor with a direct interaction with the environment and the content, allow exhibits and collection spaces to incorporate large-scale realizations. Multimedia advantages are that information is presented transparently and in a variety of modes, without the "distance" that characterizes traditional interfaces. NUIs offer effective, experiential and idea-driven ways to visit exhibitions. This is similar to what occurs in science centers and children's museums, but also in traditional "object-based" museums the desire to touch and manipulate artworks is very frequent in visitors. Thanks to NUIs, the limits due to preservation and conservation issues of the exhibited objects could be overcome. NUIs are also crucial to facilitate disabled visitors who could not enjoy traditionally designed exhibits.

2.1 Gesture-Based Interaction

It is already common to interact with a new class of devices by using natural movements and gestures [15]. As mentioned, motion and gesture detection technology creates an experience that closely mimics the real world common experience. Motion-sensing input devices are becoming more and more common and are radically transforming the visitor's engagement with collections. Immersive and interactive installations combine tools and software that react naturally to human behavior and activity.

An important distinction in the context of gesture-based interaction is that between *explicit* and *implicit* gestures [16]. The purpose of explicit (or control) gestures is to provide some forms of input to the computer, such as commands. Conversely, implicit gestures are exploited to obtain indirect information about users and their environments, for instance to recognize the activity that is being carried out. A further distinction is that between dynamic gestures, which imply body movements, and static gestures, that are characterized by body positions and/or postures.

Gesture based interaction implies two steps: (i) the user's hand must be recognized and tracked over time; (ii) a dictionary of gestures must be defined which codes the evoked actions. The set of interactive gestures must be intuitive, comfortable and meanwhile easy to remember for the user.

The first possibility, in a growing array of alternative devices, that recognizes and interprets natural physical gestures is Microsoft Kinect. This is a low-cost, non-intrusive sensor originally conceived for gaming; it supplies an RGB camera, a 3D depth image sensor that measures the distance between an object and the sensor, a data sensitive microphone, and skeleton video streams. Its use is convenient because: (i) it is widespread and already used for many purposes; (ii) it does not require camera calibration or environment setup; (iii) it allows markerless tracking; (iv) it is quite cheap; (v) it provides powerful libraries for creating custom applications [17]. For all these reasons, its impact has extended far beyond the gaming industry. In 2012, Microsoft has produced a new sensor expressively devoted to scientific research. Today researchers and practitioners are leveraging the technology for applications in robotics, electronic engineering, etc. [18].

In the exhibition of the Battle of Pavia, we selected the Microsoft Kinect sensor as a widespread and cheap technology able to monitor body motion and gestures, so as to allow users to control the computer by their hand and finger gestures (Fig. 3). An analogous example is the exhibition "Spotlight on The Antinoe Veil", at the Louvre in Paris, that deployed Kinect technology to enable visitors to interact with

Fig. 3 User gesture interaction to browse the artwork

an ancient artifact. Visitors could manipulate a veil dating back to the 4th century AD, and could explore different narratives painted on the fabric, without physically touching it [19].

2.2 Avatar

An avatar is the representation of the user, the user's alter ego or the user in third person. Avatar in historical exhibitions is often used in computer generated virtual experiences in which visitors play an online role in a story-driven approach. The visitor's avatar is full immersed into the virtual world of the exhibition and can directly control the actions performed in the stage. The virtual simulation of the human body is often made based on high quality characters related to the exhibition (in gaming, avatars are often bought off-the-shelf models). There is a growing secondary industry devoted to the creation of products that can be used to build and manage avatars, and which provide tools for changing shapes, hair, skin, gender, etc.

A characteristic avatar project for museums and exhibitions is TOURBOT, a RTD Project funded by the EU-IST Programme [20]. The goal of this project is the development of an interactive TOUr-guide RoBOT that operates as the visitor's avatar to provide individual access over the Internet to visit cultural heritage specific exhibits. The virtual reconstruction of the museum or exhibit is delivered over the Internet to a remote visitor, so that he/she can enjoy a personalized tele-presence experience, being able to choose the preferred viewing conditions (point of view, distance, resolution, etc.).

In the exhibition of the Battle of Pavia some characters mimic in real-time facial and head movements of the visitor (Fig. 4).

2.3 Gaze-Based Interaction

Techniques for measuring gaze have been an important part of cognitive psychology and many other fields of study since the early researches carried out in the 1960s. Alfred Lukyanovich Yarbus, in particular, can be considered the father of eye tracking studies. In a famous experiment, Yarbus discovered that eye movements of different subjects observing a painting were similar but not identical; furthermore, he found that repeating the experiments at different times with the same subjects produced very similar eye movements, but never the same [22]. In the late 70s, the focus of research was on the link among eye tracking, visual attention and cognition. Of particular interest is the so-called Eye-Mind Hypothesis [10, 23], according to which there is a direct correspondence between the user's gaze and the point of attention. As demonstrated by several experiments, while it is possible to shift the focus of attention without moving the eyes, the opposite is

Fig. 4 Avatar: facial and head expressions and movements replicated in real time by the 3D model of a character. The application has been developed using the web service and the facial capture and animation technology of *Face Plus* [21]. Testing in the CVML laboratory

much more difficult [24]. During the 1980s, the Eye-Mind Hypothesis was often questioned in light of covert attention, i.e. the attention to something that is not in the fixation restrict neighborhood. According to Hoffman [25], "current consensus is that visual attention is always slightly (100–250 ms) ahead of the eye". However, as soon as the attention moves to a new position, the eyes will want to follow. The Eye-Mind Hypothesis, due to its simplicity, is widely exploited in the evaluation of user interfaces, but also in other areas, such as in studies on reading or online advertising (e.g. [26]).

The gaze-contingency paradigm is a general framework to change the information presented on the monitor depending on what the user is looking at [27]. Due to an imperfect coupling between overt and covert attentions, it is not possible to exactly know which visual information the viewer is processing based on the fixation locations. However, in gaze-contingent paradigms the stimulus on the display is continuously updated as a function of the observers' current gaze position: by controlling the information feeding, it is possible to eliminate any ambiguity about what is fixated and what is processed.

Milekic [28, 29] gives an overview of gaze interaction and its potential applications for museums and exhibitions. Eye-tracking technology offers to museums the possibility to study and measure the visitor's attention to artworks, to investigate the correlation between suggested interpretation and visitor's comprehension and to inspire interpretive content delivery. In fact, this technology offers the opportunity to explore and model various interaction modalities that can be used to improve visitor experience.

Eye tracking has the potential to change the ways we relate to arts, and offers a direct way of studying several important features of an exhibition visit, reporting on conditions and factors affecting it. For example, it allows to reliably measure the amount of time a visitor spends looking at an art-work.

The study of the visual behavior of an exhibition visitor requires the solution of two sub-problems: gaze estimation and gaze analysis. Eye tracking has the purpose to find the user's gaze direction, through the identification of the eyes pupils and corneal reflections (usually produced by infrared light). Gaze coordinates may then be analyzed over time. Fixations and scanpaths allow the assessment of the user's interest. For example, the time spent by users fixating an object provides information on how much they are interested in the object itself. Furthermore, gaze interaction can exploit 'augmented' content; for instance, the user may zoom in and see different parts of the image in more detail just by looking at it.

To determine the feasibility of these techniques in museums, the first problem is the selection of the hardware to be used: of course head-mounted devices cannot be considered, as the visitor's experience should be as natural and "transparent" as possible. A number of different eye tracking equipments and methods have been proposed, but currently the most widespread and commercially available technique is that exploiting infra-red illumination. The gaze direction is determined by analyzing the position of the pupil and of corneal glints (i.e. spots of reflected infra-red light on the cornea [30]). Prices of eye trackers are constantly decreasing, and very cheap devices are now available. In particular, for the exhibition of the Battle of Pavia, we chose the EyeTribe [31] eye tracker.

In Fig. 5 one of the three workstations realized for the Pavia exhibition. The eye tracker is positioned at the bottom of the display, in front of which the user sits

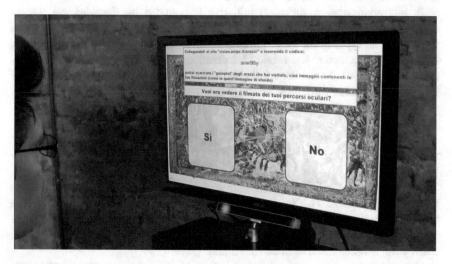

Fig. 5 An example of gazeplot; the scan path is *yellow* and fixation points are *red* (circles size explicit the fixation time). A code is proposed to user to download data recorded during his/her exploration

during the interaction. The seat needs to be adjusted for each visitor, so that his/her eyes can be properly detected. Infrared light is used to find the gaze direction. After a short calibration phase (in which the user has to look at a moving target point), the eye tracker is able to compute gaze coordinates. Moreover, additional data that can be obtained are fixation time, saccade length, pupil size, and number of blinks. At the end of each session, a comprehensive report is produced so that the interaction experience can be monitored. The most common output is the visitor's scanpath, that is a visual representation of the sequence of fixations (usually indicated as circles whose area is proportional to their duration). A second common output is the "heat map", i.e. a visual depiction, through different colors, of the time spent by users while watching the different areas of the screen.

Logging and analyzing eye tracking data provides rich information on the exhibit attractiveness and about the interaction with the multimedia content set-up by cultural heritage professionals. Not only the final resulting performance is important, but also the understanding of all processes involved [32].

3 Services to Visitors and Accessibility

It is interesting to investigate experiences involving 3D touch and how people actually approach touch interaction [33]. For some people it seems unnatural being only able to watch, and they get frustrated at exhibitions where it is not allowed to touch and feel objects and materials. Actually, touch is extremely important particularly for blind and partially sighted visitors, who exploit tactile sensing to access the external world information through contours and textures.

In this connection a structural initiative is actuated by the Museum Access Consortium (MAC) that strives to enable people with disabilities to access cultural facilities of all types in the New York metro area. The aim is to share experiences, learn from one another, and refine best practices to reach advancing accessibility and inclusion [34]. In the sequel the activities of the exhibition of Pavia in this framework are shortly summarized.

3.1 Tactile Images

Tactile images—in relief, are made in such a way that they can be read and interpreted by touch. The implementation is not immediate; to be suitable to a tactile interpretation, a figure must be simplified and transferred to a relief image that presents distinct and logically homogeneous forms, so that each component can convey the original content in an intelligible way.

The exhibition proposes a tactile reading of an artwork, experimenting the transposition of the information that the tapestries depict in a version accessible to partially sighted and blind visitors. The aim is to allow to see with your hands:

Fig. 6 The tapestry "The Advance of the Imperial army and the attack of the French gendarmerie led by Francis I" has been split into 23 layers. The figure represents three layers related to the French king Francis I

Fig. 7 The tactile images at the exhibition of the Pavia Battle

the tapestry images are digitized, modified, adapted (Fig. 6), reconstructed as three-dimensional models and finally printed in 3D to be read with the fingertips (Figs. 7 and 8).

Fig. 8 Original version, re-elaboration for tactile version, and 3D printing of the tapestry depicting the defeat of the French cavalry and the Imperial infantry reigning over the enemy artillery. In the re-elaboration all layers are shown. In the 3D print, *white* h = 0: bottom and inner contours in full figures; *red* h = 1: external contours of all, *black* h = 1: internal boundaries and Braille in hollow figures, and full figures; *green* h = 2: Braille in full figures)

3.2   3D Collections

People partially sighted or blind cannot enjoy all the treasures that cultural heritage offers. However, new technologies are now available which can make art accessible to everybody. 3D printing is an affordable, practical and effective solution to create reproductions of artwork objects that usually are not, or not always, easily available, manageable or accessible [35].

In particular, 3D models can be profitably exploited with different targets: (i) *to reconstruct*: from fragments and picture documents or other sources, 3D models of destroyed artifacts can be built; (ii) *to replicate*: for preservation of originals, that are too fragile to move, or just for making commercially available replicas; (iii) *to interpret*: for understanding the past of an object, an architecture or a place map, through the analysis of their reconstructions; (iv) *to investigate*: building various items to compare artifacts from different sources; (v) *to share*: grouping 3D resources for extended integration of parts and models [36].

Figures 9, 10, 11, 12 and 13 show the Battle of Pavia 3D collection made by the students of the "Computer Vision" course of the Master in Computer Engineering of the University of Pavia, in details: tapestry characters and models; tapestry scenes and models; a French cannon and the Pavia landscape from tapestry to model to 3D print; character models and 3D prints.

Fig. 9 Rendering of 3D models produced by *Fuse Character Creator* [37] and *Cinema 4D* [38]: Charles de Lannoy; Charles III of Bourbon; Duke of Alençon on the bridge made of boats; Guillaume Gouffier de Bonnivet; Great Lord

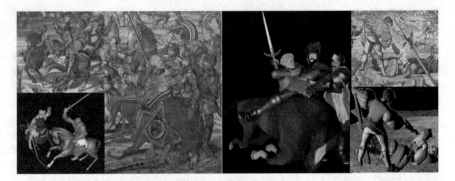

Fig. 10 Rendering of 3D scenes produced by *Fuse Character Creator* [37] and *Cinema 4D* [38]: Ferrante Castriotta and Francesco I; capture of the French king; people coming out from shelters

Fig. 11 French cannon: tapestry; 3D model; 3D object printed by *ProJet 460Plus*

Fig. 12 The city of Pavia: tapestry; 3D model; 3D city printed by *Sharebot NG PLA*

Fig. 13 Characters models and 3D print made by the *ProJet 460Plus* printer; a lady wearing a *red dress*, following the French army; Francesco Fernando d'Avalos; Georg von Frundsberg; Jean De Diesbach; stableman of the lady in *red*

4 Impact of New Technologies in Exhibitions: Issues and Visitors' Feedback

The purpose is to measure the value of digital media opportunities in exhibitions, in order to develop strategies, processes, and technologies to maximize their benefits for exhibitions and visitors. A structured approach has been followed at the Metropolitan Museum of Art, in which the full strategy focuses on three phases of interaction: (i) before the visit, by presenting the exhibition content online; (ii) during the visit, by developing solutions to enrich the visitor's experience; (iii) after the visit, by using visitor-generated data to tune plans and expectations and to assure positive memories by visitors [39].

Museums and exhibitions are increasingly leveraging emerging technologies, such as NUIs. However, the emphasis on these digital tools often prevails on the assessment of the impact of the deployed technologies. This evaluation should occur before and after technologies are implemented, in order to establish a framework to measure the impact and to validate if the adopted devices and tools are/have been suited to the exhibition's mission and goals. At last, if museums share the results with other institutions, the risk factor will be reduced in the future. It is important for museums to develop benchmarks before deploying new technologies to avoid retro-fit a solution in a post-implementation phase.

Test and evaluation sessions have so far occurred in museums and galleries. Some of these sessions had as goal to advance our understanding about specific system components. A list of recommended general guidelines for program eval-uations, project planning tools, and evaluation methods already exists [39, 40]. As more and more visitors will get acquainted with the use of these tools and methods, we may effectively assess the impact of the interactive experiences that have been tailored so far. Adaptive strategies shaped by the visitor interest and curiosity can be used to personalize an interactive process and to promote a positive emotional and memorable visitor experience in the Cultural Heritage context.

5 Conclusions

Discovery-based learning at museums and exhibitions are steadily increasing [9] even targeting subversive and non-traditional tour [41]. However, participatory strategy involves innovative approaches in an experimental territory [42]. Therefore, it is important to be able to evaluate successes and failures, because, despite many experiments, the risk that the average visitor gets lost for the abun-dance of information and technology still remains high. Some studies have revealed that the design properties of those exhibition and realization conditions have a significant impact on success or failure rates.

Crowdsourcing, a term coined in 2005, has been defined (Merriam-Webster) as the process of obtaining needed services, ideas, or content by soliciting contributions from a large group of people, and especially to draw thorough conclusions about effective and ineffective models of implementation. Driven by such practical issues and societal general trends, crowdsourcing seems likely to become a necessary workflow of heritage institutions [43]. Nevertheless, despite the widespread preva-lence of crowdsourcing projects in the cultural heritage domain, not all initiatives to date have been universally successful. Several have failed to recruit adequate numbers of participants or those with the requisite skills to complete the desired task.

Specific attributes for project success are: (i) size of the institution, and in particular the familiarity of the institution with projects that involve digital stimuli and multimedia; (ii) type of the collection: there is a correlation between the content and the type of crowd that the project might attract; (iii) complexity of the target task and its specificity, opportunity to adapt the content to be delivered to the visitor accordingly; (iv) crowd and people involved, factors shaping the amount of crowd involvement and how to keep crowd interested and motivated; (v) project site infrastructure and its suitability to the target; (vi) team able to identify and monitor progress, and how they can be trained; (vii) capability to evaluate the project successes and failures and the value of the experience [11].

The monitoring of the impact of all physical and digital stimuli is exploited to properly adapt the content to the visitor. In this connection, the future for NUIs is very promising and some exhibitions begin now to build museum inventories of the employed devices. NUIs can be used in many imaginative ways, in order to guide visitors through satisfying experiences, encouraging them to feel, digest, reflect, touch, and pursue sensorial involvement, memorable and easily achievable.

References

1. Hein, G.E.: Learning in the Museum. Routledge, London (1998)
2. Smith, J.K., Smith, L.F.: Spending time on art. Empirical Stud. Arts **19**(2), 229–236 (2001)
3. Worts, D.: On the brink of irrelevance? Art museums in contemporary society. In: Tickle, L., Sekules, V., Xanthoudaki, M. (eds.) Researching Visual Arts Education in Museums and Galleries: An International Reader, pp. 215–233. Kluwer Academic Publishers, Dordrecht (2003)
4. Johnson, L., Adams Becker, S., Freeman, A.: The NMC Horizon Report: 2013 Museum Edition. The New Media Consortium, Austin, Texas (2013)
5. Lohr, S.: Museums Morph Digitally: The Met and Other Museums Adapt to the Digital Age. The New York Times (2014)
6. Bautista, S.S.: Museums in the Digital Age: Changing Meanings of Place, Community, and Culture. Lanham (MD), USA (2014)
7. Chen, C., Wactlar, H.D., Wang, J.Z., Kiernan, K.: Digital imagery for significant cultural and historical materials. Int. J. Digit. Libr. **5**(4), 275–286 (2005)
8. Din, H., Wu, S.: Digital Heritage and Culture: Strategy and Implementation. World Scientific Publishing Co Pte Ltd, Singapore (2014)
9. Johnson, L., Adams Becker, S., Witchey, H., Cummins, M., Estrada V., Freeman, A., Ludgate, H.: The NMC Horizon Report: 2012 Museum Edition. The New Media Consortium, Austin, Texas (2012)
10. Just, M.A., Carpenter, P.A.: A theory of reading. From eye fixation to comprehension. Psychol. Rev. **87**, 329–354 (1980)
11. Johnson, L., Adams Becker, S., Estrada, V., Freeman, A.: NMC Horizon Report: 2015 Museum Edition. The New Media Consortium, Austin, Texas (2015)
12. http://ec.europa.eu/culture/library/publications/2014-heritage-communication_en.pdf. Accessed 2 Nov 2015
13. Damala, A., Schuchert, T., Rodriguez, I., Moragues, J., Gilleade, K., Stojanovic, N.: Exploring the affective museum visiting experience: adaptive augmented reality (A^2R) and cultural Heritage. Int. J. Heritage Digital Era **2**(1), 117–142 (2013)
14. Myers Brad A.: A brief history of human computer interaction technology. ACM Interact. **5** (2), 44–54 (1998)
15. Dondi, P., Lombardi, L., Porta, M.: Development of gesture-based HCI applications by fusion of depth and colour video streams. IET Comput. Vision **8**(6), 568–578 (2014)
16. Porta, M.: Vision-based user interfaces: methods and applications. Int. J. Hum Comput Stud. **57**, 27–73 (2002)
17. Gianaria, E.: Methodologies for automatic detection and analysis of human body. Ph.D. Thesis, University of Turin (2015)
18. Zhengyou, Z.: Microsoft kinect sensor and its effect. IEEE MultiMedia **19**(2), 4–10 (2012)
19. http://circanews.com/news/louvre-adds-microsoft-kinect-exhibit. Accessed 2 Nov 2015
20. http://www.ics.forth.gr/tourbot/. Accessed 2 Nov 2015
21. https://www.mixamo.com/faceplus. Accessed 26 Nov 2015
22. Yarbus, A.L.: Eye Movements and Vision. Plenum Press, NY (1967)
23. Just, M.A., Carpenter, P.A.: Eye fixations and cognitive processes. Cogn. Psychol. **88**, 441–480 (1976)
24. Poole, A., Ball, L.J.: Eye tracking in human-computer interaction and usability research: current status and future. In: Ghaoui C. (ed.) Encyclopedia of Human-Computer Interaction, pp. 211–219. Idea Group, Inc., PA (2005)
25. Hoffman, J.E.: Visual attention and eye movements. In: Pashler H. (ed.) Attention, pp. 119–154. Psychology Press, Hove, UK (1998)
26. Porta, M., Ravarelli, A., Spaghi, F.: Online newspapers and ad banners: an eye tracking study on the effects of congruity. Online Inf. Rev. **37**(3), 405–423 (2013)

27. Duchowski, A.T.: Eye Tracking Methodology—Theory and Practice, 2nd edn. Springer-Verlag, London (2007)
28. Milekic, S.: Designing digital environments for art education/exploration. J. Am. Soc. Inf. Sci. **51–1**, 49–56 (2000)
29. Milekic, S.: Gaze-Tracking and museums: current research and implications. In: Trant J., Bearman D. (eds.) Proceedings of the Museums and the Web 2010, Archives & Museum Informatics, Toronto, 31 March 2010
30. Cantoni, V., Porta, M.: Eye tracking as a computer input and interaction method. In: Proceedings of the 15th International Conference on Computer Systems and Technologies (CompSysTech 2014), pp. 1–12. ACM Press, Ruse, Bulgaria, 27 June 2014
31. https://theeyetribe.com/. Accessed 2 Nov 2015
32. Bachta, E., Stein, R.J., Filippini-Fantoni, S., Leason, T.: Evaluating the Practical Applications of Eye Tracking in Museums. In: Proceedings of Museums and the Web 2012, The Annual Conference of Museums and the Web, San Diego, CA, USA (2012). http://www.museumsandtheweb.com/mw2012/papers/evaluating_the_practical_applications_of_eye_t.html. Accessed 2 Nov 2015
33. http://www.openculture.com/2015/03/prado-creates-first-art-exhibition-for-visually-impaired.html. Accessed 2 Nov 2015
34. http://www.cityaccessny.org/mac.php. Accessed 2 Nov 2015
35. Mazura, M., Horjan, G., Vannini, C., Antlej, K., Cosentino, A.: eCult vademecum a guide for museums to develop a technology strategy & technology providers to understand the needs of cultural heritage institutions, (2015). http://www.ecultobservatory.eu/sites/ecultobservatory.eu/files/documents/Vademecum_PDF_V2.0.pdf. Accessed 2 Nov 2015
36. D-COFORM Tools & Expertise for 3D Collection Formation. http://www.3d-coform.eu/index.php. Accessed 2 Nov 2015
37. https://www.mixamo.com/fuse. Accessed 24 Nov 2015
38. http://www.maxon.net/. Accessed 2 Nov 2015
39. http://www.kurtsalmon.com/en-us2/Telecoms/vertical-insight/1219/Metropolitan-Museumof-Art%3A-adapting-its-organization-to-a-digital-offering. Accessed 2 Nov 2015
40. http://www.zdnet.com/article/digital-transformation-at-the-metropolitan-museum-of-art/. Accessed 2 Nov 2015
41. http://www.museumhack.com/. Accessed 2 Nov 2015
42. Simon, N.: The participatory museum. http://www.participatorymuseum.org/. Accessed 2 Nov 2015
43. Noordegraaf, J., Bartholomew, A., Eveleigh, A.: Modeling crowdsourcing for cultural Heritage. In: Museums and the Web 2014, The Annual Conference of Museums and the Web. Baltimore, MD, USA (2014). http://mw2014.museumsandtheweb.com/paper/modeling-crowdsourcing-for-cultural-heritage/98/. Accessed 2 Nov 2015

Mining Clinical Events to Reveal Patterns and Sequences

Svetla Boytcheva, Galia Angelova, Zhivko Angelov
and Dimitar Tcharaktchiev

Abstract This paper presents results of ongoing project for discovering complex temporal relations between disorders and their treatment. We propose a cascade data mining approach for frequent pattern and sequence mining. The main difference from the classical methods is that instead of applying separately each method we reuse and extend the result prefix tree from the previous step thus reducing the search space for the next task. Also we apply separately search for diagnosis and treatment and combine the results in more complex relations. Another constraint is that items in sequences are distinct and we have also parallel episodes and different time constraints. All experiments are provided on structured data extracted by text mining techniques from approx. 8 million outpatient records in Bulgarian language. Experiments are applied for 3 collections of data for windows with size 1 and 3 months, and without limitations. We describe in more details findings for Schizophrenia, Diabetes Mellitus Type 2 and Hyperprolactinemia association.

Keywords Medical informatics · Big data · Text mining · Temporal information · Data mining

S. Boytcheva (✉) · G. Angelova
Institute of Information and Communication Technologies,
Bulgarian Academy of Sciences, Sofia, Bulgaria
e-mail: svetla.boytcheva@gmail.com

G. Angelova
e-mail: galia@lml.bas.bg

Z. Angelov
Adiss Lab Ltd., Sofia, Bulgaria
e-mail: angelov@adiss-bg.com

D. Tcharaktchiev
Medical University Sofia, University Specialised Hospital for Active
Treatment of Endocrinology, Sofia, Bulgaria
e-mail: dimitardt@gmail.com

© Springer International Publishing Switzerland 2016
S. Margenov et al. (eds.), *Innovative Approaches and Solutions in Advanced Intelligent Systems*, Studies in Computational Intelligence 648,
DOI 10.1007/978-3-319-32207-0_7

95

1   Motivation

Analyzing relations between temporal events in clinical narratives has high importance for proving different hypotheses in healthcare: in risk factors analysis, treatment effect assessment, comparative analysis of treatment with different medications and dosage; monitoring of disease complications as well as in epidemiology for identifying complex relations between different disorders and causes for their co-existence—so called comorbidity. A lot of research efforts were reported in the area of electronic health records (EHR) visualization and analysis of periodical data for single patient or searching patterns for a cohort of patients [1–5]. The work [6] proposes a method for temporal event matrix representation and a learning framework that discovers complex latent event patterns or Diabetes Mellitus complications. Patnaik et al. [1, 2] report one of the first attempts for mining patients' history in big data scope—processing over 1.6 million of patient histories. They demonstrate a system called EMRView for mining the precedence relationships to identify and visualize partially ordered information. Three tasks are addressed in their research: mining parallel episodes, tracking serial extensions, and learning partial orders.

Mining frequent event patterns is a major task in data mining. It filters events with similar importance and features; this relationship can be specified by temporal constraints. There are two major tasks in data mining related to the temporal events analysis: (i) frequent patterns mining and (ii) frequent sequence mining. The difference between them is that in the first case the event order does not matter.

In *frequent patterns mining* the events are considered as sets—collections of objects called *itemsets*. We investigate how often two or more objects co-occur. Usually they are considered as a database of transactions presented like tuples (*transaction, itemset*), the sets of transaction identifiers are called *tidsets*. Several methods are proposed for solving this task that vary from the naive BruteForce and Apriori algorithms, where the search space is organized as a prefix tree, to Eclat Algorithm that uses tidsets directly for support computation, by processing prefix equivalence classes [7]. An improvement of Eclat is dEclat, it reduces the space by keeping only the differences in the tidsets as opposed to the full tidsets. Another efficient algorithm is Frequent Pattern Tree Approach—FPGrowth Algorithm. Using the generated frequent patterns by all these methods we can later generate association rules.

For *frequent sequence mining* the order does matter [7]. The Level-wise generalized sequential pattern (GSP) mining algorithm searches the prefix tree using breadth-first search. SPADE algorithm applies vertical sequence mining, by recording for each symbol the position at which it occurs. PrefixSpan algorithm uses projection-based sequence mining by storing only the suffix after the first occurrence of each symbol and removing infrequent symbols from the suffix. This algorithm uses depth-first search only for the individual symbols in the projection database.

There are different mining approaches for temporal events, for instance we can consider sequences leading to certain target event [8]. Gyet and Quiniou [9] propose recursive depth-first algorithm QTIPrefixSpan that explores the extensions of temporal patterns. Further they extract temporal sequences with quantitative temporal intervals with different models using a hyper-cube representation and develop a version of EM algorithm for candidates' generation [10]. Patnaik et al. present the streaming algorithm for mining frequent episodes over a window of recent events in the stream [2]. Monroe et al. [11] presents a system with visual tools that allows the user to narrow iteratively the process for mining patterns to the desired target with application in EHRs. Yang et al. [12] describe another application of temporal event sequence mining for mining patient histories. They develop a model-based method for discovering common progression stages in general event sequences.

2 Project Setup

The main goal of our research is to examine comorbidity of diseases and their relationship/causality with different treatment, i.e. how the treatment of a disease can affect the co-existing other disorders. This is a quite challenging task, because the number of diagnoses (more than 10,000) and of medications (approx. 6,500) is huge. Thus the theoretically possible variations of diagnoses and corresponding treatments are above 10^{500} for one patient. That is why we shall examine separately chronic versus acute diseases [13] and afterwards shall combine the patterns into more complex ones. Chronic diseases constitute a major cause of mortality according to the World Health Organization (WHO) reports and their study is of higher importance for healthcare.

In order to solve this challenging task we split it down into two subtasks:

- **Task 1**: To find frequent patterns of distinct chronic diseases. Afterwards for each frequent pattern of chronic diseases—to find frequent patterns of treatment and investigate their relationship in order to identify complex relations.
- **Task 2**: To search for causality and risk factors for chronic diseases by sequence mining. In this task several experiments are explored—with no limitations for the distance between events, only the order matters, and with different window limitations between events—1, 3 months, etc. In this task also more complex sequences are considered like parallel (simultaneous) episodes/disorders.

Each of these tasks needs to be investigated in the general case and also for gender and age specific conditions.

3 Materials

We deal with a repository of pseudoanonymous Outpatient Records (OR) in Bulgarian language provided by the Bulgarian National Health Insurance Fund (NHIF) in XML format. The majority of data necessary for the health management are structured in fields with XML tags, but there are still some free-text fields that contain important explanations about the patient: "Anamnesis", "Status", "Clinical examinations" and "Therapy".

From the XML fields with corresponding tags we know the Patient ID, the code of doctors' medical specialty (SimpCode), region of practice (RZOK), Date/Time and ID of the outpatient record (NoAL). XML tags also point to the main diagnose and additional diagnoses with their codes according to the International Classification of Diseases, 10th Revision (ICD-10) [14]. Each OR contains a main diagnosis ICD-10 code and up to 4 additional disorders ICD-10 codes, i.e. in total from 1 to 5 ICD-10 codes. ORs describe events that can be represented in structured format (Fig. 1).

Our experiments for pattern search are made on three collections of outpatient records that are used as training and test corpora, see Table 1. They contain data about patients suffering from Schizophrenia (ICD-10 code F20), Hyperprolactinaemia (ICD-10 code E22.1), and Diabetes Mellitus (ICD-10 codes E10–E15). These collections are of primary interest for our project because they contain cases with high diversity of chronic disorders. Schizophrenia and Diabetes Mellitus are chronic diseases with a variety of complications that are also chronic

Fig. 1 Structured event data

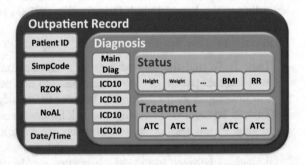

Table 1 Characteristics of data collections

Characteristics	Collections		
	S1	S2	S3
Outpatient records	1682429	288977	6327503
Patients	45945	9777	435953
Main diagnose (ICD-10)	F20	E22.1	E10–E15
Period (year)	3	3	1
Size (GB)	4	1	18

diseases. The collections are extracted by using a Business Intelligence tool (BITool) [15] from the repository of about 120 millions ORs for approx. 5 million patients for 3 years period. The size of the repository is approx. 212 GB.

4 Methods

We designed a system for exploring complex relationships between disorders and treatments, see Fig. 2. It contains two main modules—for text mining and for data mining, and two repositories—for XML documents (ORs) and for structured data (temporal events sequences).

Text mining module is responsible for the conversion of the raw text data concerning treatment and status to structured event data and in addition for the "translation" of the structured data in the XML document to event data. For extraction of information about the treatment we use a Text mining tool because the ORs contain free texts discussing drugs, dosage, frequency and route mainly in the "Therapy" section. Sometimes the "Anamnesis" also contains sentences that discuss the current or previous treatment. We developed a drug extractor using regular expressions to describe linguistic patterns [16]. There are more than 80 different patterns for matching text units to ATC drug names/codes [17] and NHIF drug codes, medication name, dosage and frequency. Currently, the extractor is elaborated and handles 2,239 drug names included in the NHIF nomenclatures. For extraction of clinical examination data we designed a Numerical value extractor [18] that processes lab and test results described in "Anamnesis", "Status", and

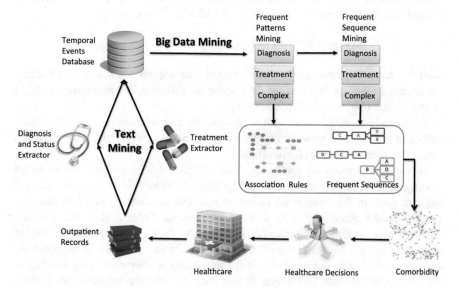

Fig. 2 System architecture

"Clinical examinations"—for instance body mass index (BMI), weight (W), blood pressure (Riva Roci—RR), etc.

Data mining module uses a cascade approach for solving the two main tasks. The process will be shown in more details in the following subsections. Briefly the idea is that Task 1 can be solved by applying modification of the classical frequent itemsets mining algorithms and association rules generation. The solution of Task 2 is based on frequent sequence mining expanding the prefix tree generated as a result from the previous task solution. There are a lot of efficient algorithms for solving each task separately. However, in our project we are interested in a single algorithm that solves both tasks. That is why we propose a cascade method that uses the obtained results from the previous task and in such a way reduces the search space for the next task. Another constraint is that items in sequences in Task 2 are distinct and we have also parallel (simultaneous) episodes and different time constraints.

4.1 Formal Presentation

Each collection $S \in \{S1, S2, S3\}$ is processed independently from the other two collections. For the collection S the set of all different patient identifiers $P = \{p_1, p_2, \ldots, p_N\}$ is extracted. This set corresponds to transaction identifiers (tids) and we call them *pids* (patient identifiers). For each patient $p_i \in P$ events sequence of tuples $\langle event, timestamp \rangle$ is generated: $E(p_i) = (\langle e_1, t_1 \rangle, \langle e_2, t_2 \rangle, \ldots, \langle e_{k_i}, t_{k_i} \rangle)$, $i = \overline{1, N}$ where timestamps $t_{n-1} \leq t_n, n = \overline{2, N}$. Let $\mathcal{E}$ be the set of all possible events and $\mathcal{T}$ be the set of all possible timestamps. Let $\mathcal{C} = \{c_1, c_2, \ldots, c_p\}$ be the set of all chronic diseases [13], which we call *items*. Each subset of $X \subseteq \mathcal{C}$ is called an *itemset*. We define a projection function $\pi : (\mathcal{E} \times \mathcal{T})^n \to (\mathcal{C} \times \mathcal{T})^n$:

$$\pi(E(p_i)) = C(p_i) = (\langle c_1, t_{1i} \rangle, \langle c_2, t_{2i} \rangle, \ldots, \langle c_{m_i}, t_{m_i} \rangle)$$

such that for each patient $p_i \in P$ the projected time sequence contains only the first occurrence (onset) of each chronic disorder of the patient p_i that is recorded in $E(p_i)$ in the format $\langle chronic\ disease, timestamp \rangle$.

To investigate both cases: with no limitations and with different window limitations of the distance between events, we store two versions of the temporal event sequences database for each collection.

In the first version all timestamps are substituted with consecutive numbers starting from 0. In this case the particular dates of events do not matter, only the order matters. In this section we introduce a simple synthetic example to illustrate the proposed method (Table 2). For the example on Table 2, the corresponding database with normalized timestamps is shown on Table 3a. In the second version all timestamps are replaced with relative time—to the first event in the sequence we assign time 0, and for all other events the timestamp is converted to the number of days distance from the first event. In this case the distance between events does

Table 2 Example database for chronic diseases with timestamps

pid				
1.	$\langle A, 12/01/2012 \rangle$	$\langle B, 12/01/2012 \rangle$	$\langle C, 01/02/2012 \rangle$	$\langle D, 16/05/2012 \rangle$
2.	$\langle B, 27/01/2012 \rangle$	$\langle C, 06/02/2012 \rangle$	$\langle D, 13/09/2012 \rangle$	–
3.	$\langle B, 03/02/2012 \rangle$	$\langle C, 08/02/2012 \rangle$	$\langle A, 08/02/2012 \rangle$	$\langle D, 27/06/2012 \rangle$
4.	$\langle B, 10/02/2012 \rangle$	$\langle A, 10/02/2012 \rangle$	$\langle D, 19/06/2012 \rangle$	–
5.	$\langle A, 22/02/2012 \rangle$	$\langle C, 22/02/2012 \rangle$	$\langle D, 20/08/2012 \rangle$	–
6.	$\langle B, 28/02/2012 \rangle$	$\langle A, 14/03/2012 \rangle$	$\langle C, 14/03/2012 \rangle$	$\langle D, 25/09/2012 \rangle$

Table 3 Example database for chronic diseases with normalized timestamps

pid	(a)				pid	(b)			
1.	$\langle A, 0 \rangle$	$\langle B, 0 \rangle$	$\langle C, 1 \rangle$	$\langle D, 2 \rangle$	1.	$\langle A, 0 \rangle$	$\langle B, 0 \rangle$	$\langle C, 20 \rangle$	$\langle D, 125 \rangle$
2.	$\langle B, 0 \rangle$	$\langle C, 1 \rangle$	$\langle D, 2 \rangle$	–	2.	$\langle B, 0 \rangle$	$\langle C, 10 \rangle$	$\langle D, 130 \rangle$	–
3.	$\langle B, 0 \rangle$	$\langle C, 1 \rangle$	$\langle A, 1 \rangle$	$\langle D, 2 \rangle$	3.	$\langle B, 0 \rangle$	$\langle C, 5 \rangle$	$\langle A, 5 \rangle$	$\langle D, 145 \rangle$
4.	$\langle B, 0 \rangle$	$\langle A, 0 \rangle$	$\langle D, 1 \rangle$	–	4.	$\langle B, 0 \rangle$	$\langle A, 0 \rangle$	$\langle D, 230 \rangle$	–
5.	$\langle A, 0 \rangle$	$\langle C, 0 \rangle$	$\langle D, 1 \rangle$	–	5.	$\langle A, 0 \rangle$	$\langle C, 0 \rangle$	$\langle D, 180 \rangle$	–
6.	$\langle B, 0 \rangle$	$\langle A, 1 \rangle$	$\langle C, 1 \rangle$	$\langle D, 2 \rangle$	6.	$\langle B, 0 \rangle$	$\langle A, 15 \rangle$	$\langle C, 15 \rangle$	$\langle D, 210 \rangle$

matter. The corresponding database with normalized timestamps is shown on Table 3b. Note that ORs always contain dates.

Additionally we generate time sequences for treatment. For each medication we can find the corresponding diseases with which treatment it is associated. Similarly to disorders we define a projection function that, applied over the event sequence, results a treatment sequence.

4.2 Frequent Patterns Mining

This module applies a modification of the classical Eclat algorithm [7] and generates a prefix-based search tree for chronic diseases itemsets (Fig. 4). In our task frequent patterns are itemsets. Then the main difference is that instead of using pids intersection we apply projection of items in the database and at each level we merge the projection vectors. Let $x \in C$ is an item and $\mathcal{D}$ is the database. We define a vector projection $v(x, \mathcal{D}) = \langle x_1, \ldots, x_n \rangle$, where:

$$x_i = \begin{cases} k, & \exists k : \langle p_i, X \rangle \in D \text{ and } \langle x, k \rangle \in X \\ -1, & otherwise \end{cases}$$

For the database on Table 3a we obtain the projection vectors shown in Fig. 3.

	1	2	3	4	5	6
A	<0,	-1,	1,	0,	0,	1>
B	<0,	0,	0,	0,	-1,	0>
C	<1,	1,	1,	-1,	0,	1>
D	<2,	2,	2,	1,	1,	2>

Fig. 3 Projection vectors of the chronic diseases A, B, C, D (Table 3a)

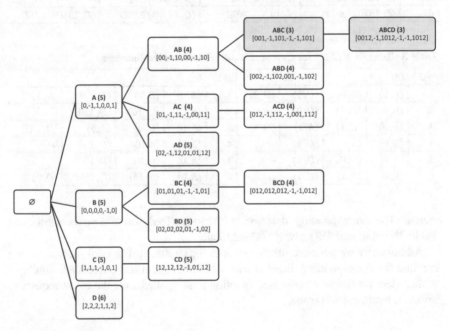

Fig. 4 Prefix tree with projection vectors for *minsup* = 4. *Shaded boxes* indicate infrequent itemsets. For simplicity parentheses are omitted. Next to the itemset is shown its support

Each itemset is considered as a sequence of items in lexicographical order, according to the ICD-10 indices. Let $a = p + d', b = p + d''$ be two sequences of chronic diseases with the same prefix p and corresponding vectors $\langle a_1, \ldots, a_n \rangle$ and $\langle b_1, \ldots, b_n \rangle$. If $d' < d''$, then we generate the result sequence $ab := p + d' + d'' = ad''$. Let the vector for item d'' be $\langle d''_1, \ldots, d''_n \rangle$. For the sequence ab with length k we generate the vector of ordered k-tuples $\langle m(a_1, d''_1), \ldots, m(a_n, d''_n) \rangle$, where the function m is defined as follows:

$$m\left(a_i, d''_i\right) = \begin{cases} -1, & a_i = -1 \text{ or } d''_i = -1 \\ \langle a_i, d''_i \rangle, & a_i \neq -1 \text{ and } d''_i \neq -1 \end{cases}$$

Fig. 5 Association rules generated with *minconf* = 0.9

Association rule	confidence
$AB \rightarrow D$	1.0
$AC \rightarrow D$	1.0
$AD \rightarrow B$	1.0
$AD \rightarrow C$	1.0
$BC \rightarrow D$	1.0
$A \rightarrow D$	1.0
$B \rightarrow D$	1.0
$C \rightarrow D$	1.0

We define a support for sequence a with vector $\langle a_1, \ldots, a_n \rangle$ as $sup(a) = |\{a_i | a_i \neq -1, i = \overline{1,n}\}|$. We are looking for itemsets with frequency above given *minsup*.

From the collection $\mathcal{F}$ of frequent itemsets, using the classic algorithm we generate association rules in the form: $\alpha \xrightarrow{s,c} \beta$, where $\alpha, \beta, \alpha\beta \in \mathcal{F}$, $s = sup(\alpha\beta) \geq minsup$, and the confidence c is defined as follows:

$$c = \frac{sup(\alpha\beta)}{sup(\alpha)}$$

We are looking for strong rules, i.e. with confidence above a given minimum confidence value *minconf*.

Although the collection $\mathcal{F}$ gives us some information about the chronic disorders coexistence necessary for Task 1 solution, association rules show more complex relations between disorders association. In Fig. 5 are shown association rules generated from the prefix tree in Fig. 4.

For example from association rules in Fig. 5 we can conclude that existence of anyone or two of the diseases A, B, C is a risk factor for presence of disorder D as well. In addition the coexistence of diseases A, and D is a risk factor for presence also either of disorder B, or of disorder C.

4.3 Frequent Sequence Mining

For Task 2 we need to study the sequences of chronic disorders and their treatment. We apply breadth-first search in the prefix-tree generated from the previous module and map on each node a substitution σ that converts vectors of k-tuples to vectors of patterns. Let $v = \langle v_1, v_2, \ldots, v_n \rangle$ be a vector for some frequent itemset $X \in \mathcal{F}$, and $v_i = \langle n_{i1}, n_{i2}, \ldots, n_{ik} \rangle$, where $k = |X|$, and $n_{ij}, j = \overline{1,k}$ are numeric values corresponding to the positions of items in the database event sequences. Let the sorted sequence of distinct values in v_i be $n_{ia_0} < n_{ia_1} < \cdots < n_{ia_l}$, where $l \leq k - 1$. We apply normalization of v_i by substitution σ of all occurrences of n_{ia_q} by q, $0 \leq q \leq l$.

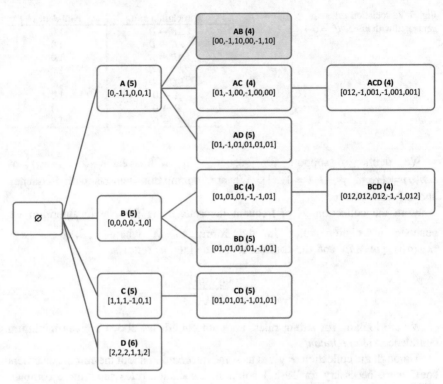

Fig. 6 Prefix tree with patterns mapped vectors for *minsup* = 3. *Shaded boxes* indicate infrequent itemsets

Fig. 7 Sequence patterns.
Parallel events are presented
in *brackets*

Sequence pattern	support
[AC]	3
[AC]D	3
AD	5
BC	4
BCD	4
BD	5
CD	5

The prefix tree in Fig. 4 is transformed to the tree in Fig. 6 after the normalization. For the itemset {A, B} we have support 2 only for both patterns [0, 0] and [1, 0], hence this set will be pruned. The resulting frequent patterns are listed in Fig. 7, where parallel (simultaneous) events are presented in brackets and the leftmost event is the start, and rightmost event is the final event in the sequence.

5 Experiments and Results

To cope with big data, we use tabulation in the implementation of both methods. Thus each level of the tree is stored in separate table as a file. During the tree generation we deal only with the last generated table and the table corresponding to the level 1. For the projection we need to store in the memory only the vectors for three nodes.

We applied both methods for each collection separately and obtained the following results for chronic diseases and treatment (Table 4). These methods are applied also for age and gender specific constraints for each collections. The figures below present the extracted frequent patterns for chronic diseases in total and for age 15–44 years, with $minsup = 80$ for S1 (Fig. 8), and $minsup = 45$ for S2 (Fig. 9). For both collections the minimal support is chosen as 0.5 % of the number of patients, respectively.

We also experimented with different window lengths—1 and 3 months. These intervals are with high importance for healthcare, because they denote the minimal and the optimal interval for which the medication treatment can show effect. In our experiments for window 1 month the results are almost similar to the results for window 0, because the frequency of patient visits usually exceeds 1 month. For window 3 months the obtained results decline dramatically, because of the rather short period of our retrospective study—only 3 years. And therefore the maximal size of the frequent sequence is 2 chronic diseases only.

Table 4 Characteristics of structured event data

Characteristics	Collections			
	S1		S2	
	Total	Age 15–44	Total	Age 15–44
Total number of extracted ICD-10 codes	782,448	288,625	248,067	198,588
Total number of distinct ICD-10 codes	5,790	4,530	4,697	4,240
Total number of extracted chronic diseases	107,789	36,180	31,151	22,194
Total number of distinct chronic diseases	227	216	228	215
Total number of patient with chronic diseases	37,921	16,059	8,414	6,933
Avg length k of chronic diseases sequence	2.843	2.253	3.702	3.201

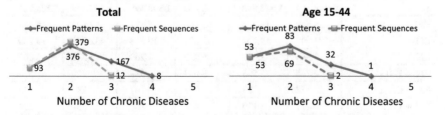

Fig. 8 Total number of extracted frequent patterns and sequences for collection S1

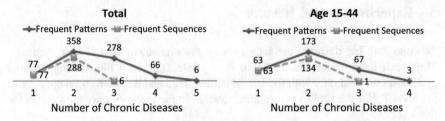

Fig. 9 Total number of extracted frequent patterns and sequences for collection S2

Together with a bunch of well known associations in each collection we found some patterns that are not well studied in the medical literature and further study will be needed to investigate their significance in healthcare management.

5.1 Case Study

Initially we started processing the collection $S3$ for patients with Diabetes Mellitus Type 2 (T2D—E11) and extracted relatively high number of frequent patterns containing different mental disorders—ICD10 codes F00–F99. This result motivated us to process collection $S1$ for patients with Schizophrenia (SCZ—F20). Mining complex itemsets including both chronic and acute diseases we obtained as a result high frequency of some acute diseases as well. And we included in our study collection $S2$ for patients with Hyperprolactinemia (E22.1) to investigate in more details the association of these three diseases (Table 1).

On the other side SCZ develops in relatively young age 18–35 years and the peak ages of onset are 25 years for males and 27 years for females. In contrast T2D develops relatively later and the peak ages of onset are 45–65 years. This motivates us to study in more details age specific event sequences (Table 5). These co-morbidity facts together with the administrated medications were interpreted as

Table 5 Statistics for collections S1, S2 and S3

Year	2012		2013		2014	
ICD10	Total	Age 15–44	Total	Age 15–44	Total	Age 15–44
Total	5,048,165	1,583,980	5,132,403	1,598,595	5,147,648	1,585,024
F20	38,560	16,226	38,464	15,871	37,921	15,241
E11	403,048	18,704	419,237	19,304	432,705	19,200
E22.1	3,663	3,018	5,273	4,269	5,347	4,267
E11 + F20	2,775	505	3,093	538	3,209	576
F20 + E22.1	158	73	251	107	237	183
E11 + E22.1	271	120	472	206	534	231
F20 + E22.1 + E11	19	7	33	18	30	20

temporal events and the event sequences were processed by the mining tool for pattern search. The data are extracted from more than 8 millions ORs (Table 1).

It is well known that patients with SCZ are at an increased risk of T2D [19], therefore a better understanding of the factors contributing to T2D is needed. SCZ is often treated with antipsychotic agents but the use of antipsychotics has been associated with Hyperprolactinemia, or elevated prolactin levels (a serious hormonal abnormality). Thus, given the large repository of ORs, that covers more than 5 million citizens of Bulgaria, it is interesting to study associations and dependencies among SCZ, T2D and Hyperprolactinemia in the context of the treatment prescribed to the patients.

Regarding the treatment it is well known that the classical antipsychotics, blocking D2 dopamine receptors, lead to extrapyramidal effects related to antagonism in the nigrostriatal pathway, and Hyperprolactinemia is due to antagonism in the tuberoinfundibular pathway. In the early 1990s a new class of antipsychotics was introduced in the clinical practice with the alleged advantage of causing no or minimal extrapyramidal side effects, and the resulting potential to increase treatment adherence. However, there are data, that some of these antipsychotics can induce Diabetes, Hyperlipidaemia and weight gain.

Our study considers the presence of:

- Hyperprolactinemia in the patients with Schizophrenia,
- T2D and Schizophrenia in the patients with Hyperprolactinemia,
- T2D and Hyperprolactinemia in the patients with Schizophrenia and
- T2D in the patients with Schizophrenia and Hyperprolactinemia.

We also combine diagnosis patterns with treatment sequences patterns that include first generation of antipsychotics (ATC codes: N05AF01, N05AB02, N05AD01, N05AF05) and/or second generation of antipsychotics (ATC codes: N05AL05, N05AX12, N05AH02, N05AH03, N05AX13, N05AH04, N05AX08, N05AE03, N05AE04).

We found an increased rate of Hyperprolactinemia and T2D in patients with Schizophrenia, compared to presence of these diseases in patients without Schizophrenia. Table 6 presents extracted frequent sequence patterns for SCZ (F20), T2D (E11) and Hyperprolactinemia (E22.1). We can observe that in these sequences dominates the relation SCZ (F20) precedes T2D (E11). In Fig. 10 is presented the temporal relation between the onset of T2D (E11) after the onset of SCZ (F20) measured in months. Table 7 shows the prevalence of T2D (E11) in the entire population and among patients suffering from SCZ (F20). Although there is no significant difference for the total collection, the statistics for age 15–44 years show that for patients with SCZ (F20) there is about three times higher risk for development of T2D (E11). The finding is explicated in relation with the administrated treatment.

Table 6 Extracted sequence patterns with *minsup* = 3 from S1

Sequence pattern	Total	Age 15–44 years
$F20 \rightarrow E11 \rightarrow E22.1$	19	6
$F20 \rightarrow E22.1 \rightarrow E11$	11	6
$F20 \rightarrow [E22.1, E11]$	21	8
$[F20, E11] \rightarrow E22.1$	8	4
$[F20, E11, E22.1]$	5	3
$F20 \rightarrow E22.1$	314	264
$[F20, E22.1]$	25	22
$E22.1 \rightarrow F20$	27	23
$E11 \rightarrow F20$	785	142
$[F20, E11]$	1,512	231
$F20 \rightarrow E11$	2,201	491

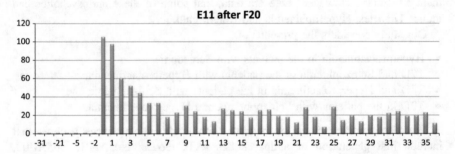

Fig. 10 Total number of patients with E11 after diagnosis F20, grouped by period measured in months

Table 7 Prevalence of E11

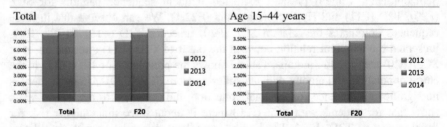

6 Conclusion and Further Work

The proposed approach presents an elegant and efficient solution for the task of frequent patterns and sequences mining in the scope of big data applied over real medical datasets. Due to the constraints for distinct diagnoses in our patterns the proposed solution allows to reuse the frequent patterns tree generated for the Task 1 for frequent sequences mining. This is due to the fact that the tree from Task 1 can only shrink on Task 2, i.e. not all possible permutations of the itemsets have sufficient support above the *minsup*. It also allows checking different hypotheses for treatment effect and risk factors, by using different size for search window. In contrast to the other classical methods for frequent sequence mining the proposed solution allows parallel (simultaneous) events in sequences to be recognized and grouped. This approach can be applied also for other domains where heterogeneous events need to be mined.

In contrast to other approaches for temporal events mining in EHR the proposed approach is not so expensive like EMRView [1, 2] which requires offline preliminary generation of frequent patterns over which later the system applies filtering and projection. Also we process the entire collection with no limitation like interactive systems [5] that reduce the task by initially selecting events of interest for analysis and visualization queries or sentinel events [3].

The main advantage of the proposed approach is that it takes into account the data specificity and enables flexible parameterization of (i) the set of the diagnoses in the research interest, and (ii) the time window size.

The contribution of the paper is that it demonstrates a powerful, sufficiently generalized technology for discovering correlations in the medical domain.

Our future plans include experiments with collections of ORs for patients with Cardio-vascular Disorders and Malignant neoplasm. We also plan to process the complete sequence of diseases—including both chronic and acute diseases to investigate more complex causalities.

Integrating mining of status data for patients will further elucidate the risk factors and causality for some acute and chronic disorders. Moreover, changing patient states brings new events in the whole picture and hopefully longer sequences will be revealed that will point to new interesting medical facts.

Acknowledgments The research work presented in this paper is supported by the FP7 grant 316087 AComIn "Advanced Computing for Innovation", funded by the European Commission in the FP7 Capacity Programme in 2012–2016. The authors also acknowledge the support of the Bulgarian Health Insurance Fund, the Bulgarian Ministry of Health and Medical University—Sofia for providing the experimental data.

References

1. Patnaik, D., Parida, L., Butler, P., Keller, B.J., Ramakrishnan, N., Hanauer, D.A.: Experiences with mining temporal event sequences from electronic medical records: initial successes and some challenges. In: Proceedings of the 17th ACM SIGKDD International Conference on Knowledge Discovery and Data Mining (KDD'11), San Diego, August 2011, pp. 360–368 (2011)
2. Patnaik, D., Laxman, S., Chandramouli, B., Ramakrishnan, N.: Efficient episode mining of dynamic event streams. In: Proceedings of the IEEE International Conference on Data Mining (ICDM'12), Brussels, Belgium, December 2012, pp. 605–614 (2012)
3. Wang, T., Plaisant, C., Quinn, A.J., Stanchak, R., Murphy, S., Shneiderman, B.: Aligning temporal data by sentinel events: discovering patterns in electronic health records. In: Proceedings of the SIGCHI Conference on Human Factors in Computing Systems (CHI '08), ACM, New York, NY, USA, 2008, pp. 457–466 (2008)
4. Gotz, D., Wang, F., Perer, A.: A methodology for interactive mining and visual analysis of clinical event patterns using electronic health record data. J. Biomed. Inf. **48**, 148–159 (2014)
5. Rind, A., Wang, T.D., Aigner, W., Miksch, S., Wongsuphasawat, K., Plaisant, C., Shneiderman, B.: Interactive information visualization to explore and query electronic health records. J. Found. Trends® Human–Comput. Interact. **5**(3), 207–298 (2013)
6. Lee, N., Laine, A.F., Hu, J., Wang, F., Sun, J., Ebadollahi, S.: Mining electronic medical records to explore the linkage between healthcare resource utilization and disease severity in diabetic patients. In: Proceedings of the First IEEE International Conference on Healthcare Informatics, Imaging and Systems Biology (HISB), pp. 250–257 (2011)
7. Zaki, M.J., Wagner Jr., M.: Data Mining and Analysis: Fundamental Concepts and Algorithms. Cambridge University Press (2014)
8. Sun, X., Orlowska, M., Zhou, X.: Finding event-oriented patterns in long temporal sequences. In: Proceedings of the 7th Pacific-Asia Conference on Knowledge Discovery and Data Mining (PAKDD 2003), Seoul, South Korea, April 2003, Springer LNCS 2637, pp. 15–26 (2003)
9. Guyet, T., Quiniou, R.: Mining temporal patterns with quantitative intervals. In: Zighed, D., Ras, Z., Tsumoto, S. (eds.) Proceedings of the 4th International Workshop on Mining Complex Data, IEEE ICDM Workshop, pp. 218–227 (2008)
10. Guyet, T., Quiniou, R.: Extracting temporal patterns from interval-based sequences. In: Proceedings 22nd International Joint Conference on Artificial Intelligence, pp. 1306–1311 (2011)
11. Monroe, M., Lan, R., Lee, H., Plaisant, C., Shneiderman, B.: Temporal event sequence simplification. IEEE Trans. Vis. Comput. Graph. **19**(12), 2227–2236 (2013)
12. Yang, J., McAuley, J., Leskovec, J., LePendu, P., Shah, N.: Finding progression stages in time-evolving event sequences. In: Proceedings of the 23rd International Conference on World Wide Web (WWW '14), ACM, New York, NY, USA, pp. 783–794 (2014)
13. Chronic diseases, WHO. http://www.who.int/topics/chronic_diseases/en/
14. International Classification of Diseases and Related Health Problems 10th Revision. http://apps.who.int/classifications/icd10/browse/2015/en
15. Nikolova, I., Tcharaktchiev, D., Boytcheva, S., Angelov, Z., Angelova, G.: Applying language technologies on healthcare patient records for better treatment of Bulgarian diabetic patients. In: Agre, G. et al. (eds.) Artificial Intelligence: Methodology, Systems, and Applications, Lecture Notes in Artificial Intelligence, vol. 8722, pp. 92–103. Springer (2014)
16. Boytcheva, S.: Shallow medication extraction from hospital patient records. In: Koutkias, V., Nies, J., Jensen, S., Maglaveras, N., Beuscart, R. (eds.) Studies in Health Technology and Informatics, vol. 166, pp. 119–128. IOS Press (2011)
17. Anatomical Therapeutic Chemical (ATC) Classification System. http://atc.thedrugsinfo.com/
18. Boytcheva, S., Angelova, G., Angelov, Z., Tcharaktchiev, D.: Text mining and big data analytics for retrospective analysis of clinical texts from outpatient care. Cybern. Inf. Technol. **15**(4), 58–77 (2015)

19. Marder, S., Essock, S., Miller, A., Buchanan, R., Casey, D., Davis, J., Kane, J., Lieberman, J., Schooler, N., Covell, N., Stroup, S., Weissman, E., Wirshing, D., Hall, C., Pogach, L., Pi-Sunyer, X., Bigger Jr., J.T., Friedman, A., Kleinberg, D., Yevich, S., Davis, B., Shon, S.: Physical health monitoring of patients with schizophrenia. Amer. J. Psychiat. **161**(8), 1334–1349 (2004)

Emerging Applications of Educational Data Mining in Bulgaria: The Case of UCHA.SE

Ivelina Nikolova, Darina Dicheva, Gennady Agre, Zhivko Angelov, Galia Angelova, Christo Dichev and Darin Madzharov

Abstract As part of the EC FP7 project "AComIn: Advanced Computing for Innovation", which focuses on transferring innovative technologies to Bulgaria, we have applied educational data mining to the most popular Bulgarian K-12 educational web portal, UCHA.SE. UCHA.SE offers interactive instructional materials— videos and practice exercises—for all K-12 subjects that can be used in schools and for self-learning. Currently it offers more than 4,150 videos in 17 subjects and more than 1,000 exercises. The goal of the project is to study how educational data mining can be used to improve the quality of the educational services and revenue generation for UCHA.SE. In this paper we describe the conducted study and outline the machine learning methods used for mining the log data of the portal as well as the problems we faced. We then discuss the obtained results and propose measures for enhancing the learning experiences offered by UCHA.SE.

I. Nikolova (✉) · G. Agre · G. Angelova
Institute of Information and Communication Technologies,
Bulgarian Academy of Sciences, 1113 Sofia, Bulgaria
e-mail: iva@lml.bas.bg

G. Agre
e-mail: agre@iinf.bas.bg

G. Angelova
e-mail: galia@lml.bas.bg

D. Dicheva · C. Dichev
Winston-Salem State University, 601 S. Martin Luther King Jr. Drive,
Winston Salem, NC 27110, USA
e-mail: dichevad@wssu.edu

C. Dichev
e-mail: dichevc@wssu.edu

Z. Angelov
ADISS Lab. Ltd, Sofia, Bulgaria
e-mail: angelov@adiss-bg.com

D. Madzharov
UCHA.SE, 1113 Sofia, Bulgaria
e-mail: darin@ucha.se

© Springer International Publishing Switzerland 2016
S. Margenov et al. (eds.), *Innovative Approaches and Solutions in Advanced Intelligent Systems*, Studies in Computational Intelligence 648,
DOI 10.1007/978-3-319-32207-0_8

1   Introduction

Web-based learning is becoming a significant part of contemporary education. Supported by online learning environments spanning from online digital libraries and encyclopedias, such as Wikipedia, to educational websites, such as Khan Academy, to Massive Open Online Courses (MOOCs), such as Coursera, Udacity and edX, it is easily accessible and ubiquitous. The intrinsically interactive nature of the online environments results in the generation of huge volumes of information about the online activities of the learners using them. This is a gold mine of educational data that can be used to improve users' learning experiences. To extract useful relevant information from it, data mining methods are typically used. Data mining is concerned with the automatic extraction of implicit and interesting patterns from large data collections. Applying it in the context of education is known as Educational Data Mining (EDM).

EDM is an emerging discipline "concerned with developing methods for exploring the unique types of data that come from educational settings, and using those methods to better understand students, and the settings which they learn in."[1] Typical questions addressed by EDM are: What student actions are associated with better learning? What actions indicate satisfaction and engagement? What features of an online learning environment lead to better engagement and learning? Educational data mining is typically applied to student information systems, learning management systems (LMS), web-based courses, adaptive intelligent educational systems, personal learning environments, open data sets, and social media. The targeted data are of two main categories: student-performed actions with a given outcome (e.g. content viewing during a time span, completing an activity with a given result, etc.) and student profile (incl. age, gender, interests, etc.).

EDM leverages data mining techniques to analyze education-related data. The primary categories of methods used for educational data mining are classification, clustering, regression, relationship mining, and discovery with models. The most popular are classification and clustering—together they reach 69 % of the DM tasks used by EDM approaches [16]. Prediction is one of the important categories of EDM. In prediction, the goal is to develop a model which can infer a single aspect of the data from some combination of other aspects of the data. Prediction models are used to either predict future, or to predict variables that are not feasible to directly collect in real-time, such as data on affect or engagement [6].

Educational data mining is largely used for constructing and or improving student models to enable personalized learning experiences and recommendations for students, providing feedback for supporting instructors, predicting student performance, detecting undesirable student behaviors including disengagement within educational software, grouping students, social network analysis, planning and

---

[1]www.educationaldatamining.org.

scheduling, and scientific research into learning and learners [5, 17]. EDM is one of the most efficient approaches for automated adaptation of educational software. It can be used for identifying a problem or a need and automatically focusing on it thus personalizing the learner's experience (cf. [1, 4, 11]). Automated detectors of a range of disengaged behaviors have been developed using prediction methods or knowledge engineering methods [6], including detectors of gaming the system [2, 7, 15, 20], off-task behavior [3, 8], and carelessness [18].

For online educational websites and learning environments, educational data mining can provide valuable information on user behavior on the site pages, such as how often students use the site, what they are interested in, do they find what they are looking for, what learning materials are used the most, etc. These findings can be effectively used to improve customer experience and increase the engagement and the rate at which people learn.

In this paper we present our effort to apply educational data mining to the most popular Bulgarian educational site, UCHA.SE ('ucha' is the Bulgarian word for 'learn'). Similarly to Khan Academy, UCHA.SE offers instructional videos and practice exercises but in Bulgarian language. The overall goal of the project is to improve the quality of educational services and the subscription rate for the site. To this end we employed two approaches: usability study and educational data mining, with the second one emphasized here. The paper is organized as follows. In the next section we discuss the educational site UCHA.SE and the results of our usability analysis of it. Section 3 describes the procedure which we have used for mining the educational data. In Sect. 4 we present the data miming results received so far, which are then discussed in Sect. 5. Section 6 concludes and describes our future plans.

2　Case Study: Educational Site UCHA.SE

In this section we present briefly UCHA.SE and some recommendations related to the structure of the site and the services it provides based on our usability analysis.

2.1　UCHA.SE—Structure and Services

UCHA.SE is an online learning environment, aimed at supporting the K-12 National Bulgarian Curricula through offering interactive instructional materials—videos and practice exercises—for all subjects. The site can be used to support formal and informal education, that is, in schools and for self-learning. Currently it offers more than 4,150 videos in 17 subjects, including the basic K-12 subjects, as well as Introductory level English, German, French, and Spanish, and Introduction to Programming. Figure 1 shows a screenshot from the site's homepage.

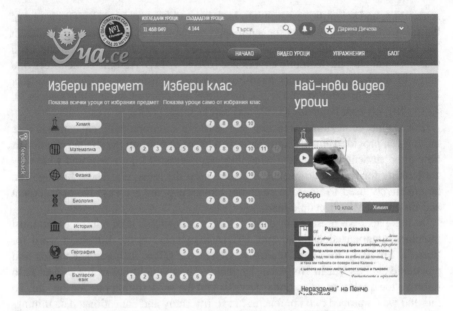

Fig. 1 A screenshot from the home page of the educational site UCHA.SE

To facilitate browsing and resource finding on the site, the learning resources are structured using the following classification: *Subject—Class—Section—Topic/Lesson* (see Fig. 2). The metadata of the video resources include: video title, description, keywords, length of the video, date of uploading, subject, class, and section.

UCHA.SE employs two main instructional strategies: memorizing, in which students watch videos, and practicing, in which students take quizzes consisting of questions related to the videos. After submitting a quiz, the student may ask for help for particular questions, or for the correct answers with corresponding explanations. When asking for help, the student is sent to the corresponding fragment in the related video.

The site implements some gamification elements [14] to engage and motivate the users, including points, levels, badges, progression and status, as well as social features such as commenting, voting and liking.

As of September 2015, the site has more than 240,000 registered users—students (including college students), teachers, and parents.

2.2 UCHA.SE—Evaluation and Recommendations

Our evaluation of the functionality and usability of the site resulted in several recommendations, which can be grouped as follows:

Fig. 2 A screenshot demonstrating the navigational structure in the site UCHA.SE and the topic "Vowels in the Bulgarian Language"

- Personalization of user experiences in the portal based on comprehensive user modeling.
- Enriching the semantic annotation of the learning content for enabling recommendation.
- Enhancing portal gamification for more effective motivation and meaningful engaging of its users.
- Improving portal user interface, including the use of icons, links, breadcrumbs menus, scrollable areas, etc.

The initial interface design and functionality of the website bears a resemblance to a digital collection of learning resources (videos and exercises linked to them), rather than a learning environment. As a result of this, little attention has been paid to the learning process and to the adaptation and personalization of user experiences. The latter however is vital for building an effective educational environment. In the context of UCHA.SE, it would be useful to distinguish between at least two categories of users, students and teachers, which have different goals. It is critical for students to be informed about their learning progress and presented with a view summarizing and visualizing their achievements. For example, upon their entering the portal, students could be presented with a personalized view showing which videos they have already watched (even which they have completely finished), how many/which exercises they have attempted (how many times and with what results), etc. To this end the system should be extended with a user modeling component responsible for building and maintaining comprehensive models of its users.

The recommendation about a richer semantic annotation of the learning content targets the facilitation of more efficient navigation, search and content recommendation. For example, "knowledge maps" can be built to semantically organize the metadata of the learning resources in the portal. Such maps can represent not only knowledge relationships among the various learning resources in UCHA.SE but also pedagogical relations and indicators such as prerequisite relationships, level of difficulty, etc. These can be used for both navigation and the provision of feedback. The later can be implemented as a personalized recommender service that uses the semantic annotation of the resources and is based on the user model. Such a recommender can suggest to the student what would be the most appropriate activity for them to complete next, for example, to watch a particular video, to complete a particular exercise linked to the last watched video, or to go back and attempt another exercise linked to a previously watched video (in case the student hasn't demonstrated sufficient knowledge of a prerequisite topic). In the implementation of this functionality open source software, such as the search server Apache Solr (lucene.apache.org/solr/), may be useful.

Another recommendation is related to the gamification of UCHA.SE and the engagement of learners in meaningful interactions with the site. Gamification, when properly used, can create an environment stimulating students to actively involve themselves through challenge, competition, and fun [13]. While a few of the most popular game mechanisms, including points, badges/achievements, leaderboard and encouraging messages have been incorporated in the site, there are problems with their definition and interpretation. For example, the meaning of the user's score (total accumulated points) is not evident; there is no clear difference between points and achievements (all points are used for achievements). It is recommendable to separate points given for learning-related activities (e.g. watched videos, highly rated comments, etc.) and such given for site usage-related activities (e.g. site visits and recommending the site to a friend). The former points can then be used as a correct measure of user learning progress, while the later can be used for determining other user achievements. In addition, such a separation will reduce the possibility of 'gaming' the system.

Given the recognized ability of gamification to increase users' interest and motivation, the possibility of including additional gamification elements and incentives in UCHA.SE should be considered. An example of a suitable game mechanism is 'virtual currency'. Another line to pursue is the integration of game techniques with semantic technologies. For example, the above mentioned "knowledge maps" can be used for the provision of feedback, freedom of choice, leveling-up, and other game mechanisms.

3   Mining the Site

The second goal of the study, the one which is in the focus of this paper, is the improvement of the revenue generation for the educational site UCHA.SE. For this purpose we explored the relationship between users' behavior and their subscription to the site employing educational data mining.

3.1   Data Preparation

The user-related information needed for exploring the relations between users' behavior and their subscription to the site was extracted from the system logs and students' performance data stored in the relational database of UCHA.SE, using the business intelligence software BITool.[2] We defined a *user model* consisting of 27 attributes that reflect four types of information:

- General information about the user. These attributes include user ID, *user type* (pupil, student, teacher, parent, other), *school grade, favorite subject(s)* and *type of first subscription*. This information is entered by the user during registration.
- Information related to the user activities. These attributes reflect the user engagement with the site by describing the relative intensity of site usage, the variety of subjects and grades the user accesses, the average time per day he/she spends in watching videos, exercising or making comments, etc. Most of the attributes are aggregations of values extracted from the user logs over the whole period of site use or for a certain time span (e.g. the last subscription period). The attributes include *period between the first and last access to resources by the user, the most watched subject, number of accessed distinct subjects, number of accessed distinct grades, average time watched videos per day (in seconds), average number of attempted exercises per day, average number of accesses to distinct exercises per day, average number of accessed resources per day, average number of user comments per day, percent videos for which the user also attempted corresponding exercises out of the total number watched videos, percent watched distinct videos the user also commented on*, etc.
- Information related to the user knowledge and skills. In this category we included attributes reflecting users' ability to be persistent, to learn efficiently and to complete successfully the tests. The attributes are *percent of videos watched for more than 90 % of their length out of all watched videos, percent of videos watched more than once, percent of distinct videos watched more than once, percent of exercises completed with more than 75 points out of all exercise attempts, percent of exercises retaken out of all exercises attempts, percent of distinct exercises which were retaken*. When calculating the attributes related to

---

[2]http://www.adiss-bg.com/en/bitool.

watched videos we took into account only videos watched for at least 10 s. We included two distinct attributes for repeated use of exercises and videos—the first one was used to measure the percentage of overall time spent in repeating the activity, and the second one—to measure the percentage of distinct resources which were used more than one time.

- Information about the social behavior of the user on the site. These attributes represent the *level* achieved by the user with regard to demonstrated: *motivation* (defined by the site developers as the number of consecutive days of using the site), *leadership* (define as the number of user comments with rating higher than 3), *charisma* (defined as the number of friends invited to the portal), *influence* (defined as the number of lessons shared in Facebook), etc. The numbers determine the level for each attribute; all levels are from 1 to 6. For example, if the user has accessed the site for 6 or more consecutive days, her level of motivation is 2.

Most of the attributes, especially in the second group, were derived from attributes in the relational database of UCHA.SE, some by applying non-trivial calculation. This made the data transformation stage very labor-intensive. In constructing the user model we also faced several problems related to the quality of available data. For example, personal information is typically not provided by users; the points in the system are not separated into knowledge level-related (e.g. videos watched, high rating on a comment, etc.) and site usage-related (e.g. site visits, resource share, etc.), etc. In addition, we had to find a way to approximate missing values. For example, the registration date of some users was not recorded in the system database immediately after releasing the site. For such users, we decided to use the first date the user accessed a video or an exercise as a registration date.

3.2 Formulating the Data Mining Task

As it has been already mentioned, the general desire of the site developers was to understand the relationship between users' behavior and their use and subscription to the site. The most significant aspect of such formulated task was *to increase the site subscription rate*. In order to do this, we reformulated this task as a classification task for predicting whether a user will renew her current subscription in a certain period of time (for example 3 months) after its expiration.

However, in order to solve this classification task by means of a machine learning approach we had to construct a training data set. Each example from that set is a description of a site user represented by the user model attributes and extended by a value of the target (class) attribute describing whether the user has renewed her subscription or not. The construction of such training data set requires the following decisions to be made:

- Are there restrictions on the available data to be included into the training set?
- What is the reference point (in the timeline) for creating the training data set?
- What is the period for aggregating data values?
- How to calculate the values of the target attribute?

3.2.1 Restrictions on Data Collection

UCHA.SE offers free access to some demo videos and paid access to all other resources. The user could pay either for separate visits or for a certain subscription period in which she can use the resources unlimited number of times. For the purpose of this study we have decided to include into the training set only users who had at least one subscription to the site (no matter of its length).

3.2.2 Reference Point of Time for Training Set Creation

The academic year in the Bulgarian schools has 2 semesters: from mid September to mid February and from mid February to mid June or end of June. This impacts the use of UCHA.SE because the students need the site mostly during the academic year. During the vacation periods there is a lower number of subscriptions and number of views. Therefore, in order to be able to predict the behavior of the students, it is very important to consider the time of the year for which the data is being analyzed.

In the beginning of the academic year there are many subscriptions awaiting renewal and in the end of the year there are too many subscriptions which will not be renewed before the beginning of the next academic year. That is why we decided to construct the training set from data collected as of the beginning of the second semester when there is a sufficient number of available subscriptions. Thus our data consists of records within the time span 14 February 2014–28 February 2015. Based on these records we constructed the training set to be used for learning predictive models for solving the formulated above classification task. To be more precise, the selected reference point of 28 February 2015 defines the moment in time in which we calculate whether a user has renewed her subscription to the site or not.

3.2.3 Data Aggregation Period

Since the site proposes different types of subscriptions (varying from 3 days to 12 months), the values for some attributes describing the training dataset are calculated per day, e.g. *average number of accessed distinct exercises per day, average number of accessed resources per day,* etc. Moreover, we calculated these values only from the last subscription period which, in our opinion, is the most significant for a subscription renewal. The values of the attributes which represent percentages or raw numbers (e.g. *period between the first and last access to resources by the user, percent of videos for which the user also attempted corresponding exercises out of the total number watched video,* etc.) were calculated over the whole period since the registration, except for *the number of accessed grades,* which is also calculated for the last subscription period only.

3.2.4 Calculating Values of the Target Attribute

We defined a 'target' function to present whether a user renewed her subscription in the period of 90 days from the expiration of her previous subscription and the selected reference point. Such a definition led to a 3-class problem (see Fig. 3).

According to this definition:

- An example is considered as positive example if either the current subscription has been started before ending the previous one or the period between these subscription is less or equal to 90 days.
- An example is considered as negative example if the user did not renew his/her subscription in the period of 90 days after its expiration.
- The value of the target attribute of an example can not be defined (i.e., is "unknown") if either the current subscription is the first one for the user (related to the selected reference point) or less than 90 days have passed after the date of expiration of the previous subscription but the user still has not renewed her subscription in the moment of time defined by the selected reference point.

The distribution of the training examples constructed with the reference point of 28 February 2015 is shown in Fig. 4.

After removing the examples for which the value of the target attribute was *unknown,* we got a two-class problem with 11,263 "negative" and 2,430 "positive" examples.

4 Experiments and Results

For solving the formulated above classification task we used WEKA[3] and Orange[4]—open source data mining toolboxes that implement numerous machine learning algorithms and provide tools for data preprocessing, classification, regression, clustering, association rules, and visualization.

We were looking for a predictive model that solves the task and satisfies the following requirements:

---

[3]http://www.cs.waikato.ac.nz/ml/weka/.

[4]http://orange.biolab.si/.

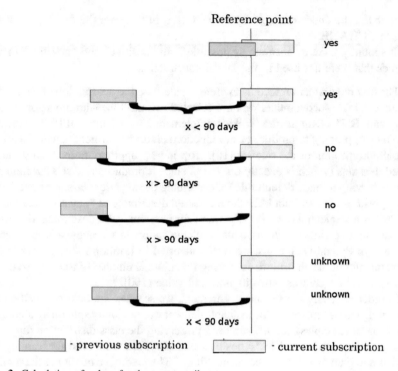

Fig. 3 Calculation of values for the target attributes

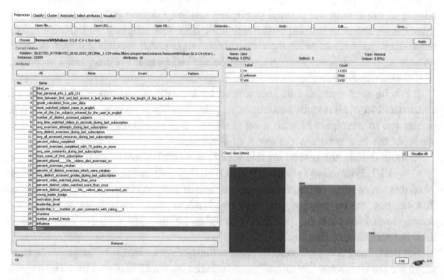

Fig. 4 Distribution of the training examples: 11,263—"no" (*blue*), 2,430—"yes" (*light blue*), 8,966—"unknown" (*red*)

- It should be understandable by the end-user, in this case the developers of the site UCHA.SE.
- It should predict as accurate as possible both "positive" and "negative" examples that were not used in the model construction.

The first requirement forced us to prefer a rule-based representation of the model. In order to learn such a model we have selected two machine learning algorithms—JRip and CN2. JRip is the Weka's implementation of the RIPPER (repeated incremental pruning to produce error reduction) algorithm, which includes heuristic global optimization of the rule set [10]. JRip is very appropriate for highly imbalanced data sets (which is exactly our case), since it produces rule sets for minority classes leaving a single default rule for describing the majority class. However, with this algorithm we could not learn the rule-based description of "negative" examples.

CN2 is a sequential covering algorithm for inducing rules describing all classes [9]. In our experiments we used its implementation in Orange—a Data Mining Toolbox in Python [12]. Since we were interested in learning rather general rules, the parameter of the algorithm, describing the minimal number of training examples a rule should cover, was set to its maximal value (100).

In order to satisfy the second requirement, we selected the hold-out method for evaluating learned models—the available data set was randomly split on a training set (90 % of all examples) and a testing set, preserving the class distribution (approximately 82 % negative and 18 % positive examples) that were used for learning the models and then their testing, correspondingly. The predictive quality of the models was measured by the classification accuracy and the F-measure (for the minority class).

Since the training set is highly imbalanced, we carried out a set of experiments aiming at improving the learned predictive models via manipulating the class distribution in the set. We experimented with both subsampling and oversampling techniques, as well as with their combination [19]. All models learned on such manipulated training set were evaluated on the independent testing set. The results of the experiments are shown in Table 1.

It can be seen that the rule set induced by the JRip algorithm from the data set constructed from all negative examples and duplication of all positive examples (the proportion of examples in such a set is about 70 % negatives and 30 % positives) achieves the best accuracy (in terms of F-measure).

In more details, the results for this set of rules are: Precision (yes) = 0.58, Recall (yes) = 0.70, F-measure (yes) = 0.634, Overall classification accuracy = 85.69 %. Thus this set of rules (without the last one, which assigns class "no" to all examples that do not match any of the "positive" rules) was selected as a rule-based description of the class of "positive" examples, i.e. the class of users who renewed their subscription in due time.

For describing users who did not renew their subscriptions (i.e. "negative" examples), we have selected "negative" rules induced by the CN2 algorithm. The experiments showed that the best results were achieved for the rules learned from a data set constructed from all positives examples and randomly selected 67 % of all negative examples: Precision (no) = 0.893, Recall (no) = 0.944, F-measure (no) = 0.918,

Table 1 Experiments with JRip and CN2

		All data (82–18 %)	All yes + 67 % no (76–24 %)	All yes + 50 % no (70–30 %)	All yes -33 % no (60–40 %)	2 × All yes and All no (70–30 %)	3 × All yes and All no (60–40 %)	2 × All yes and 67 % no (62–38 %)	2 × All yes and 50 % no (54–46 %)	Majority vote
JRIP	Clas. Acc. (%)	85.47	**85.84**	85.04	81.75	85.69	85.26	84.45	78.90	82.26
	F-Meas. (yes)	0.530	0.589	0.592	0.595	**0.634**	0.631	0.62	0.578	–
	№ rules	14	16	15	8	19	49	28	21	1
CN2	Clas. Acc. (%)	85.18	**86.06**	85.84	84.16	85.40	85.84	85.26	84.16	82.26
	F-Meas. (yes)	0.418	0.546	0.587	0.556	0.552	0.586	**0.599**	0.580	–
	№ rules (no)	77 (43)	58 (32)	54 (29)	45 (22)	95 (51)	107 (47)	76 (36)	66 (29)	1

Table 2 Experiments with J48 and SVM

		All data (82–18 %)	All yes + 67 % no (76–24 %)	All yes + 50 % no (70–30 %)	All yes -33 % no (60–40 %)	2 × All yes and All no (70–30 %)	3 × All yes and All no (60–40 %)	2 × All yes and 67 % no (62–38 %)	2 × All yes and 50 % no (54–46 %)	Majority vote
J48	Clas. Acc. (%)	**87.08**	86.13	84.67	83.21	85.04	85.47	84.74	82.34	82.26
	F-Meas. (yes)	0.576	0.583	0.582	0.615	0.602	**0.632**	0.619	0.602	–
SVM	Clas. Acc. (%)	87.08	**87.81**	87.74	87.08	87.45	86.50	86.35	85.18	82.26
	F-Meas. (yes)	0.526	0.628	0.650	**0.679**	0.646	0.654	0.646	0.656	–

Overall classification accuracy = 86.06 %. It should be mentioned that these results are better than the corresponding results for the "default" negative JRip rule (Precision (no) = 0.932, Recall (no) = 0.891, F-measure (no) = 0.911). Thus, we may expect that the rule set constructed by a combination of the mentioned above "positive" rules induced by JRip and "negative" rules induced by CN2 (and extended by the additional default rule classifying all unmatched examples as "positive" ones) will have F-measure (yes) ≥ 0.634 and Classification accuracy ≥ 86.06.%.

In order to evaluate the relative predictive quality of this easy understandable to the end-users rule-based model, we conducted a set of experiments with two other very effective but non rule-based machine learning algorithms—J48 and SVM (see Table 2). J48 is the Weka's implementation of the decision tree learner C4.5 [21]. For the experiments with the Support Vector Machine (SVM) algorithm we used its implementation in Orange with the linear kernel. The models were induced on the same variants of the training data set as in the experiments with JRip and CN2 and tested on the same independent test set.

As one can see, the predictive quality of our rule set is close to that of the model produced by one of the best classification algorithm—SVM (87.08 % and 0.679), however, the SVM model is not understandable by the end-user.

5 Discussion

The final model predicting whether a user would renew her subscription or not contains two sets of rules—19 of them describe examples of the minority class "*yes*" (the users who would renew their subscription) and 32—the majority class "no". Among the "positive" rules, 15 rules have precision above 75 %, while 25 of the "negative" rules have precision higher than 90 %. Below are given examples of positive and negative rules:

```
IF motivation_level = 1
and percent_video_watched_more_than_once >= 7
and time_between_first_and_last_access_in_last_subscr <= 0.22
and number_of_distinct_accessed_subjects = 8
and type_name_of_first_subscription = 3 months
and percent_distinct_played_videos_also_commented_om <= 1
THEN class=YES

IF type_name_of_first_subscription = 15_days) and
and percent_video_watched_more_than_once >= 37) and
and percent_distinct_video_watched_more_than_once <= 39)
THEN class=YES

IF motivation_level = 0
and avg_time_watched_videos_in_secs_during_last_subscription<= 14.0
and percent_exercises_completed_with_75_points_or_more <= 0.00
and percent_video_watched_more_than_once <= 8.00
THEN class=NO

IF motivation_level = 0
and percent_videos_completed <= 42.00
and avg_time_watched_videos_in_secs_during_last_subscription<= 79.0
and avg_exercises_attempts_during_last_subscription <= 0.00
THEN class=NO
```

Table 3 Most dominant attributes in the rules for classes "YES" and "NO"

	Attribute	# in YES rules	Values in YES rules	# in NO rules	Values in NO rules
YES & NO	**Motivation level**	15	**1** (2 consecutive days); **2** (6 consecutive days)	13	**0** (no visits in consecutive days); 1 (2 consecutive days)
	Percent video watched more than once	13	MEDIUM and HIGH	10	Various
	First subscription type	11	**15 days;** 1 month; 3 months	12	**3 days;** 1 month; 3 months
	Time between first and last access in last subscription divided to the length of the last subscription	9	Various	10	**<=0; >0**
	Average of all accessed resources during last subscription	7	Around MEDIUM	10	MEDIUM and LOW
	Percent distinct video watched more than once	6	MEDIUM and HIGH	6	**HIGH**
YES only	Distinct accessed grades during last subscription	7	1, 2	–	–
	User kind	3	Pupil	–	–
NO only	Percent videos completed	–	–	11	Various
	Average time watched videos in seconds during last subscription	–	–	10	Around MEDIUM
	Number of distinct accessed subjects	–	–	8	1, 2, 3
	Average exercises attempts during last subscription	–	–	6	**LOW**
	Young leader badge	–	–	6	0
	Percent exercises completed with 75 points or more	–	–	5	**LOW**

A further analysis of the extracted rules showed that there are several attributes which appear with higher frequency in the rules and deserve to be discussed here (see Table 3). In the first section are listed the attributes which appear in both sets of rules (positive and negative), in the second section—only the attributes which appear in the positive rules, and in the third one—only the attributes appearing in the negative rules. The second and the fourth columns of the table show correspondingly the number of the positive (negative) rules in which the given attribute appears. The third and the fifth columns present some concrete values (for nominal attributes) or comment on the attribute values (for numeric attributes) occurring in positive and negative rules, correspondently. The values and comments are given in order to facilitate the understanding of the extracted rules. The comments describing the values of a numeric attribute in a rule are based on its mean and standard deviation values. For an attribute A they are as follows:

- LOW, if value(A) < mean(A) − standard deviation(A)
- MEDIUM, if mean(A − standard deviation(A) ≤ value(A) ≤ mean (A) + standard deviation(A)
- HIGH, if value(A) > mean(A) + standard deviation(A)

For example, in Table 3, MEDIUM means that an attribute takes values within the standard deviation from its mean value in all rules. MEDIUM and HIGH means that the attribute takes mostly MEDIUM values but also have HIGH values in a few rules.

The analysis of the rules showed that the attribute *motivation level*, which indicates the number of consecutive days of using the portal, seems to have major importance for both classes, as it appears in 15 out of 19 positive rules and 13 out of 25 negative rules. Table 3 shows that the users who renewed their subscriptions have entered the portal for at least 2 consecutive days, whereas users who did not— often haven't ever accessed the site for two consecutive days (motivation level = 0).

The nominal attribute *first user subscription type* has also interesting 'behaviour' in the rules: while its values "1 month" and "3 months" occur in both positive and negative rules, its 'smaller' values are different for the different types of rules. For example, the positive rules describe users who subscribed for the period of at least 15 days, whereas in 50 % of the cases where this attribute is present in negative rules, the users subscribed for 3 days only. Our understanding of this phenomenon is that 3 days is either too short period of time for the user to get used to the portal and appreciate its true advantages or that these are cases where the user has urgent one-time need to access certain resources and once she did that, she would stop using the site.

Another attribute which plays an important role, especially in the negative rules, is *time between first and last access in last subscription normalized by the length of the subscription*. With this attribute we try to measure how fast the user looses interest to the site. While the attribute takes various values in the positive rules, in the negative examples there is an isolated value <=0 (meaning that the user did not access resources at all or that first and last accesses were done within the same day). Users who practically do not use the resources obviously would not renew their subscription in the future.

Two more attributes also worth mentioning. They appear only in the negative rules and are related to taking exercises: one is *average number of attempted exercises during last subscription,* which values are all below the standard deviation from the mean and the other one is *percent of exercises completed with 75 points or more,* which values are also below the standard deviation from the mean. This indicates that the users who tend to interrupt their subscription are such who do not take exercises after watching videos or if they do their scores are below the average of all users.

Thus, from the analysis of the extracted rules we can conclude that in general the users who do not use the website in consecutive days and those who don't attempt the exercises along with the videos, tend to lose their interest in the portal and do not renew their subscription.

6 Conclusion

This paper presents a study of the educational web portal UCHA.SE aiming to improve the quality of the provided educational services. The main recommendations to the developer team of the portal are to extend the system with more opportunities for learning about the offered topics, including a variety of exercises and quizzes with different level of difficulty for practicing and leveled (need driven) help for completing them. This would allow *personalization, resource recommendation, and tracking and visualizing learner progress and achievements.* In addition, enhanced gamification will increase users' interest, engagement and motivation which, as empirical research demonstrates, can lead to improved student learning. This certainly will enhance significantly the educational services provided by the portal and thus the customer appreciation and loyalty.

We employed machine learning algorithms to extract educational data from the system log that can contribute to the evaluation of the portal and to identifying gaps that can be remediated. By applying subsamplng and oversampling methods for manipulating the available training data set with highly imbalanced classes, we constructed a quite accurate and understandable for the end-user rule-based model for predicting whether a user would renew her subscription to the site or not. The model was analyzed in order to find significant factors, which would signal whether a user will continue or discontinue their subscription. The analysis has shown that such important factors are the length of the first subscription, the speed with which the user looses interest to the portal and the completion of corresponding exercises after watching videos.

We are going to continue our work for studying and mining the educational site UCHA.SE in several directions. One of them is to design a more elaborate user model allowing pedagogically-based clustering of the users by reevaluating the importance of characteristics included in the existing model and by extending it by other characteristics. We will also try to explore other methods for constructing training data sets, which are independent of the selection of a reference point in time and reflect in a more adequate way the time-structure of the academic year in the Bulgarian schools.

Another direction of our future research is to assess the public interest to the resources from the different subject categories and level in order to inform the planning of the future production of learning resources for UCHA.SE. This will be done by a thorough analysis of the resource model and the resource category model as well as with mining dependencies between them and the user model.

Acknowledgments This work was partially supported by the project "AComIn: Advanced Computing for Innovation" grant 316087 funded by the European Commission in FP7 Capacity (2012–2016).

References

1. Arroyo, I., Ferguson, K., Johns, J., Dragon, T., Meheranian, H., Fisher, D., Barto, A., Mahadevan, S., Woolf. B.P.: Repairing disengagement with non-invasive interventions. In: Proceedings of the 13th International Conference on Artificial Intelligence in Education, pp. 195–202 (2007)
2. Baker, R.S., Corbett, A.T., Koedinger, K.R.: Detecting student misuse of intelligent tutoring systems. In: Proceedings of the 7th International Conference on Intelligent Tutoring Systems, pp. 531–540 (2004)
3. Baker, R.S.J.D.: Modeling and understanding students' off-task behavior in intelligent tutoring systems. In: Proceedings of ACM CHI 2007: Computer-Human Interaction, pp. 1059–1068 (2007)
4. Baker, R.S.J.D., Corbett, A.T., Koedinger, K.R., Evenson, S.E., Roll, I., Wagner, A.Z., Naim, M., Raspat, J., Baker, D.J., Beck, J.: Adapting to when students game an intelligent tutoring system. In: Proceedings of the 8th International Conference on Intelligent Tutoring Systems, pp. 392–401 (2006)
5. Baker, R.: Data mining for education. In: McGaw, B., Peterson, P., Baker, E. (eds.) International Encyclopedia of Education, 3rd edn. Elsevier, Oxford, UK (2010)
6. Baker, R., Siemens, G.: Educational data mining and learning analytics. In: Sawyer, K. (ed.) Cambridge Handbook of the Learning Sciences, 2nd edn. pp. 253–274 (2014)
7. Beal, C.R., Qu, L., Lee, H.: Mathematics motivation and achievement as predictors of high school students guessing and help-seeking with instructional software. J. Comput. Assist. Learn. **24**, 507–514 (2008)
8. Cetintas, S., Si, L., Xin, Y., Hord, C.: Automatic detection of off-task behaviors in intelligent tutoring systems with machine learning techniques. IEEE Trans. Learn. Technol. **3**(3), 228–236 (2010)
9. Clark, P., Niblett, T.: The CN2 induction algorithm. Mach. Learn. **3**(4), 261–283 (1989)
10. Cohen, W.W.: Fast effective rule induction. In: Machine Learning: Proceedings of the 12th International Conference on Machine Learning, Lake Tahoe, pp. 115–123. Morgan Kaufmanm, California (1995)
11. Corbett, A.T., Anderson, J.R.: Knowledge tracing: modeling the acquisition of procedural knowledge. User Model. User-Adap. Inter. **4**, 253–278 (1995)
12. Demsar, J., Curk, T., Erjavec, A., Gorup, C., Hocevar, T., Milutinovic, M., Mozina, M., Polajnar, M., Toplak, M., Staric, A., Stajdohar, M., Umek, L., Zagar, L., Zbontar, J., Zitnik, M. Zupan, B.: Orange: data mining toolbox in python. J. Mach. Learn. Res. **14**(Aug), 2349–2353 (2013)
13. Dichev, C., Dicheva, D., Angelova, G. Agre, G.: From gamification to gameful design and gameful experience in learning. Cybern. Inf. Technol. **14**(4), 80–100 (2014). ISSN 1311-9702 (Sofia)
14. Dicheva, D., Dichev, C., Agre, G., Angelova, G.: Gamification in education: a systematic mapping study. Educ. Technol. Soc. **18**(3), 75–88 (2015). ISSN 1176-3647
15. Muldner, K., Burleson, W., Van de Sande, B., VanLehn, K.: An analysis of students' gaming behaviors in an intelligent tutoring system: predictors and impacts. User Model. User-Adap. Inter. **21**(1–2), 99–135 (2011)
16. Peña-Ayala, A.: Educational data mining: a survey and a data mining-based analysis of recent works. Expert Syst. Appl. **41**, 1432–1462 (2014)

17. Romero, C., Ventura, S.: Educational data mining: a review of the state-of-the-art. IEEE Trans. Syst. Man Cybern. Part C Appl. Rev. **40**(6), 601–618 (2010)
18. San Pedro, M.O.C., Baker, R. Rodrigo, M.M.: Detecting carelessness through contextual estimation of slip probabilities among students using an intelligent tutor for mathematics. In: Proceedings of 15th International Conference on Artificial Intelligence in Education, pp. 304–311 (2011)
19. Tang, P-N., Steinbach, M., Kumar, V.: Introduction to Data Mining. Pearson Education Inc (2006)
20. Walonoski, J.A., Heffernan, N.T.: Detection and analysis of off-task gaming behavior in intelligent tutoring systems. In: Ikeda, Ashley, Chan (eds.) Proceedings of the 8th International Conference on Intelligent Tutoring Systems, Jhongli, Taiwan, pp. 382–391. Springer, Berlin (2006)
21. Witten, I.H., Frank, E., Hall, M.A.: Data Mining—Practical Machine Learning Tools and Techniques. 3rd edn. Elsevier (2011)

About Sense Disambiguation of Image Tags in Large Annotated Image Collections

Olga Kanishcheva and Galia Angelova

Abstract This paper presents an approach for word sense disambiguation (WSD) of image tags from professional and social image databases without categorial labels, using WordNet as an external resource defining word senses. We focus on the resolution of lexical ambiguity that arises when a given keyword has several different meanings. Our approach combines some knowledge-based methods (Lesk algorithm and Hyponym Heuristics) and semantic measures, in order to achieve successful disambiguation. Experimental results and performance evaluation are 95.16 % accuracy for professional images and 87.40 % accuracy for social images, for keywords included in WordNet. This approach can be used for improving machine translation of tags or image similarity measurement.

Keywords Word sense disambiguation · Semantic similarity measures · Social tagging · Lexical semantics · Lexical ambiguity · Tag sense disambiguation

1 Introduction

Word sense ambiguity (WSA) is a central problem for many established Human Language Technology applications (Machine Translation, Information Extraction, Question Answering, Information Retrieval, Text Classification and Text Summarization). This is also the case for the associated subtasks (e.g., reference resolution, acquisition of subcategorization patterns, parsing and, obviously, semantic interpretation) [1, 2].

WSA concerns also tagging: in blogs, videos, images and digital objects in general; website authors attach keyword descriptions (also called tags) as a category

O. Kanishcheva (✉) · G. Angelova
Institute of Information and Communication Technologies,
Bulgarian Academy of Sciences, 25A Acad. G. Bonchev Street, 1113 Sofia, Bulgaria
e-mail: kanichshevaolga@gmail.com

G. Angelova
e-mail: galia@lml.bas.bg

© Springer International Publishing Switzerland 2016
S. Margenov et al. (eds.), *Innovative Approaches and Solutions in Advanced Intelligent Systems*, Studies in Computational Intelligence 648,
DOI 10.1007/978-3-319-32207-0_9

or topic to identify sub-units within their websites. Digital resources with identical tags can be linked together allowing users to search for similar or related content. Moreover, multimedia platforms supporting photo archives, due to the prevalence of digital cameras and the growing practice of photo sharing in community-contributed image websites like Flickr (www.flickr.com) and Zooomr (www.zooomr.com). All such services are based on tagging for images and video. Tags also support Image Retrieval, Classification and Clusterization.

Tags ambiguity leads to inaccuracy and misunderstandings. For example in Fig. 1 no distinction is made between images related to *birds* and those related to *machines* since they share the same tag *crane*. Resolving the ambiguity of tags can help improve the accuracy of machine translation of keywords and image classification.

The paper is organized as follows: Sect. 2 discusses related work and summarizes different knowledge-based methods for word sense disambiguation (WSD) and

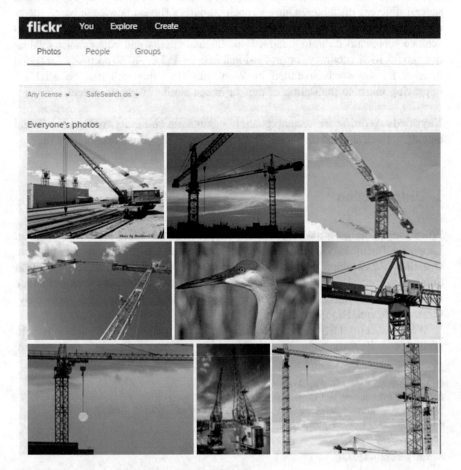

Fig. 1 The results from Flickr for the query "crane"

similarity measures applied in our experiments. In Sect. 3 we describe our approach for tag sense disambiguation of annotated images. The results and evaluation using different component configurations are reported in Sect. 4. We also describe experiments in the comparison of image tags coming from different sources. Finally, in Sect. 5 we briefly sketch future work and present the conclusion.

2 Related Works

Research on WSD started decades ago; here we provide an overview only of articles related to tag sense disambiguation for the social web.

An unsupervised method for disambiguating tags in blogs is presented in [3]. Initially the authors clustered the tags by their context words, using Spectral Clustering. Afterwards they compared a tag with these clusters to find the most suitable meaning. The Normalized Google Distance is used to measure word similarity instead of manual labeling and dictionary search. Evaluation was done on 27,327 blog posts, for 36 non-ambiguous tags and 16 pseudo-tags. The micro average precision is 84.2 %.

Ching-man Au Yeung et al. presented a method for discovering the different meanings of ambiguous tags in a folksonomy [4]. Their approach is based on the tripartite structure of folksonomies. They made experiments with four ambiguous tags, and the results showed that the method is pretty efficient in tag disambiguation. However, some issues, such as the fact that a particular meaning of a tag can be observed in more than one cluster, remain to be investigated. Wu et al. [5] proposed a global semantic model to disambiguate tags and group synonymous tags.

In [6] the authors proposed a method, called Tag Sense Disambiguation (TSD), which is applied in the social tagging environment. TSD builds mappings from tags to Wikipedia articles. The authors considered the tags of the top-10 most popular URLs at del.icio.us and mapped 10 randomly selected of them to the corresponding Wikipedia articles (i.e. a total of 100 mappings were provided for each subject) thereby clarifying the vocabulary of tags. The method presented in this work has 80.2 % accuracy.

In the literature WSD of image tags is usually considered as a component of different tasks related to image processing like tag translation and relevant image retrieval. Furthermore, when calculating image similarity, one might define closeness measures between the keywords within one tagset or between different tagsets, thus going beyond the core image recognition phase.

In our work we consider only image tags as sets of keywords. Their disambiguation is quite a challenging as no textual context is available in the image annotation. We use knowledge-based methods for WSD to disambiguate nouns by matching tags in the context to the entries of a specific lexical resource—in this case WordNet because it combines the characteristics of both a dictionary and a structured semantic network, provides definitions for the different senses of English words and defines groups of synonyms by means of synsets, which represent distinct lexical concepts. WordNet also organizes words into a conceptual structure by representing a number of semantic relationships (hyponymy, hypernymy, meronymy, etc.) among synsets.

2.1 Lesk Algorithm

This is a classical approach to WSD where the text contexts of an ambiguous word w are matched to the sense definitions of w [7]. The goal is to find the most probable sense referred to at the particular occurrence of w. When WordNet senses are used, each match increases the synset weight by one. One varsion of the Lesk algorithm is currently applied which identifies the sense of a word w whose textual definition has the highest overlap with the words in the context of w. Formally, given a target word w, the following score is computed for each sense S of w:

$$scoreLeskVar(S) = |context(w) \cap gloss(S)|, \tag{1}$$

where $context(w)$ is the bag of all content words in a context window around the target word w [8]. For our task, given an ambiguous keyword w, all other tags will be viewed as the context of its use.

Example 1 We consider the tags of an image from the photo hosting site Flickr shown in Fig. 2, annotated by five keywords:

   Shetland, Collie, dog, Atherstone, Smorgasbord

   Three keywords have only one sense (as no linguistic resource offers more): *Collie, Atherstone,* and *Smorgasbord.*

   Two tags are ambiguous in WordNet: *Shetland* and *dog.*

   Below we apply the Lesk algorithm for disambiguation of the tag *Shetland.*

Fig. 2 A Flickr photo

Context: *Shetland, Collie, dog, Atherstone, Smorgasbord*
Words to be disambiguated: *Shetland* and *dog*
Senses for *Shetland* in WordNet: Shetland#1, Shetland#2 defined as

- Shetland#1: (n) Shetland, Shetland Islands, Zetland (an archipelago of about 100 islands in the North Atlantic off the north coast of Scotland)
- Shetland#2: (n) Shetland sheepdog, Shetland sheep **dog**, Shetland (a small sheepdog resembling a **collie** that was developed in the Shetland Islands)

For Shetland#1 *weight* = 0 (no matches between the definition words and the other image tags).

For Shetland#2 *weight* = 2 (two tags *dog* and *collie* appear in the definition (gloss), we emphasized these words in bold and underline).

So the sense Shetland#2 is chosen as the meaning referred to in the photo annotation, because it has the largest weight.

2.2 Gloss Hypernym/Hyponym Heuristic

This method extends the Lesk approach by using glosses of the hypernym/hyponym synsets of the ambiguous word [9]. To disambiguate a given word, all the glosses of the hypernym/hyponym synsets are checked looking for words occurring in the context. Coincidences are counted. As in the Lesk algorithm, the synset having the largest weight is chosen:

$$score_{ExtLesk}(S) = \sum_{S':S\xrightarrow{rel}orS\equiv S'} |context(w)\cap gloss(S')|, \qquad (2)$$

where *context*(w) is, as above, the bag of all content words in a context window around the target word *w* and *gloss*(S') is the bag of words in the textual definition of a sense S' which is either S itself or related to S through a relation *rel*(hypernym/hyponym). The overlap scoring mechanism is also parametrized and can be adjusted to take into account gloss length (i.e. normalization) or to include function words [8].

Example 2 Below we illustrate the application of this approach for the image in Fig. 2 but now for the disambiguation of the tag *dog*.
Context: *Shetland, Collie, dog, Atherstone, Smorgasbord*
Words none disambiguated: *Shetland* and *dog*
Senses for *dog* in WordNet: dog#1, dog#2, dog#3, dog#4, dog#5, dog#6, dog#7.

- dog#1: (n) dog, domestic dog, Canis familiaris (a member of the genus Canis (probably descended from the common wolf) that has been domesticated by man since prehistoric times; occurs in many breeds)

→ hyponym S: (n) **Shetland** sheepdog, **Shetland** sheep dog, **Shetland** (a small sheepdog resembling a **collie** that was developed in the **Shetland** Islands)
→ hyponym S: (n) **collie** (a silky-coated sheepdog with a long ruff and long narrow head developed in Scotland)
→ hyponym S: (n) Border **collie** (developed in the area between Scotland and England usually having a black coat with white on the head and tip of tail used for herding both sheep and cattle)
→...

- dog#2: (n) frump, dog (a dull unattractive unpleasant girl or woman)
- dog#3: (n) dog (informal term for a man)
- dog#4: (n) cad, bounder, blackguard, dog, hound, heel (someone who is morally reprehensible)
- dog#5: (n) frank, frankfurter, hotdog, hot dog, dog, wiener, wienerwurst, weenie (a smooth-textured sausage of minced beef or pork usually smoked; often served on a bread roll)
- dog#6: (n) pawl, detent, click, dog (a hinged catch that fits into a notch of a ratchet to move a wheel forward or prevent it from moving backward)
- dog#7: (n) andiron, firedog, dog, dog-iron (metal supports for logs in a fireplace)

For the sense dog#1 we match all hyponym definitions to the image tagset and find that the word *Shetland* is met four times and the word *collie*—three times. Therefore, the weight for the sense dog#1 equals seven. The other senses' weights equal zero. So, the sense dog#1 is chosen, because it has the largest weight.

3 Our Approach to TSD

In our experiment we used 4,221 professional images from Professional image marketplace (www.stockpodium.com) and 5,118 social images from Flickr. The images have no categorial tags, they have only tags provided by the Imagga's auto-tagging platform (www.imagga.com). Table 1 presents features of the test datasets by mapping tags to WordNet.

The test sets contain about 22 % more social images; however the comparative analysis of the figures in Table 1 shows that: (i) the social images have fewer tags

Table 1 Some statistical information about the test datasets

	Number of files	Number of tags	Number of tags with one senses	Number of tags with many senses	Number of tags outside WordNet
Professional images	4,221	292,418	37,285	250,439	4,694
Social images	5,118	277,065	37,305	238,972	788

than the professional ones; (ii) the professional data set contains more ambiguous tags; (iii) the number of "unknown" tags which are not found in WordNet is significant for both groups but not equal (4,694—professional; 788—social).

Finding (i) can be explained as follows: social images are often low-quality photos without clearly focused objects, which is somewhat problematic for the auto-tagging platform and therefore fewer tags are assigned on average. To understand more deeply finding (iii) we analyzed the "unknown" tags. These are: (a) phrases (the auto-tagging program assigns phrasal annotations and many of them are outside WordNet, e.g. "*one person*", "*low fat*"); and (b) rather new words (for example, "*clipart*", "*keywords*", "*eco*", "*bqq*" etc.).

Given a conceptual hierarchy, the *depth(C)* of a concept node C is defined as the number of edges on the shortest path from the top of the hierarchy to the concept node C. The least common subsumer $LCS(C_1,C_2)$ of two concept nodes C_1 and C_2 is the lowest node that can be a parent for C_1 and C_2. A semantic similarity measure, called WordNetPath (WUP), was introduced by Wu and Palmer [10]: "within one conceptual domain, the similarity of two concepts is defined by how closely they are related in the hierarchy, i.e., their structural relations".

Definition 1 [10] The conceptual similarity *ConSim* between C_1 and C_2 is
$ConSim(C_1, C_2) = 2 \cdot N_3/(N_1 + N_2 + 2 \cdot N_3)$ where

- $C_3 = LCS(C_1,C_2)$;
- N_1 is the number of nodes on the path from C_1 to C_3;
- N_2 is the number of nodes on the path from C_2 to C_3; and
- N_3 is the number of nodes on the path from C_3 to the top.

Thus $ConSim(C_1,C_2) \in [0,1]$.

Using the notion of *depth*, the definition is transformed to formula (3) (Fig. 3):

$$sim_{WUP}(C_1, C_2) = \frac{2 \cdot depth(LCS(C_1, C_2))}{depth(C_1) + depth(C_2)}. \qquad (3)$$

WUP returns a score between 0 and 1, where 1 indicates the highest possible similarity score for a pair of words and 0 means that they are completely dissimilar. Due to the limitation of *is-a* hierarchies, we only work with "noun-noun" and

Fig. 3 A conceptual hierarchy

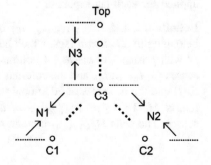

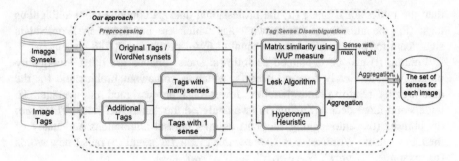

Fig. 4 Pipeline for tag sense disambiguation

"verb-verb" parts of speech. Thus the similarity score between different parts of speech or in "adj-adj" pairs is zero [11].

Our approach is illustrated in Fig. 4. The input to our system are annotated images. The keywords are assigned to the images (but in the case of our present experiment, they come from the Imagga's auto-tagging program). Let the tagset for a given image be denoted as $T = \{t_1, t_2, \ldots, t_p\}$, where p is the total number of keywords. At the preprocessing phase, for each image we split the tags into "*original*" (assigned automatically by the original Imagga's image annotation platform, where the annotation contains an WordNet synset as well and thus the tag's sense is fixed) and "*additional*" ones (borrowed from similar images available in public hosting sites in order to extend the description of the image content). Our main goal is to perform WSD for the *additional* tags because they might be ambiguous. We note that having (original) tags with one sense is helpful as they are used as context for disambiguation.

We work with sets of senses for the tags t_j, $1 \leq k \ll p$ that annotate an image. Let $S_o(t) = \{S_{o_1}, S_{o_2}, \ldots, S_{o_k}\}$ be the set of the original tags' senses in WordNet and $S_a(t) = \{S_{a_{k+1}}, S_{a_{k+2}}, \ldots, S_{a_p}\}$ be the set of the additional tags' senses in WordNet. For each additional tag we determine the number of senses by means of the WordNet concept hierarchy. If the tag has only one sense we put this sense into the set of results. Let $S'_a(t) = \{S'_{a_1}, S'_{a_2}, \ldots, S'_{a_m}\}$ be the set of senses of additional tags which have one sense. Otherwise, if a tag has multiple senses, we use a special algorithm that is introduced hereafter. This tag sense disambiguation algorithm is applied for each tag separately.

Definition 2 Let $T = \{t_1, t_2, \ldots, t_k\}$ be the tags of a given image. For each tag t_x belonging to *additional* tags, which has p senses t_{x1}, $t_{x2}, \ldots, t_{xp}$, we build the matrix M with p rows and $k + m + 1$ columns, where the rows correspond to the various senses t_{x1}, $t_{x2}, \ldots, t_{xp}$, and the columns are juxtaposed to the j senses of original tags $S_o(t)$ and the m senses of additional tags with one sense $S'_a(t)$. Each cell M_{ij} of the matrix M, $1 \leq i \leq p$, $1 \leq j \leq k + m$, contains the WUP similarity score between senses. The cells $M_{i(k+m+1)}$ contain the sums of all cell values in the i-th row.

Fig. 5 Oranges

Table 2 The set of senses for the Fig. 5 tagset

Tag	Senses
fruit	S: (n) fruit (the ripened reproductive body of a seed plant)
food	S: (n) food, solid food (any solid substance (as opposed to liquid) that is used as a source of nourishment)
plate	S: (n) plate (dish on which food is served or from which food is eaten)
tasty	S: (adj) tasty (pleasing to the sense of taste)
closeup	S: (n) closeup (a photograph taken at close range)
vitamin	S: (n) vitamin (any of a group of organic substances essential in small quantities to normal metabolism)

Example 3 The image in Fig. 5 has tags {*orange*, *fruit*, *plate*, *food*, *tasty*, *closeup*, *vitamin*}. Let us consider the task how to find which sense of *orange* is referred to in the annotation of this image. Or, more practically, we want to design algorithms that suggest with certain accuracy the most probable sense that is meant in the image annotation. The TSD procedure will be performed in the context of *fruit*, *food*, *plate*, *tasty*, *closeup* and *vitamin* senses

The tags *fruit*, *food* and *plate* belong to the group of original tags and the tags *tasty*, *closeup* and *vitamin* belong to the group of additional tags with one sense in WordNet. The senses for these tags are presented in Table 2.

Table 3 lists the senses of *orange* in WordNet—there are six senses, S_1–S_5 correspond to nouns (*n*) and S_6 is an adjective (*adj*).

Table 4 illustrates the matrix M for the image in Fig. 5, built following Definition 2. Its rows correspond to the six senses for *orange* in WordNet. The senses of *fruit*, *food*, *plate*, *tasty*, *closeup* and *vitamin* are arranged on matrix columns. The 7th column contains sums of similarity coefficients.

Let the 7th column for a given image be denoted as the vector:

$$V_{matrix}(orange) = (\ 2.91, \quad 1.13, \quad 1.72, \quad 1.94, \quad 1.79, \quad 0\ ). \qquad (4)$$

In the 7th column we see that the sense S_1 has maximal weight. It is the correct result, because the tag *orange* has the meaning of "fruit", not of "color". In Table 4

Table 3 The set of WordNet senses for the tag *orange*

	Senses
S_1	S: (n) orange (round yellow to orange fruit of any of several citrus trees)
S_2	S: (n) orange, orangeness (orange color or pigment; any of a range of colors between red and yellow)
S_3	S: (n) orange, orange tree (any citrus tree bearing oranges)
S_4	S: (n) orange (any pigment producing the orange color)
S_5	S: (n) Orange, Orange River (a river in South Africa that flows generally westward to the Atlantic Ocean)
S_6	S: (adj) orange, orangish (of the color between red and yellow; similar to the color of a ripe orange)

Table 4 The matrix M constructed for the tag *orange* in the context of *fruit*, *food*, *plate*, *tasty*, *closeup* and *vitamin*

	S(*fruit*)	S(*food*)	S(*plate*)	S(*tasty*)	S(*closeup*)	S(*vitamin*)	Sum
S_1	0.87	0.75	0.42	0	0.43	0.44	**2.91**
S_2	0.21	0.27	0.20	0	0.21	0.24	**1.13**
S_3	0.40	0.28	0.38	0	0.40	0.26	**1.72**
S_4	0.32	0.53	0.30	0	0.32	0.47	**1.94**
S_5	0.33	0.43	0.32	0	0.33	0.38	**1.79**
S_6	0	0	0	0	0	0	**0**

we also see zero values for row 6th and column 4th. This is due to the fact that the sense S_{tasty} denotes an adjective while the senses of *orange* are either nominal or adjectival. The 6th row corresponds to *orange* as an adjective. In fact the matrix M reflects a WordNet-based, structural point of view regarding the senses of the noun tags under consideration.

As shown in Fig. 4 (the TSD pipeline), in order to obtain a more holistic perspective on lexical ambiguity, we also calculate the similarity scores using the Lesk algorithm (1) and Hyponym Heuristic (2) for each sense of the tag *orange*.

We use the Lesk algorithm and match all six definitions of senses (Table 2) to the image tagset (*fruit*, *food*, *plate*, *tasty*, *closeup* and *vitamin*). Thus we obtain a Lesk-similarity vector for all senses. We find that the word *fruit* is met once in the gloss of the sense S_1 (Table 3). Therefore, the weight of the sense S_1 is 1. The other senses have weights zero. For the hyponym heuristic we build another vector with elements juxtaposed to senses; we find that the word *fruit* is found two times for the sense S_1 and the word *plate* is found two times for the sense S_3. Thus, for the tag *orange* we receive the following vectors:

$$
\begin{aligned}
V_{Lesk}(orange) &= (1, \quad 0, \quad 0, \quad 0, \quad 0, \quad 0\ ), \\
V_{Hyponym}(orange) &= (2, \quad 0, \quad 2, \quad 0, \quad 0, \quad 0\ ),
\end{aligned}
\tag{5}
$$

where each vector element is the weight of the respective sense. After we calculate both vectors, we sum up the vector elements and receive the aggregated vector:

$$V_{Lesk + Hyponym}(orange) = (\ 3,\quad 0,\quad 2,\quad 0,\quad 0,\quad 0\ ). \tag{6}$$

The vectors V_{matrix} and $V_{Lesk + Hyponym}$ have an identical dimension which corresponds to the number of multiple tag meanings. Thus, after we calculated the vectors V_{matrix} and $V_{Lesk + Hyponym}$ for *orange* in Fig. 5, we sum up these vectors component wise:

$$V_{matrix} + V_{Lesk + Hyponym} = (\ 5.91,\quad 1.13,\quad 3.72,\quad 1.94,\quad 1.79,\quad 0\ ). \tag{7}$$

Finally, we find the maximum component in the vector (7). For our example it has a value of 5.91. So we obtain the result that the sense S_1 is the correct one: (*n*) *orange* (*round yellow to orange fruit of any of several citrus trees*), and this result is supported by the WUP measure as well. Thus the ranking score of the correct sense S_k is determined by two parameters: the WordNet similarity score and the similarity score based on knowledge-based methods (Lesk algorithm and Hyponym Heuristic).

4 Experiments and Results

Our approach was evaluated manually on 60 images from Professional image marketplace and 60 images from Flickr. Information about tags is presented in Table 5.

All professional images were classified manually by human annotators into three categories with certain informal names ("*Food*", "*Scene*", "*Children*") and social images from Flickr were classified into three categories with certain informal names ("*Scene*", "*Flowers*", "*Animals*"). Table 6 summarizes some data about all tags.

All results of our experiment were analyzed manually by annotators, where the humans were asked to classify the suggested tag sense as "correct", "acceptable" and "irrelevant/wrong". This is necessary because the WordNet senses might often be quite close (in meaning), so sometimes it is not easy for a human (being) to identify only one correct sense in the context of all image tags. We performed experiments with the similarity calculation. For the first experiment we used only the senses of original tags $S_o(t)$ and for the second experiment—senses of original

Table 5 Tags in the test datasets

Tags	Professional image marketplace	Flickr
Original tags	350	505
Additional tags	4,334/569 with one sense	2,462/264 with one sense
Total	4,684	2,967

Table 6 The statistic information about test datasets of evaluation

Categories		Number of images	Number of tags	Number of tags with 1 sense	Number of tags with many senses	Number of tags outside WordNet
Prof. images	*"Food"*	20	1,315	188	1,111	16
	"Scene"	17	1,109	128	968	13
	"Children"	23	1,910	253	1,637	20
	Total	**60**	**4,334**	**569**	**3,716**	**49**
Social images	*"Scene"*	18	1,088	105	972	11
	"Flowers"	24	1,104	130	973	1
	"Animals"	18	270	29	237	4
	Total	**60**	**2,462**	**264**	**2,182**	**16**

Table 7 Manual evaluation of tag sense disambiguation

Category	Correct	Acceptable	Irrelevant
Professional images			
"Food"	1,073	2	36
"Scene"	883	4	81
"Children"	1,580	17	40
Total	3,536	23	157
Social images			
"Scene"	899	11	62
"Flowers"	796	44	133
"Animals"	212	4	21
Total	1,907	59	216

tags $S_o(t)$ together with those of additional tags with one sense $S'_a(t)$. The accuracy in the second experiment is much higher; these results are presented in Table 7.

The assessment of word sense disambiguation systems is usually performed in terms of evaluation measures borrowed from the field of information retrieval: Recall, Precision and F-measure [8]. Since our system makes an assignment for every tag and thus precision and recall are the same, we used a specific measure of accuracy (6):

$$Accuracy = \frac{correct\ answers\ provided}{total\ answers\ to\ provide}. \tag{8}$$

Figure 6 presents the results of our experiments for professional images (with 3,716 ambiguous tags) and social images (with 2,182 ambiguous tags). The blue columns present the accuracy A_1 (only irrelevant tags are considered wrong) and the red columns present the accuracy A_2 (both irrelevant and acceptable tags are viewed as mistakes). The result *"Correct"* according to A_2: 95.16 % for professional images and 87.4 % for social images, seems to be rather good.

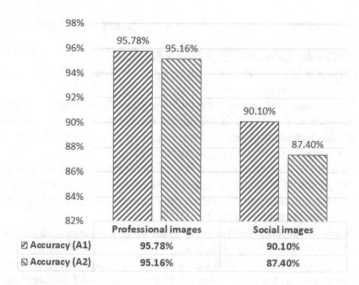

Fig. 6 The evaluation of our approach

Further, we calculate the percentage of "*Acceptable*" senses: 0.62 % for professional images and 2.7 % for social images, as well as the percentage of "*Irrelevant*" senses: 4.22 % for professional images and 9.9 % for social images. These results are promising too.

Here we illustrate the results by some examples. For the category "*Correct*" we have good results in tag sense disambiguation. For example, the tag *bowl* has 12 senses in WordNet (nine as a noun and three as a verb). For the category "*Food*" our system defined the following sense for the tag *bowl*:

Synset('bowl.n.03') a dish that is round and open at the top for serving foods

This is the correct intended meaning. Many examples prove that the neighboring tags provide reliable context for tag sense disambiguation.

In the category "*Acceptable*" we considered the senses that are not quite relevant to the image category, at least according to the human annotator. For example, for the tag *sweet* from the category "*Children*" our system proposed the following sense:

Synset('sweet.n.03') a food rich in sugar

The tag *sweet* has 16 senses in WordNet (five as a noun, ten as an adjective and one as an adverb). But a more relevant sense for this category would be:

S: (adj) angelic, angelical, cherubic, seraphic, sweet (having a sweet nature befitting an angel or cherub) "an angelic smile"; "a cherubic face"; "looking so seraphic when he slept"; "a sweet disposition".

This sense is not identified in the particular case as we do not deal with adjectives.

The last column of Table 5—"*Irrelevant*"—contains the numbers for the real mistakes. For example, for the tag *france* from the category "*Scene*" the system determined the sense:

Synset('france.n.02') French writer of sophisticated novels and short stories *(1844–1924)*

But in this case the correct suggestion is the sense:

S: (n) France, French Republic (a republic in Western Europe; the largest country wholly in Europe)

We also calculated the accuracy for each category. The accuracy for professional images is given in Fig. 7a; it is lowest for the category "*Scene*" because these images contain numerous objects and are tagged by many keywords with multiple senses. The accuracy for social images is given in Fig. 7b: in general it is lower than the accuracy for professional images. As we said before, social images are often low-quality photos without clearly focused objects, which is somewhat problematic for the auto-tagging platform. Therefore these images are annotated by fewer tags with lower correct accuracy (cf. Table 1).

We studied the dependences between: (i) the accuracy (green dash line in Figs. 8 and 9), (ii) the number of tags with a single sense—which for us are the original tags $S_o(t)$ plus additional tags with one sense (orange graph in Figs. 8 and 9), and (iii) the number of all tags for the images in the experimental dataset (blue line in Figs. 8 and 9). The horizontal ax in Figs. 8 and 9 corresponds to images from the test data set; the left vertical ax in Figs. 8 and 9 gives number of tags for the blue and orange lines and the right vertical ax shows percentage of accuracy (green dash line). Figure 8 presents the results for social images and Fig. 9—the results for professional.

Figure 9 shows for professional images that the increase in tags (their number is shown by the blue line) influences TSD accuracy (green line), but for social images it is not an essential change (Fig. 8). In Figs. 8 and 9 we observe some dependence between the green line and the orange space, it shows the influence of the similarity matrix on TSD accuracy (more senses—an orange peak—imply in general higher accuracy). This would mean that the similarity matrix contributes more (or always) to the TSD accuracy compared to the Lesk algorithm and the Hyponym Heuristic.

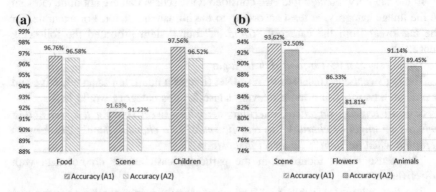

Fig. 7 Evaluation of accuracy per category

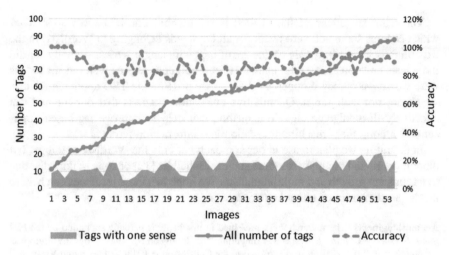

Fig. 8 The dependence of accuracy and amount of tags for social images

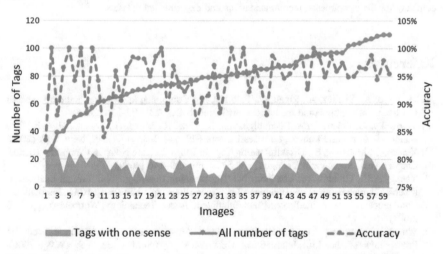

Fig. 9 The dependence of accuracy and amount of tags for professional images

5 Conclusions and Future Works

In this paper we present an integrated method for tag disambiguation of images, and show its effectiveness for arbitrary tags obtained using an auto-tagging program. We concentrate on nominal and verbal senses because the automatic image annotation usually recognizes material objects (nouns) and disambiguation is needed for the auto-tagging process. In general adjectives are assigned as tags in manual annotation.

The main advantage of the method is the combination of known approaches which ensures some non-zero result for all keywords belonging to WordNet. In the experiments we strongly rely on the set of tags with single senses. Actually we did more tests and constructed the matrix M for ambiguous tags as well but the results were worse—accuracy dropped by 4 %, so we propose the TSD approach for tags which are not ambiguous. Quantitative evaluation suggests that our method can effectively disambiguate tags. The method can help to improve tag-based applications, among them machine translation and information retrieval.

In the future we plan to use in our approach not only the WordNet ontology, but other modern and complex resource such as BabelNet [12] or even Wikipedia, due to its constant growth and multilingualism. We hope that these resources can help to tackle the dynamics of novel keywords.

Acknowledgments The research work presented in this paper is partially supported by the FP7 grant 316087 AComIn "Advanced Computing for Innovation", funded by the European Commission in 2012–2016. It is also related to the COST Action IC1307 "Integrating Vision and Language (iV&L Net): Combining Computer Vision and Language Processing for Advanced Search, Retrieval, Annotation and Description of Visual Data". The authors are thankful to Imagga company for their comments, recommendations and experimental datasets.

References

1. Montoyo, A., Su´arez, A., Rigau, G., Palomar, M.: Combining knowledge- and corpus-based word-sense-disambiguation methods. J. Artif. Intell. Res. **23**, 299–330 (2005)
2. Yarowsky, D.: Word Sense Disambiguation. In: Dale, R., Moisl, H., Somers, H. (eds.) The Handbook of Natural Language Processing, pp. 629–654. Marcel Dekker, New York (2000)
3. Xiance, S., Maosong, S.: Disambiguating Tags in Blogs. Lecture Notes in Computer Science, Chapter Text, Speech and Dialogue, vol. 5729, pp. 139–146 (2009)
4. Yeung, A.C., Gibbins, N., Shadbolt, N.: Tag meaning disambiguation through analysis of tripartite structure of Folksonomies. In: Proceedings of IEEE/WIC/ACM International Conferences on Web Intelligence and Intelligent Agent Technology Workshops, pp. 3–6 (2007)
5. Wu, X., Zhang, L., Yu, Y.: Exploring Social Annotations for the Semantic Web. In: Proceedings of the 15th International Conference on World Wide Web (WWW2006), pp. 417–426 (2006)
6. Lee, K., Kim, H., Shin, H., Kim, H.-J.: Tag sense disambiguation for clarifying the vocabulary of social tags. In: Proceedings of International Conference on Computational Science and Engineering, CSE'09, vol. 4, pp. 729–734 (2009)
7. Lesk, M.: Automatic sense disambiguation using machine readable dictionaries: how to tell a pine cone from an ice cream cone. In: SIGDOC'86: Proceedings 5th Annual International Conference on Systems Documentation, pp. 24–26. ACM, USA (1986). doi:10.1145/318723. 318728
8. Navigli, R.: Word sense disambiguation: a survey. ACM Comput. Surv. **41**(2) (2009) (Article 1)
9. Banerjee, S., Pedersen, T.: Extended gloss overlaps as a measure of semantic relatedness. In: Proceedings of the 18th International Joint Conference on Artificial Intelligence (IJCAI, Acapulco, Mexico), pp. 805–810 (2003)
10. Wu, Z., Palmer, M.: Verb Semantics and Lexical Selection, pp. 133–138. Las Cruces, Mexico (1994)

11. Pedersen, T., Patwardhan, S., Michelizzi, J.: WordNet: Similarity—measuring the relatedness of concepts. In: Proceedings of Fifth Annual Meeting of the North American Chapter of the Association for Computational Linguistics, Boston, MA, pp. 38–41 (2004)
12. Navigli, R., Ponzetto, S.: BabelNet: The automatic construction, evaluation and application of a wide-coverage multilingual semantic network. Artificial Intelligence, vol. 193, pp. 217–250. Elsevier (2012)

Knowledge Graph Extension for Word Sense Annotation

Kiril Simov, Alexander Popov and Petya Osenova

Abstract One of the most successful approaches to Word Sense Disambiguation (**WSD**) in the last decade has been the knowledge-based approach, which exploits lexical knowledge sources such as wordnets, ontologies, etc. The knowledge encoded in them is typically used as a sense inventory and as a relations bank. However, this type of information is rather sparse in terms of senses and the relations among them. In this paper we present a strategy for the enrichment of WSD knowledge bases with data-driven relations from a gold standard corpus (annotated with word senses, valency information, syntactic analyses, etc.). We focus on Bulgarian and English as use cases, but our approach is scalable to other languages as well. The results show that the addition of new knowledge improves accuracy on the WSD task in a statistically significant way.

Keywords Knowledge-based WSD · Inference of semantic relations

1 Introduction

The recent success of *knowledge-based Word Sense Disambiguation* (**KWSD**) approaches depends on the quality of the *knowledge graph* (**KG**)—whether the knowledge represented in terms of nodes and relations (arcs) between them is sufficient for the algorithm to pick the correct senses of the ambiguous words. Several extensions of the KG constructed on the basis of WordNet have been proposed and already implemented. The solutions to Word Sense Disambiguation

K. Simov (✉) · A. Popov · P. Osenova
Institute of Information and Communication Technologies,
BAS, Akad. G. Bonchev. 25A, 1113 Sofia, Bulgaria
e-mail: kivs@bultreebank.org

A. Popov
e-mail: alex.popov@bultreebank.org

P. Osenova
e-mail: petya@bultreebank.org

© Springer International Publishing Switzerland 2016   151
S. Margenov et al. (eds.), *Innovative Approaches and Solutions in Advanced
Intelligent Systems*, Studies in Computational Intelligence 648,
DOI 10.1007/978-3-319-32207-0_10

(**WSD**) related tasks usually employ lexical databases, such as wordnets and ontologies. However, lexical databases suffer from sparseness with respect to the availability and density of relations. One approach towards remedying this problem is the BabelNet [9], which relates several lexical resources—WordNet[1] [5], DBpedia,[2] Wiktionary,[3] etc. Although such a setting takes into consideration the role of lexical and world knowledge, it does not incorporate contextual knowledge learned from actual texts (such as collocational patterns, for example). This happens because the knowledge sources for WSD systems usually capture only a fraction of the relations between entities in the world. Many important relations are not present in the ontological resources but could be learned from texts.

Here we present an approach towards the enrichment of WSD knowledge bases with relations from gold standard corpora. We focus on Bulgarian (*BTB* [11]) and English (*SemCor* [7]) as use cases, but our approach is scalable to other languages as well. Such an approach is justified by the fact that the lexical databases are sparse with respect to the available knowledge, its density and appropriateness. Also, the predominance of paradigmatic knowledge (synonymy, hypernymy, etc.) is balanced by the addition of syntagmatic relations (valency). Analysis is provided on the impact of the various semantic relations, which lead to different variants of the KG. More precisely, in our current work we perform an analysis on the various semantic relations in WordNet[4] and Extended WordNet [6] knowledge graphs. The investigation is performed via experiments with different subgraphs that include only part of the semantic relations in these resources. Some of the relation types allow for inference to be applied over them. Thus, inferred semantic relations have been included in some of the KGs as well.

The structure of the paper is as follows: the next section discusses the related work on the topic. Section 3 presents the resources—the Bulgarian sense annotated treebank *BTB* and *SemCor*. Section 4 introduces the knowledge-based tool for WSD. Section 5 describes: the experiments performed, based on the semantic relations in WordNet; the additional inference over WordNet relations; the semantic relations in the Extended WordNet; as well as syntax-based relations. Section 6 concludes the paper.

2 Related Work

Knowledge-based systems for WSD have proven to be a good alternative to supervised systems, which require large amounts of manually annotated data. Knowledge-based systems require only a knowledge base and no additional

---

[1]https://wordnet.princeton.edu/ .

[2]http://wiki.dbpedia.org/ .

[3]https://en.wiktionary.org/wiki/Wiktionary:Main_Page .

[4]In this work we used version 3.0 of Princeton WordNet: https://wordnet.princeton.edu/.

corpus-dependent information. An especially popular knowledge-based disambiguation approach has been the use of graph-based algorithms known under the name of "Random Walk on Graph" [3]. Most approaches exploit variants of the PageRank algorithm [4]. Agirre and Soroa [2] apply a variant of the algorithm to WSD by translating WordNet into a graph in which the synsets are represented as nodes and the relations between them are represented as arcs. The resulting graph is called a *knowledge graph* in this paper. Calculating the PageRank vector **Pr** is accomplished through solving the equation:

$$\mathbf{Pr} = cM\mathbf{Pr} + (1 - c)\mathbf{v} \tag{1}$$

where M is an $N \times N$ transition probability matrix (N being the number of nodes in the graph), c is the damping factor and v is an $N \times 1$ vector. In the traditional, static version of PageRank the values of **v** are all equal $(1/N)$, which means that in the case of a random jump each vertex is equally likely to be selected. Modifying the values of **v** effectively changes these probabilities and thus makes certain nodes more important. The version of PageRank for which the values in **v** are not uniform is called *Personalized PageRank* (PPR). The words in the text that are to be disambiguated are inserted as nodes in the KG and are connected to their potential senses via directed arcs. These newly introduced nodes serve to inject initial probability mass (via the vector **v**) and thus to make their associated sense nodes especially relevant in the *knowledge graph*. Applying the PPR algorithm iteratively over the resulting graph determines the most appropriate sense for each ambiguous word. Montroyo et al. [8] present a combination of knowledge-based and supervised systems for WSD, demonstrating that the two approaches can boost one another, due to the fundamentally different types of knowledge they utilise (paradigmatic vs. syntagmatic). They explore a knowledge-based system that uses heuristics for WSD depending on the position of potential word senses in the WordNet knowledge graph (**WN**). In terms of supervised machine learning based on an annotated corpus, it explores a Maximum Entropy model that takes into account multiple features from the context of the to-be-disambiguated word. This earlier line of research demonstrates that combining paradigmatic and syntagmatic information is a fruitful strategy, but it does so by doing the combination in a postprocessing step, i.e. by merging the output of two separate systems; also, it still relies on manually annotated data for the supervised disambiguation. Building on the already mentioned work on graph-based approaches, it is possible to combine paradigmatic and syntagmatic information in another way—by incorporating both into the KG. This approach is described here. The success of KWSD approaches apparently depends on the quality of the knowledge graph—whether the knowledge represented in terms of nodes and relations between them is sufficient for the algorithm to pick the correct senses of ambiguous words. Several extensions of the KG constructed on the basis of WordNet have been proposed. An approach similar to ours is described in [1], which explores the extraction of syntactically supported semantic relations from a manually annotated corpus: *SemCor*. *SemCor* was processed with the MiniPar parser and the subject-verb as well as object-verb relations were extracted. The new relations were represented on several levels: as word-to-class and class-to-class relations. The extracted selectional relations were then added to WordNet. The main difference with our approach is that the set of relations used here is larger (including indirect-object-to-verb relations, noun-to-noun relations, etc.). Another difference is that the new relations are not added as selectional relations, but as semantic ones. This means that the specific syntactic role of the participant is not taken into account, but only the connectedness between the participant and the event is registered in the KG.

3   The Sense Annotated Treebanks: *BTB* and *SemCor*

As it was stated above, our goal is to experiment with different kinds of semantic relations. The relations missing in WordNet are the syntagmatic ones. As sources of such types of relations we consider treebanks annotated with semantic information. In this case we are able to extract syntagmatic relations between semantic classes of syntactically related words. Thus, we need semantically and syntactically annotated corpora of Bulgarian and English. For Bulgarian we have annotated *BTB* with synsets from the BTB WordNet (**BTB-WN**). For English we use a parsebank created over the texts in *SemCor* [7]. Since these syntactically and semantically annotated corpora have been exploited for extracting relations, we divided them in test and training parts in a ratio of one-to-three. The sense annotation process over *BTB* is organized in three layers: verb valency frames [10]; senses of verbs, nouns, adjectives and adverbs; DBpedia URIs over named entities. However, in the experiment presented here, we used mainly the annotated senses of nouns and verbs, as well as the concept mappings to WordNet.

The sense annotation was organized as follows: the lemmatized words per part-of-speech (POS) from *BTB* received all their possible senses from the explanatory dictionary of Bulgarian and from BTB-WN.[5] When two competing definitions came from both resources, preference was given to the one that was mapped to the WordNet. In the ambiguous cases the correct sense was selected according to the context of usage. For the purposes of the evaluation, some of the files were independently checked by two human annotators. In total, 92,000 running words have been mapped to word senses. Thus, about 43 % of all the treebank tokens have been associated with senses. Additionally, we extended the BTB-WN with new synsets which were mapped to English WordNet. We have divided *BTB* in test and training parts by selecting three files as a test part, which contains news articles from two different newspapers and covers about a quarter of the tokens in the whole treebank. The rest of the data was selected for the training set. The training set is used for extraction of new semantics relations on the basis of syntactic information in the treebank.

---

[5]Available at http://compling.hss.ntu.edu.sg/omw/ .

The sense annotations in *SemCor* were also performed manually, using WordNet senses. It comprises texts from the Brown corpus,[6] which is a balanced corpus. In this respect, *SemCor* contains very diverse types of texts. We use *SemCor* in two ways: first, for testing the WSD for English; and second, as a source for extracting new semantic relations. To achieve this, we parsed *SemCor* with a dependency parser included in the IXA pipeline.[7] Then we divided the corpus in a proportion of one-to-three: first part comprises of 49 documents (from br-a01 to br-f44) and it was used as a test set in the experiments reported below in the paper. The rest of the documents formed the training set from which the new relations were extracted. Similarly to the case of *BTB*, the new semantic relations were extracted on the basis of the syntactic relations in the dependency parses of each sentence in the training part of *SemCor*.

4   Knowledge-Based Tool for the WSD

The experiments that serve to illustrate the outlined approaches were carried out with the UKB[8] tool, which provides graph-based methods for WSD and measuring lexical similarity. The tool uses the PPR algorithm, described in [2]. It builds a knowledge graph over a set of relations that can be induced from different types of resources, such as WordNet or DBPedia; then it selects a context window of open-class words and runs the algorithm over the graph. There is an additional module called NAF UKB[9] that can be used to run UKB with input in the NAF format[10] and to obtain output structured in the same way, only with added word sense information. For compatibility reasons, NAF UKB was used to perform the experiments reported here; the input NAF document contains in its "term" nodes the lemma and POS information, which is necessary for the running of UKB. We have used the UKB default settings, i.e. a context window of 20 words that are to be disambiguated together, and 30 iterations of the PPR algorithm.

The UKB tool requires two resource files to process the input file. One of the resources is a dictionary file with all lemmas that can be possibly linked to a sense identifier. In our case WordNet-derived relations were used for our knowledge base; consequently, the sense identifiers are WordNet IDs. For instance, a line from the dictionary extracted from WordNet looks like this:

predicate 06316813-n:0 06316626-n:0 01017222-v:0
01017001-v:0 00931232-v:0

First comes the lemma associated with the relevant word senses, after the lemma the sense identifiers are listed. Each ID consists of eight digits followed by a hyphen and a label referring to the POS category of the word. Finally, a number following a colon indicates the frequency of the word sense, calculated on the basis of a tagged corpus. When a lemma from the dictionary has occurred in the analysis of the input text, the tool assigns all the associated word senses to the word form in the context and attempts to disambiguate its meaning among them. The Bulgarian dictionary comprises all the lemmas of words annotated with WordNet senses in the *BTB*. It has 8,491 lemmas mapped to 6,965 unique word senses. The second resource file required for running the tool is the set of relations that is used to construct the knowledge graph over which PPR is run. Since the *BTB* has been annotated with word senses from WordNet 3.0, the resource files for version 3.0, distributed together with the tool, have been used in our experiments. The distribution of UKB comes with a file containing the standard lexical relations defined in WordNet, such as hypernymy, meronymy, etc., as well as with a file containing relations derived on the basis of common words found in the synset glosses, which have been manually disambiguated. The format of the relations in the KG is as follows:

u:SynSetId01 v:SynSetId02 s:Source d:w

where SynSetId01 is the identifier of the first synset in the relation, SynSetId02 is the identifier of the second synset, Source is the source of the relation, and w is the weight of the relation in the graph. In the experiments reported in the paper, the weight of all relations is set to 0. Here is one concrete example:

u:01916925-n v:02673969-a s:30glc d:0

This tool is used for performing all the experiments reported in the next section. The goal in this paper is to investigate the impact of the different sets of relations over the knowledge graph. Thus, further experiments utilizing the full functionality of the UKB tool are left for future work.

5   Experiments with Different Knowledge Graphs

All the experiments, reported here, use the same algorithm and the same test data. Only the KG differs in the different cases, as it is generated out of various sets of relations. Thus, via using this tool we performed several groups of experiments: (1) experiments with the various relations from the WordNet KG; (2) experiments with relations resulting from applying inference over the relations in the original WordNet KG; (3) experiments with relations from Extended WordNet; and (4) experiments with relations extracted from the syntactically and semantically annotated corpora. These experiments are reported here.

5.1   Experiments with Semantic Relations in WordNet

The WordNet-based KG (**WN**) has been constructed out of the relations in the Princeton WordNet (**PWN3.0**). PWN3.0 groups together words in synsets, which we consider as concepts, and thus as units. The possible relation types between the various synsets are 16. In our experiments we separated the relations in WN into 16 sets of relations corresponding to the relations in PWN3.0:

1. **WN-Hyp** (*hypernymy*) **89089**. (N-N), (V-V).[11]
2. **WN-Ant** (*antonymy*) **8689**. (A-A), (N-N), (R-R), (V-V).
3. **WN-At** (*attribute relation between noun and adjective*) **886**. (N-A), (A-N).
4. **WN-Cls** (*a member of a class*) **9420**. (A-N), (N-N), (R-N), (V-N).
5. **WN-Cs** (*cause*) **192**. (V-V).
6. **WN-Der** (*derivational morphology*) **74644**. (A-N), (N-A), (N-N), (N-V).
7. **WN-Ent** (*entailment*) **408**. (V-V).
8. **WN-Ins** (*instance*) **8576**. (N-N).
9. **WN-Mm** (*member meronym*) **12293**. (N-N).
10. **WN-Mp** (*part meronym*) **9097**. (N-N).
11. **WN-Ms** (*substance meronym*) **797**. (N-N).
12. **WN-Per** (*pertains/derived from*) **8505**. (A-N), (R-A).
13. **WN-Ppl** (*participle of the verb*) **79**. (A-V).
14. **WN-Sa** (*additional information about the first word*) **3269**. (A-A), (V-V).
15. **WN-Sim** (*similar in meaning*) **21386**. (A-A).
16. **WN-Vgp** (*similar in meaning verb synsets*) **1725**. (V-V).

These classes differ in the type of semantic relations they represent, the number of relations in each class, the parts-of-speech of the words in the synsets that are connected by the relation. Obviously, isolated nodes do not play a role in the disambiguation process. Thus, if we exploit only relations between nouns, we cannot expect that the system could select appropriate senses for other parts-of-speech. Nevertheless, we performed some experiments with only some of the relations in order to have a basis for comparison with larger combinations. As a basic relation we consider the superordinate-subordinate relation (hypernymy), because it provides relations between the biggest groups of synsets: nouns and verbs. Thus, we assume that this set of relations always has to be used in the knowledge graph.

The baseline results include **WN** relations, gloss-derived relations (**GL**) and the combination of WN and GL: **WNG**. The results for these three basic KGs are given at the top three lines of Table 1. In Table 1 we also present the results for

---

[11]Here we present the combination of synsets in each relation as POS. The POS are: A—adjective, N—noun, R—adverb, and V—verb. Also we present the number of arcs for the relations in WordNet.

Table 1 Experimental results when using the sets of relations from the WordNet KG on the two test corpora

KG	SemCor		BTB		
WN	49.37		52.97		
GL	51.66		51.15		
WNG	58.97		55.90		
KG	SemCor	BTB	KG	SemCor	BTB
WN-Hyp	33.52	45.03	WN-Hyp + WN-Mm	33.70	44.81
WN-Hyp + WN-Ant	38.63	48.41	WN-Hyp + WN-Mp	35.67	45.22
WN-Hyp + WN-At	36.97	47.91	WN-Hyp + WN-Ms	33.57	45.31
WN-Hyp + WN-Cls	34.23	46.11	WN-Hyp + WN-Per	**39.57**	48.19
WN-Hyp + WN-Cs	33.54	44.99	WN-Hyp + WN-Ppl	33.53	45.11
WN-Hyp + WN-Der	**39.03**	**50.63**	WN-Hyp + WN-Sa	38.29	48.31
WN-Hyp + WN-Ent	33.30	44.65	WN-Hyp + WN-Sim	**42.89**	**49.28**
WN-Hyp + WN-Ins	34.18	45.13	WN-Hyp + WN-Vgp	34.22	46.07

combinations between the hypernymy relation and all other relations. The biggest improvement is observed for the combination **WN-Hyp + WN-Sim** for *SemCor*. It shows 9 % of improvement over the **WN-Hyp** relation alone. In our opinion, the great difference is due to the different coverage of the relations over the synsets in WordNet. The hypernymy relation covers only noun and verb synsets, but not adjective and adverb synsets. Thus, a KG-based only on hypernymy relations does not provide any knowledge about adjectives and adverbs. Additionally, it does not contain any knowledge about the interactions between verbs and nouns. The relations that improve over hypernymy ones in fact introduce knowledge about adjectives or interaction across parts-of-speech. We have performed some more experiments in order to check whether we could exclude some relations without considerable loss. For instance, the combination of the following eight sets: **WN-Hyp + WN-Ant + WN-Der + WN-Per + WN-Sa + WN-Sim + WN-Mp + WN-Cls**, gives accuracy of **49.34 %** on the *SemCor* test corpus, which is **0.03 %** less than the accuracy obtained with the whole KG of WordNet. The same combination gives **52.36 %** for the *BTB* test corpus, which is also close (**0.61 %**) to the result for the whole set. Thus, the impact of the rest of relations is quite modest.

The results also show the differences between the corpora. *BTB* seems more compact with respect to sub-domains, while *SemCor* introduces a big variety of sub-domains. Also, *BTB* is mainly annotated with noun and verb synsets. Thus, the impact of the relations is different from the impact they have over the *SemCor* corpus.

The general conclusion from these experiments is that the addition of relations to the KG does not contribute monotonically to the accuracy of the KWSD. It shows that some of the relations in the original graph lower the accuracy.

5.2   Inference over WordNet Relations

Under inference in our experiments we consider the application of rules, given relations in the KG, which produce new relations to be added to the KG. In this section we consider some rules applicable to the relations from WordNet. Having in mind that WordNet is not a fully formalized lexical database, we cannot expect that the inferences proposed below are always correct. The main inference rule is the hypernymy hierarchy inheritance: if some relation includes a noun as an argument, then the hyponyms of the noun also could be arguments in the relation. The situation is similar for verbs. Sometimes the appropriate inference includes their hypernyms.

1. **WN-Hyp**. The hypernymy relation is transitive. Thus, we could construct its transitive closure: if **doctor** is a hypernym of **surgeon** and **professional** is a hypernym of **doctor** then **professional** is a hypernym of **surgeon**. Similarly, for the verb hierarchy. The inferred set of relations will be named by adding the suffix **Infer** to the name of the main relation. In this case: **WN-HypInfer**.
2. **WN-Ant**. Antonymy relations between adjectives and adverbs cannot partici-pate in the inference, because there is no support in WordNet. For nouns and verbs it is possible, if the antonymy relation means that corresponding synsets are disjoint. The disjointedness is preserved by the hyponymy relation: if we have two disjoint concepts, then their subconcepts are also disjoint. For example, **man** (an adult male person) and **woman** (an adult female person) do not have common instances. Then we could infer that **man** and **girl** are disjoint. Name: **WN-AntInfer**.
3. **WN-At**. The attributes of a noun usually can be inherited by its hyponyms. For example, **measure** as a quantity of something has attributes—**standard** and **nonstandard**. These attributes can be inherited by all types of measures like **time interval** and others. Name: **WN-AtInfer**.
4. **WN-Cls**. The general understanding of the relation *a member of a class* is that each hyponym of the **member** could be a member of each of the hypernyms of the **class**. For instance, **desktop publishing** is a member of **computer science** as a branch, but also it is a branch of **engineering**, which is a hypernym of computer science. Name: **WN-ClsInfer**.
5. **WN-Cs**. The *cause* relation between verbs naturally allows for inference on both arguments—each hyponym of the first argument could be a cause for each hypernym of the second argument. The sets resulting from the inference on the first and second arguments are denoted with **WN-Cs1stVInfer** and **WN-Cs2ndVInfer**.
6. **WN-Der**. The derivational relation is quite diverse, connecting adjectives and nouns, nouns and nouns, and nouns and verbs. We consider this relation as denoting an event or a state in which the noun determines a participant of the event or a state. Thus, a noun can be substituted with its hyponyms, and a verb can be substituted with its hypernyms. We divided the new set of relations by the POS of the participating words: **WN-DerNAInfer**, **WN-DerNNInfer**, **WN-DerNVInfer**, **WN-DerVNInfer**.

7. **WN-Ent**. If a verb entails another verb, then we assume that each hyponym of the first verb entails each hypernym of the second verb. The sets resulting from the inference on the first and second arguments are denoted with **WN-Ent1stVInfer** and **WN-Ent2ndVInfer**.
8. **WN-Ins**. An instance of a class is an instance of its super classes. Thus, we perform substitution of the second noun with its hypernyms. Name: **WN-InsInfer**.
9. **WN-Mm**. Each hyponym of a member of a set is a member of each hypernym of the set.
10. **WN-Mp**. The transitive closure over the part meronym relation is a feasible inference rule. However, in these experiments we do not perform it.
11. **WN-Ms**. Substitution with hyponyms of the substance noun is a feasible inference rule. Similarly to the previous relation, in these experiments we do not perform it.
12. **WN-Per**. Similarly to the derivational relation, we perform substitution with hyponyms on the noun synset.
13. **WN-Ppl**. We do not perform any inference for this relation.
14. **WN-Sa**. The additional information about the first word can be inherited by its hyponyms.
15. **WN-Sim**. We do not perform any inference for this relation.
16. **WN-Vgp**. Because the definition "verb synsets that are similar in meaning" allows for very wide interpretation, we do not perform any inference here.

Some of the above inferences produce a huge amount of new relations, which prevents us from effectively experimenting with them. We have used the inference rules only partially. We consider only combinations in which the knowledge graphs of **WN** and **WNG** are included as a basis. Table 2 presents some of the results.

Table 2 Experimental results when using some of the inferred sets of relations

KG	SemCor	BTB	KG	SemCor	BTB
WN + WN-HypInfer	**53.40**	53.70	WNG + WN-HypInfer	58.59	55.20
WN + WN-AntInfer	48.57	53.05	WNG + WN-AntInfer	**59.14**	**55.93**
WN + WN-ClsInfer	48.43	54.62	WNG + WN-ClsInfer	57.84	**56.14**
WN + WN-Cs1stVInfer	49.32	56.02	WNG + WN-Cs1stVInfer	**59.06**	**55.93**
WN + WN-Cs2ndVInfer	49.39	**57.28**	WNG + WN-Cs2ndVInfer	58.95	**56.17**
WN + WN-DerNAInfer	48.76	57.19	WNG + WN-DerNAInfer	58.49	52.13
WN + WN-DerNNInfer	47.79	56.74	WNG + WN-DerNNInfer	58.80	52.86
WN + WN-DerNVInfer	47.73	55.84	WNG + WN-DerNVInfer	55.87	52.77
WN + WN-DerVNInfer	48.72	55.99	WNG + WN-DerVNInfer	**59.00**	53.56
WN + WN-Ent1stVInfer	49.34	56.08	WNG + WN-Ent1stVInfer	**58.98**	52.55
WN + WN-Ent2ndVInfer	49.36	56.60	WNG + WN-Ent2ndVInfer	58.92	52.70
WN + WN-InsInfer	49.03	56.76	WNG + WN-InsInfer	58.38	52.89

The results that are above the baselines are *bolded*

There are few cases in which the inferred new relations add accuracy above the baselines (Notably, for the *BTB* test corpus the improvement for several combinations with **WN** is higher even with respect to the addition of relations from **WNG**). In most of the cases, however, the additional relations decrease the accuracy. With the Extended WordNet relations the improvement is quite modest for both test corpora.

5.3 Experiments with Semantic Relations in Extended WordNet

The Extended WordNet [6] is constructed through of analyses of the synset glosses. During this process, the open class words were annotated with word synsets from PWN3.0. For example, the synset {stony coral, madrepore, madriporian coral}— 01916925-n, is defined by "corals having calcareous skeletons aggregations of which form reefs and islands." After the analysis, the following synsets are selected: 02673969-a—*calcareous*, 01917882-n—*mushroom coral*, 05585383-n—*skeleton*, 07951464-n—*aggregation*, 09316454-n—*island*, 09406793-n—*reef*, and 02621395-v—*form*. Each of these synsets is related to the synset to which the gloss belongs to:

u:01916925-n v:02673969-a s:30glc d:0
u:01916925-n v:01917882-n s:30glc d:0
u:01916925-n v:05585383-n s:30glc d:0
u:01916925-n v:07951464-n s:30glc d:0
u:01916925-n v:09316454-n s:30glc d:0
u:01916925-n v:09406793-n s:30glc d:0
u:01916925-n v:02621395-v s:30glc d:0

The division of the relations in **WNG** into groups is on the basis of the parts of speech of the main synset. The four sets are: **WNG-A** (first synset is for adjectives), **WNG-N** (first synset is for nouns), **WNG-R** (first synset is for adverbs), and **WNG-V** (first synset is for verbs). In Table 3 we present the impact of each of these sets of relations on the KG of **WN**. As it can be seen, each set adds accuracy above the baseline of **WN** for *SemCor*, but not for *BTB*. For *BTB* only two are better than the **WN** baseline. One of them is even better than the baseline for **WNG** graph.

Table 3 Experimental results when using the sets of relations from WNG

KG	SemCor	BTB	KG	SemCor	BTB
WN + WNG-A	52.94	53.08	WN + WNG-R	51.76	52.85
WN + WNG-N	**57.04**	52.92	WN + WNG-V	53.01	**56.01**

Table 4 Results from experiments using the sets of relations from syntax

KG	SemCor	BTB	KG	SemCor	BTB
WNG + SC-AA	**59.20**	**55.93**	WNG + SC-RN	58.94	55.89
WNG + SC-AN	**59.30**	55.89	WNG + SC-RR	**59.07**	**55.93**
WNG + SC-AV	**59.46**	55.78	WNG + SC-RV	**59.43**	52.71
WNG + SC-NN	58.81	**56.21**	WNG + SC-VN	**59.05**	55.55
WNG + SC-NV	**59.31**	**56.21**	WNG + SC-VV	**59.26**	53.78
WNG + SC-RA	**59.52**	**56.18**			

5.4 Syntax-Based Relations

As it was mentioned above, in our experiments we have also used semantic relations from a syntactically annotated corpus.

First, we defined patterns of dependency relations. For example, we used patterns like the following: $s_1 subj s_2$, which defines a relation between a noun synset s_1 and a verb synset s_2; $s_1 mods_2$, which defines a relation between an adjective synset s_1 and a noun synset s_2; $s_1 modxpobjs_2$, which defines a relation between a noun synset s_1 and a noun synset s_2; etc. We extracted the following sets of relations: **SC-AA, SC-AN, SC-AV, SC-NN, SC-NV, SC-RA, SC-RN, SC-RR, SC-RV, SC-VN, SC-VV**, where the suffixes: **AA, AN, AV**, etc., denote the parts of speech of the related synsets. The results from the experiments performed are presented in Table 4. As it can be seen, many of the extracted new sets increase the accuracy above the baseline for **WNG**.

We have combined most of these sets in joint combinations. The combination of all the sets with the original knowledge graph: **WNG, SC-AA, SC-AN, SC-AV, SC-NN, SC-NV, SC-RA, SC-RN, SC-RR, SC-RV, SC-VN, SC-VV** gives accuracy of **60.34 %** for *SemCor* and **53.05 %** for *BTB*. It is interesting that a slight change in the combination: is **WNG, SC-AA, SC-AN, SC-AV, SC-NV, SC-RA, SC-RN, SC-RR, SC-RV, SC-VN, SC-VV** gives a much better result for *BTB*— **55.38 %** and a slightly worse one for *SemCor*—**60.33 %**.

We also preformed experiments adding sets of relations from the WordNet relations. The best result for *SemCor* was achieved for the combination: **WNG, SC-AA, SC-AN, SC-AV, SC-NV, SC-RA, SC-RR, SC-RV, SC-VN, SC-VV, WN-HypInfer, WN-AntInfer, WN-DerVNInfer, WN-Ent1stVInfer, WN-Ent2ndVInfer**. Its accuracy is **60.70 %**. This result is 1.73 % higher than the baseline for WNG. The improvement is statistically significant. The result for *BTB* is **56.39 %**.

We have performed similar experiments with extraction of syntactic relations from the Universal Dependency version of BTB.[12] The targeted dependency relations are of the types: *nsubj, nmod, amod, iobj, dobj*. We have used these relations

---

[12]http://universaldependencies.github.io/docs/ .

in a similar way as the ones extracted from the parsed *SemCor* corpus. These syntactic relations have been extended similarly to the hypernymy relations. For example, in the case of the *nsubj* relation, the hyponyms of the dependent node have been replicated in new relations of the same kind, for all hyponyms of that particular word sense encountered in the golden corpus. Thus, the relation

u:00118523-v v:00510189-n

is derived from an *nsubj* relation, where 00118523-v stands for a sense of the Bulgarian verb "prodalzha" (continue) and is the head node (the predicate in *nsubj*), and 00510189-n, corresponding to a particular word sense of "veselba" (revelry), is the dependent node (the subject). The dependent node has a number of hyponyms in the WordNet hierarchy, therefore all these (and their hyponyms, too) have been added into a relation with the node 00118523-v. For instance, 00510723-n (the synset for particular word senses of the words "binge", "bout" and "tear") has been entered analogously in the same slot as 00510189-n.

Our motivation for using the hyponyms to infer new relations is based on the intuition that these syntactic relations connect an entity to an event[13] in which the entity participates or connects two participants of an event. We assume that if a class of entities contains possible participants in an event, then the instances of all sub-classes are possible participants in the same kind of event. The original relations are trusted to be valid, because they were annotated manually in the semantically annotated treebank. Another important assumption is that the relations found in the treebank are not the most general ones, which means that there is room for generalization over the participants in these events.

Thus, in addition to the extension of the dependency relations outlined above, we did a further enrichment of the knowledge graph by taking the hypernym of the node of interest in the syntactic relation and then taking all nodes beneath it in the hypernym hierarchy, and inserting them in the relevant relation attested in the golden corpus. Returning to the example from above in order to illustrate this strategy, we identify the "revelry" node ("unrestrained merrymaking") as subject of the "continue" node, then we go one level up to its hypernym, which is "merry-making" ("a boisterous celebration; a merry festivity"), and extend the *nsubj* relation from there downwards the hierarchy. Thus, the hyponym sense "jinks" ("noisy and mischievous merrymaking") is also inserted in the *nsubj* relation with the relevant sense of the verb "continue". This extension leads to an additional significant increase in the size of the KG. Figure 1 illustrates the described hierarchy as a simple tree. The bolded term ("revelry") is the node we want to use to expand the *nsubj* relation. The expanding procedure finds the hypernym of that node ("merrymaking"), then takes all the nodes below it and inserts them in the same type of relation, in place of "revelry". In this way, multiple relations can be derived from the initial *nsubj* relation.

---

[13]Here we interpret the concept of "event" in a wider sense that also includes states.

Fig. 1 Traversing the
hypernymy hierarchy, an
example

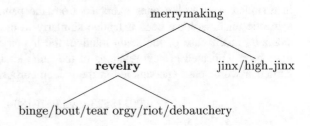

Finally, we have used information about WordNet domains, e.g. biology, linguistics, time_period, etc. An initial experiment was run whereby all synsets in a
given domain were entered in a relation with the domain. Unique WordNet-style
IDs were generated for all domains and the relevant synsets were connected to those
nodes. This approach yielded poor results, possibly due to the fact that in the PPR
algorithm the contribution of a node weakens the more outgoing edges it has, and
the artificial domain nodes have hundreds of outgoing links. Thus, an alternative
strategy was adopted of connecting all synsets within a domain to each other. In
order to avoid generating many millions of new relations, only the synsets in the
Bulgarian dictionary were connected in this fashion. This resulted in 132,596 new
relations. The following is a short description of the different configurations for the
new knowledge graphs. The basic graph is WNG:

- **WNG + I** : **WNG + WN-HypInfer**
- **WNG + ID** : **WNG + WN-HypInfer** + domain relations
- **WNG + IS** : **WNG + WN-HypInfer** + dependency relations
- **WNG + ISE** : **WNG + WN-HypInfer** + dependency relations + extended
  dependency relations
- **WNG + ISED** : **WNG + WN-HypInfer** + dependency relations + extended
  dependency relations + domain relations
- **WNG + ISEUD** : **WNG + WN-HypInfer** + dependency relations + extended
  dependency relations starting from one level up + domain relations

Table 5 represents the results from these experiments. The first two lines repeat
the baseline results from Table 1. Several interesting facts can be observed. The
addition of domain relations causes significant improvement for *BTB* test corpus.

Table 5 Results from adding syntactic relations from *BTB*

KG		*SemCor*		*BTB*	
WN		49.37		52.97	
WNG		58.97		55.90	
KG	*SemCor*	*BTB*	*KG*	*SemCor*	*BTB*
WNG + I	**58.59**	55.20	WNG + ISE	57.38	63.14
WNG + ID	58.46	55.82	WNG + ISED	57.41	63.54
WNG + IS	58.51	57.56	WNG + ISEUD	55.14	**66.71**

Adding the dependency relations extracted from the gold corpus results in close to 4.5 % improvement, while the addition of the downward extended set adds a further improvement of 5 %; extending the set by starting from one level above the original nodes in the dependency relations helps even more. Contextual information accounts for more than 11 % higher accuracy in the experiment done with the last configuration. However, the results over the *SemCor* corpus are rather negative.

When combining the *BTB* relations with relations from *SemCor* and inferred relations from WordNet, we received the following combination: **WNG + ISEUD, SC-AA, SC-AN, SC-AV, SC-NV, SC-RA, SC-RR, SC-RV, SC-VN, SC-VV, WN-HypInfer, WN-AntInfer, WN-DerVNInfer, WN-Ent1stVInfer, WN-Ent2ndVInfer**. The result for the *BTB* test corpus is **67.44 %**, which is the best result for this corpus. The same combination for *SemCor* gives a result of **57.65 %**.

6 Conclusion

The experiments with adding various bundles of relations from WordNet and from syntactically and semantically annotated corpora for Bulgarian and English have shown several directions to be considered in future work.

First of all, the addition of syntagmatic syntactic-based relations improve the results of KWSD task, since they balance the paradigmatic lexical relations. Then, the accuracy depends also on the integrity of the domain—in more homogeneous domains the accuracy is more stable and increases, while in more heterogeneous domains the accuracy drops. We consider the accuracy as a measure of quality of the knowledge graph with respect to the KWSD task. The conclusion is that adding important language and world knowledge in the form of relations between lexical concepts does not necessarily improve the quality of the knowledge graph.

Another issue is the different impact of the various relation types on the knowledge graph. Since the quantity of the added information is huge, our idea was to reduce it through the selection of the contributing relations, without losing the quality of the result. This strategy is not trivial. It requires a lot experimentation, as well as new mechanisms for evaluating the graph and optimizing the algorithm.

Acknowledgments This research has received partial support by the EC's FP7 (FP7/2007–2013) project under grant agreement number 610516: "QTLeap: Quality Translation by Deep Language Engineering Approaches" and FP7 grant 316087 AComIn "Advanced Computing for Innovation", funded by the European Commission in 20122016. The AComIN project kindly supported the scientific visits of Kiril Simov and Petya Osenova in Amsterdam, at Vrije Universiteit in 2014 and 2015. The authors would like to thank prof. Piek Vossen for hosting these visits as well as for the fruitful discussions in his group.

References

1. Agirre, E., Martinez, D.: Integrating selectional preferences in WordNet. In: Proceedings of First International WordNet Conference, Mysore, India. (2002)
2. Agirre, E., Soroa, A.: Personalizing PageRank for word sense disambiguation. In: Proceedings of the 12th Conference of the European Chapter of the ACL (EACL 2009), pp. 33–41. Association for Computational Linguistics (2009)
3. Agirre, E., Lacalle, O.L., Soroa, A.: Random walks for knowledge-based word sense disambiguation. Comput. Linguist. **40**(1), 57–84 (The Association for Computational Linguistics, MIT Press) (2014)
4. Brin, S., Page, L.: Reprint of: the anatomy of a large-scale hypertextual web search engine. Comput. Netw. **56**(18), 3825–3833 (2012)
5. Fellbaum, C. (ed.): WordNet an Electronic Lexical Database. The MIT Press, Cambridge (1998)
6. Mihalcea, R., Moldovan, D.: Extended Wordnet: progress report. In: Proceedings of NAACL Workshop on WordNet and Other Lexical Resources, pp. 95–100 (2001)
7. Miller, G.A., Leacock, C., Tengi, R., Bunker, R.T.: A semantic concordance. In: Proceedings of HLT '93, pp. 303–308 (1993)
8. Montoyo, A., Suárez, A., Rigau, G., Palomar, M.: Combining knowledge-and corpus-based word-sense-disambiguation methods. J. Artif. Intell. Res. (JAIR) **23**, 299–330 (2005)
9. Navigli, R., Ponzetto, S.P.: BabelNet: the automatic construction, evaluation and application of a wide-coverage multilingual semantic network. Artif. Intell. **193**, 217–250 (Elsevier) (2012)
10. Osenova, P., Simov, K., Laskova, L., Kancheva, S.: A treebank-driven creation of an ontovalence verb lexicon for Bulgarian. In: Proceedings of the Eight International Conference on Language Resources and Evaluation, pp. 2636–2640 (2012)
11. Popov, A., Kancheva, S., Manova, S., Radev, I., Simov, K., Osenova, P.: The sense annotation of BulTreeBank. In: Proceedings of the Thirteenth International Workshop on Treebanks and Linguistic Theories, pp. 127–136 (2014)

Part III
Signal and Image Processing

Multi-model Ear Database for Biometric Applications

Atanas Nikolov, Virginio Cantoni, Dimo Dimov, Andrea Abate
and Stefano Ricciardi

Abstract We present the 3DEarDB, a multi-model ear database, characterized by
different types of ear representation, either 2D or 3D, depending on the acquisition
device used. The main objective is to provide the biometrics community with a
unified tool for testing and comparing of classification algorithms not only on 2D
intensity and/or depth images, or videos, but also on detailed 3D mesh models of
human ears. The 3DEarDB features accurate 3D mesh models of right ear captured
from more than 100 subjects, with a resolution of 1 mm and an accuracy of
0.05 mm, collected via the VIUscan 3D laser scanner, available at the Smart Lab of
IICT-BAS, in the AComIn project frames. Two more ear acquisition modalities are
also included: 3D Kinect ear depth maps and 2D high-definition video clips,
associated to the basic mesh models. To extend 3DEarDB compatibilities with
known methods for 2D/3D ear detection and/or recognition, we provide two more
ear model types. Namely, a set of 2D ear intensity projections (of different orientations and/or lightening directions), and a set of 2D depth map projections can be
generated by demand from the basic 3D ear models. Finally, we report about
preliminary experiments conducted by means of Extended Gaussian Image
approach that confirm the consistency of the proposed 3D-Ear-Data-Base.

A. Nikolov · D. Dimov (✉)
Institute of Information and Communication Technologies (IICT),
Bulgarian Academy of Sciences (BAS), Sofia, Bulgaria
e-mail: dtdim@iinf.bas.bg

A. Nikolov
e-mail: a.nikolov@iinf.bas.bg

V. Cantoni
Department of Industrial and Information Engineering, Pavia University, Pavia, Italy
e-mail: virginio.cantoni@unipv.it

A. Abate · S. Ricciardi
Department of Computer Science, Salerno University, Salerno, Italy
e-mail: abate@unisa.it

S. Ricciardi
e-mail: sricciardi@unisa.it

© Springer International Publishing Switzerland 2016
S. Margenov et al. (eds.), *Innovative Approaches and Solutions in Advanced
Intelligent Systems*, Studies in Computational Intelligence 648,
DOI 10.1007/978-3-319-32207-0_11

Keywords 2D and 3D ear database · 3D mesh ear models · Ear biometrics · Extended gaussian image (EGI)

1 Introduction

The usage of biometric identifiers as a reliable and convenient way to verifying a person's identity has become common worldwide in the last decade, with particular regard to the most established ones like fingerprint, face and, more recently, iris. A key factor in diffusion of a biometric entity is its acceptability, since this characteristic directly affects the range of applications and the extent of the provided advantages in the context of both validation and identification [1]. In addition, aspects like stability over time and reduced intra class variations have been proved relevant in determining the success of biometrics-based id-check solutions. To these regards, ear seems to be a convenient biometric feature since it combines good distinctiveness, as indirectly proved by the high recognition accuracy achieved [2–4], with high acceptability (since is captured without the need for a physical contact) and permanence. The human ear was first hypothesized as a salient identifier in the end of XIX century by the French criminologist A. Bertillon [5], but only in 1949 A. Iannarelli proposed, with a more scientific approach, a set of twelve measurements characterizing the ear geometry [6]. The clear advantages in using ear biometrics are related to its tridimensional (3D) structure protruding from the overall head surface/profile (when observed frontally) that allows for simple and contactless capture by means of 2D and 3D techniques. Ear is characterized by easily recognizable ridges and valleys, whose configuration is relatively immune to variation due to ageing [7]. The almost complete absence of shape changes represents another advantage of this biometrics whose main intra-class variations derive by occlusions caused by hair, hats, earrings, etc., [8].

Though the number of contributions delivered by the research community on the topic of ear recognition are not comparable to the effort produced so far for face, fingerprint or even iris, many different methods and algorithms have been proposed with both 2D and 3D approaches over the last 15 years. 2D methods have exploited a variety of descriptors, including Principal Component Analysis (PCA) [9, 10], Independent Component Analysis (ICA) [11], Active Shape Model (ASM) [12], sparse representations [13], force fields [2, 14, 15], ear geometries [16, 17], Generic Fourier Descriptor (GFD) [18], wavelet transforms [3, 19, 20], Local Binary Patterns (LBP) [21], Gabor filters [22] and Scale-Invariant Feature Transform (SIFT) [23, 24].

The first 3D method [25] was proposed in 2004 and exploited the Local Surface Patch (LSP) representation and the Iterative Closest Point (ICP) algorithm, that was also used [4, 26, 27] for matching ears models obtained as range images or 3D mesh. A 2.5D approach was explored using surveillance videos and pseudo 3D information extracted by means of Shape-from-Shading (SFS) scheme [28]. It is worth to mention also two recent approaches to 3D ear recognition, based on the

EGI representation of 3D ear models [29], and on the 2D appearance 3D multi-view approach [30], in which additional related works are surveyed. A detailed and recent survey on Ear processing and recognition can be found in [31], as well as in [32, 33].

A crucial aspect of the research around ear biometrics is represented by the availability of public ear databases to be used as a reference to test and stress proposed methods on a common set of images captured in known conditions, and to highlight the strengths and the weaknesses of each method and/or approach in terms of recognition accuracy and robustness. To this regard, a number of ear datasets have been publicly released through the last 10 years, along with the research works that led to their creation. They typically provide 2D pictures of the ear(s) isolated or as a part of face profiles (mostly captured in laboratory), and in a limited number of cases also 3D scans of the face region near to the ear. We provide details on the existing ear datasets in Sect. 2 of this paper. Since, currently there is still a lack of a multi-model ear database, providing a full spectrum of capturing modalities for each of the enrolled subjects, in this paper we present such a kind of ear dataset that features high resolution 3D scans for each subject (both, row data and a segmented, cleaned polygonal mesh), also high resolution color pictures, high resolution video capture from variable angles, color pictures captured by last-generation mobile devices and other indirect modalities derived by the 3D data (2D intensity, and depth images).

The rest of the paper is organized as follows. Section 2 presents a description of the existing, publicly available, ear datasets. Section 3 provides a detailed description of a new dataset developed with regard to all the provided models and their capture. Section 4 presents the results of the first batch of experiments conducted on the proposed dataset and, finally, Sect. 5 draws some conclusions.

2 Publicly Available Ear-Specific Datasets—A Brief Review

As recalled in the previous section, there is a small number of publicly-available ear-specific datasets released so far, at least if we do not consider well known face database like, the FERET database [34], the CAS-PEAL database [35], the UMIST database [36], the NIST Mugshot Identification Database (MID) [37] or the XM2VTS database [38] which, though not originally aimed at ear biometrics, have been used and cited in literature mostly for testing ear detection algorithms. The ear-specific datasets are the AMI Ear Database [39], the UBEAR dataset [40], the University of Notre Dame (UND) databases [41], the University of Science and Technology Beijing (USTB) Databases [42], as well as the most recent OpenHear database [43], and the SYMARE database [44]. They are briefly described in the following lines.

AMI Ear Database [39] consists of ear images collected from students, teachers and staff of the Computer Science department at Universidad de Las Palmas de Gran Canaria (ULPGC), Las Palmas, Spain. The 700 images provided have been captured solely in an indoor environment from 100 different subjects in the age range of 19–65 years. For each individual, seven images (six right ear images and one left ear image) are taken under the same lighting conditions, at a capture resolution of 492 × 702 pixels, with the subject seated at a distance of about 2 m from the camera. Five of the captured images are right side profile (right ear) with the individual facing forward, looking up and down, and looking left and right (Fig. 1).

UBEAR Dataset [40] represents the result of a research study focused on capturing ear images on the move in uncontrolled conditions, including ample variations of posing, lighting and presence of occlusions, to the aim of providing a real-world set of samples that should result very challenging for detection and recognition algorithms. The dataset is built by means of four high-resolution (1280 × 960 pixels at 15 fps) video captures, two for each ear across two different sessions, requiring each subject to undergo the same enrollment protocol. From each video 17 frames (5 frames for stepping ahead and backwards + 12 frames for head movements in four directions, namely, 3 upwards, 3 downwards, 3 outwards, and 3 towards) are selected for each of the 126 subjects, acquired of whom 44.62 % are males and 55.38 % are females. The result database contains 4430 uncompressed gray-scale images, a few is shown in Fig. 2.

UND Databases [41] of the University of Notre Dame include a variety of biometric data in various modalities, organized in collections. The following four collections are relevant for ear biometrics:

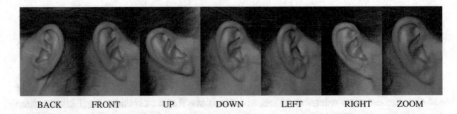

BACK   FRONT   UP   DOWN   LEFT   RIGHT   ZOOM

Fig. 1 Seven samples of two subjects captured from different directions (from AMI dataset)

Fig. 2 Samples of different posing in the UBEAR dataset

- Collection E: 464 visible-light face side profile (ear) images from 114 human subjects captured in 2002.
- Collection F: 942 3D (+corresponding 2D) profile (ear) images from 302 human subjects captured in 2003 and 2004.
- Collection G: 738 3D (+corresponding 2D) profile (ear) images from 235 human subjects captured between 2003 and 2005.
- Collection J2: 1800 3D (+corresponding 2D) profile (ear) images from 415 human subjects captured between 2003 and 2005.

USTB Databases [42] of the University of Science and Technology Beijing represent four databases dedicated to ear biometrics:

- Image Database I (dated: July–Aug 2002) contains 180 grayscale images of right ear from 60 subjects, each one photographed three times including one frontal image, another one with slight angle and one more with different lighting condition.
- Image Database II (dated: Nov 2003–Jan 2004) contains 308, 300 × 400 pixels, 24bit color images of right ear from 77 subjects, each one photographed four times with one profile image, two different form angles and one with different lighting conditions.
- Image Database III (dated: 20 Nov–30 Dec 2004) contains two ear datasets, a dataset with regular ear images and another one with occluded ear images. The first dataset includes right side profiles captured at 768 × 576 pixels, 24 bit colors from 79 subjects captured from variable rotations: 22 rotation steps to the right and 18 to the left. The second dataset contains 144 images of partially occluded ears from 24 subjects. They obey three conditions: partial occlusions (disturbance from some hair), trivial occlusions (little hair), and regular (natural) occlusions.
- Image Database IV (dated: Jun 2007–Dec 2008) contains both grayscale and color ear images, 500 × 400 pixels each, from 500 subjects acquired from multiple angles by 17 CCD cameras distributed around the volunteer at a 15° step from each other.

OpenHear, the Open head and ear database [43], is an open database of 3D surface scans of human heads and ears. Its purpose is to be used for acoustical simulation in aid design. The dataset contains head and ear 3D models of 20 subjects (10 men, 7 women, 1 baby boy, and 2 girls), see part of them in Fig. 3. The scans (available in VTK format) are acquired using a 3dMD cranial scanner, placed

Fig. 3 Samples from the current version of OpenHear dataset

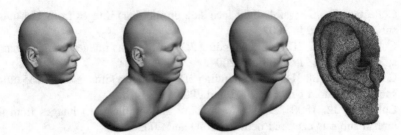

Fig. 4 Samples from SYMARE: the four types of surface meshes provided per subject

at the 3D Craniofacial Image Research Laboratory at the University of Copenhagen. The initial 3D point clouds are created via 3dMD stereo-algorithms, while surface reconstructions are obtained using the authors algorithm to create complete head and ear models from initial captured data.

SYMARE [44], the Sydney York Morphological and Acoustic Recordings of Ears database, supports acoustics research exploring the relationship between the morphology of human outer ears and their acoustic filtering properties for purpose of improving the individualization of 3D audio for personal audio devices in the future. The database includes multiple mesh models (upper torso, head and ears) at varying resolutions for 61 listeners (48 male and 13 female) in order to accommodate acoustic stimulations at different frequencies. The 3D data are collected using a Philips 3T Achieva MRI scanner. For each of the 61 subjects in the database, high-resolution (sub-millimeter) surface meshes are provided for: (i) the head and ears, (ii) the head, upper torso and ears, (iii) the head and upper torso (no ears), (iv) the separated left and right ears, see Fig. 4. The number of surface elements involved in an average head and torso mesh is about 130 K elements.

3   Overview of Our 3D Ear Database

The announced 3D Ear Database, called here 3DEarDB, was collected mainly during the middle of 2015 at the Institute of Information and Communication Technologies at Bulgarian Academy of Sciences (IICT-BAS) in the frames of AComIn[1] project. We have gathered more than 100 precise 3D mesh models of right ears of persons, who differ in gender as well as in age (25–65). A scan resolution of 1 mm between neighboring 3D points and accuracy of 0.05 mm for each 3D point was chosen for simplicity of the data gathering, considering it to be enough for near future experiments. The first version of 3DEarDB (dated May, 2014) contained 3D ear models of the same precision but for 11 persons only, and

---

[1]http://www.iict.bas.bg/acomin/.

was designed for initial experiments with both our approaches to 3D ear classification and/or recognition, [29, 30].

The recent objective of 3DEarDB is to provide, in a consistent way, many different output formats for the given human (subject, person) ear represented. These includes: (i) a raw 3D ear mesh model, (ii) a processed 3D ear mesh, (iii) Kinect 3D ear depth (range) images, (iv) accompanying 2D ear video clips, (v) generated structures of 2D ear intensity projections, and (vi) generated structures of 2D ear depth images. This consistent variety of ear capturing formats could be very useful for ear biometrics community to test and compare algorithms accuracy on possibly different input scenarios—from the ideal case of precise (and static) 3D mesh to more realistic (and dynamic) case of 2D video data and/or still images.

By our best knowledge, cf. also Sect. 2, among the existent Ear Datasets, the only DB, which provides corresponding 2D and 3D data for the same subject's ear is that of UND Collections F, G, and J2, [41]. The UND 3D ear data do not represent real polygonal 3D meshes, but only 3D range images containing depth information. Moreover, the ear video data, which could be used for performing 3D ear reconstruction as an alternative to 2D range images, are missing there. The recent 3D databases, OpenHear [43] and SYMARE [44], really concern 3D ear data, but they are not designed especially for visual ear biometrics. Besides, neither OpenHear (only 20 face models), nor SYMARE even with its 61 listeners recorded and scanned, could be considered statistically enough representative at present.

An essential requirement of the large biometrics community is that such a DB has to top 100, or more, persons represented. We also consider ear biometrics based on video data as the most realistic case according to the contemporary technology development, especially if it is intended to be build-in the portable electronics of personal use. For this reason, it is useful to provide accurate 3D ear mesh representation as reference for evaluation of 3D video reconstruction errors, and for comparing between ideal and real recognition performances of investigated descriptors and classifiers. Because of we consider colors a non-informative ear feature for classification, we do not scan it at present. Colors are kept in the accompanying 2D ear video clips.

Next section contains a more detailed description of our multi-model Ear DB, considering two main types of ear data—hardware acquired and software generated. Hardware acquired ear representations are composed by raw and post-processed 3D ear meshes (from 3D laser scanners), 3D depth maps (from Kinect cameras), and 2D Video clips (from photo cameras). The software generated ear representations from each 3D mesh model are also two types at present, namely: (i) structures of images, i.e. 2D intensity projections with different lightening and/or orientation (using MeshLab[2]); and (ii) corresponding structures of 2D depth map projections with different orientation (using Wolfram Mathematica[3]).

---

[2]http://meshlab.sourceforge.net/.

[3]https://www.wolfram.com/mathematica/.

(a)   (b)   (c)

Fig. 5 a The cartoon "helmet". **b** A person under scanning. **c** VIUscan 3D scanner

3.1 Data Acquisition

The three types of devices we use to collect ear data are described below. Only right
ears data are gathered, and only one 3D ear model per subject is represented in
3DEarDB, because of limited people resource, for the time being. For more detail
on this matter see also discussions in Sects. 4.2 and 5.

VIUscan 3D Laser Scanner. This hand-scanner of Creaform (Fig. 5c) was
bought by the AComIn project for the Smart Lab of IICT-BAS in the end of 2013.
Well computer assisted, it can reproduce a 3D mesh model of the scanned solid as
well as respective textures and/or colors. Although, we have not used the maximal
resolution (0.1 mm) and any color data, they could be very useful in other applications, where 3D objects have variable texture with fine surface details, [45].

This type of scanners require specific markers (retro-reflective targets) regularly
situated on or around the object of scanning. The scanner needs to "see" at least
four targets, which should not move in respect to the object of scan. VIUscan uses
these targets to position itself in the space. To facilitate our work, we created a
special "helmet" of cartoon with enough markers on it. The helmet is to be placed
on the subject's head around the ear before scanning (Fig. 5a, b).

Omitting of color data makes the procedure of scanning faster, up to 10 min per
ear, as well as more comfortable, because of no need of special lightening—
possible shadows do not disturb scanning.

Kinect Xbox One Sensor.[4] This motion sensor of Microsoft is an upgraded
version of its predecessor Xbox 360. Available as a standalone version since
October 2014, it has an infrared array and a 512×424 pixels time-of-flight camera
that resolves scene depth and allows for motion tracking and gesture recognition.
This new Kinect also includes a Full HD (1920×1080) video camera with
increased field of view.

We plan to use Kinect for obtaining real depth maps of ears and to apply its
accompanying software for 3D reconstruction (using video and/or depth maps).

---

[4]https://en.wikipedia.org/wiki/Kinect_for_Xbox_One#Specifications.

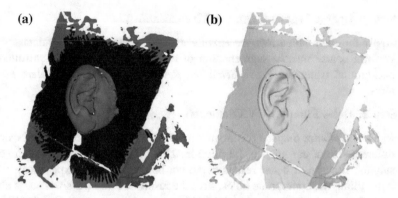

Fig. 6 **a** Raw scanned ear with color data. **b** Only the surface of the raw ear data

Olympus Photo Camera.[5] The Olympus SH-21 photo camera with its 16 MP CMOS sensor of 1/2.3″ format has been used for producing Full HD (1920 × 1080) video clips for each subject's ear, generally in a MP4 format file.

3.2 Raw (Unprocessed) Ear Data

A raw scanned ear, as shown on Fig. 6b, appears from VXelements software usually accompanying VIUscan scanners, [45]. The primary output file format is CSF, which size, in our case is about 64 MB per ear. VXelements help to convert each CSF to an OBJ format (an ASCII text) file for the ear geometry, and to an accompanying BMP file for the ear colors. In Fig. 6a we illustrate a colored ear scan, only for giving an idea of how it looks like, although not using it for now, as already mentioned. We use OBJ files at next (half-tone) post-processing, see Fig. 6b. Of course, color data could be successfully used for an automatic 3D ear segmentation, what is outside this work.

3.3 Raw Ear Data Post-processing

To create a complete and appropriately smooth 3D mesh model for each ear, we describe a post-processing of six steps using either VXelements [45] or MeshLab [46].

---

[5]http://www.olympus-global.com/en/news/2011b/nr111110sh21e.jsp.

Step 1: Coarse Segmentation (by VXelements)

- Apply the filter called *Remove Isolated Patches* on the input CSF data.
- Perform coarse manual segmentation of the ear surface from the surrounding background using the *Brush Selection*, *Reverse Selection*, and *Delete Facets* tools.

Step 2: Holes Filling (by VXelements)

- Run the *Optimize Surface* reconstruction algorithm each time when choosing a different size of ear holes to be filled-in. This procedure is the most time consuming, because of better results could not be predicted but experimented.
- After filling the appropriate holes, save the result CSF file (its size here is about 49 MB per ear). To continue with MeshLab processing, convert CSF to OBJ file that results in about 600 KB (per ear).

Step 3: Fine Editing of Mesh-Facets (by MeshLab). It includes finer background segmentation, as well as removing unpleasant sharp peaks (Fig. 7a) in the current 3D mesh model resulting from the Optimize Surface tool of the previous step. Of course, the peak facets removal leads to new holes to fill-in (Fig. 7c), but of much smaller size (Fig. 7b), that is usually no problem for MeshLab.

Step 4: Mesh Extra Smoothing (by MeshLab). After holes filling (Fig. 7c), the final step is smoothing the complete 3D object (Fig. 7d). The MeshLab function we prefer to this aim, is the *HC Laplacian Smooth*, based on the paper of Vollmer et al. [47]. At this final stage of manipulation, each ear mesh consists of about 6–8 thousands of (triangular) facets, determined by about 3–4 thousands of vertexes (3D points). Omitting the normal vectors data, considered here derivative and redundant ones for simplicity, the size of the respective OBJ file is reduced up to about 240 KB (per ear).

Step 5: Mesh Decimation and Subdivision (by MeshLab). This step is necessary for creation of test data for our EGI classification approach [29], which we use to prove experimentally the 3DEarDB functionality. The MeshLab function for increasing the facets number (Fig. 8c) is called *Subdivision Surfaces: LS3 Loop*, based on [48], and the function reducing this number (Fig. 8a) is *Quadratic Edge Collapse Decimation*.

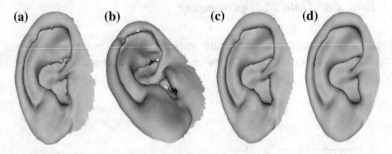

Fig. 7 **a** Sharp peaks. **b** New holes created. **c** All holes filled. **d** Final smooth

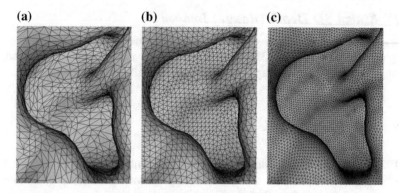

Fig. 8 **a** Decimated facets. **b** Original scan resolution. **c** Subdivided (refined) facets

Step 6: Geometric Normalization (in MATLAB). It includes translation, orientation and scale of each ear model separately:

- Translate the Cartesian origin into the model barycenter, i.e. the averaged (x, y, z) coordinates of all 3D points (vertexes) of the mesh. After subtracting it from all vertexes, the new barycenter becomes (0, 0, 0).
- Rotate Principal axes, i.e. the eigenvectors of the covariance matrix over the whole mesh (all the vertexes). To normalize by rotation, the vertexes are rotated back to the already centralized Cartesian coordinate system, see also Fig. 9.
- Scale: The three eigenvalues (associated to principal axes, they should be already rotated) are used to normalize the mesh model by scale, so that the bounding box of the model (or its equivalent ellipsoid) to reach predefined sizes, e.g. 1-s (units). The three scale coefficients (reciprocal to eigenvalues) for each model have to be saved, if the real ear size will be further essential.

Fig. 9 A normalized ear model

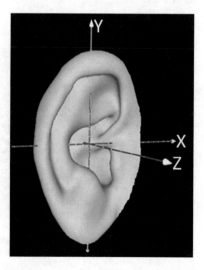

3.4   Kinect 3D Depth (Range) Images

At present, we do not give 3D ear data gathered by a Kinect camera. Instead, we have generated 2D depth-map images from 3DEarDB, as described in Sect. 3.7.

3.5   Full HD Ear Video Clips

A 1920 × 1080 video is made over each ear, uniformly filming it by azimuth from −80° to +80°, for 3 different altitude rows (upper, central, and lower ones) towards the center of the ear frontal view (Fig. 10), in the same laboratory, immediately after the 3D ear scan. Each clip is about 20 s long, at 30 fps that costs about 45 MB per clip, written in MP4 file format.

3.6   2D Intensity Projections

The 2D ear projections are produced in MeshLab, by loading a number of layers, one for each 3D rotation of an ear. Then, 2D snapshots of all these layers are made and recorded in JPEG format. The artificial lightening chosen is frontal and coherent.

The 2D intensity projections are taken according to a rotations scheme of 100 frontal view directions, uniformly distributed towards the ear barycenter, i.e. on 10 declinations and 10 azimuths uniformly chosen in the interval (−45°, +45°), cf. also Fig. 11. Of course, the angle step could be smaller or larger, in this way to manipulate the density of the resultant set of 2D projections, i.e. the size of output JPG files.

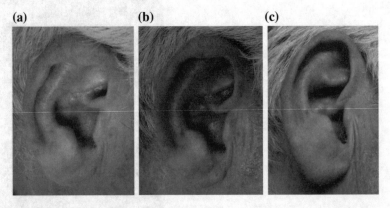

Fig. 10 Representative frames for the three horizontal rows of an ear video clip. **a** View from above. **b** A central view; and **c** view from bellow

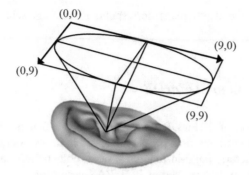

Fig. 11 A scheme of multi-view 3D modeling of a given ear

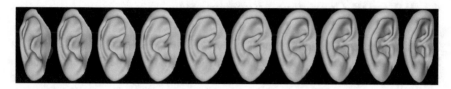

Fig. 12 2D ear images from a row of the ear model rotation scheme, cf. also Fig. 11

This type of 3D ear representation, we call it *Multi-view 3D modeling*, has been developed for our experiments in [30]. We needed there a random access to the Multi-view datasets, but the same datasets could be arbitrary ordered, e.g. top-down and left-right, like the video clips of Sect. 3.5.

An illustration of ten 2D ear images generated from a 3D ear model (for a given central row, cf. Fig. 11), is shown in Fig. 12.

3.7 2D Depth Map Images

The build-in functions of Wolfram Mathematica software was used to render 2D depth images from a 3D mesh, where instead of intensity values, the z-coordinates of the 3D points are recorded into the 2D image grid (Fig. 13). For consistence with

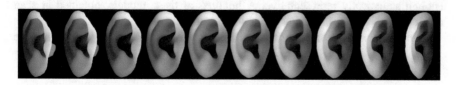

Fig. 13 Ear depth maps under orthographic projections of a given 3D ear model

previous section, the depth maps correspond to rotation scheme illustrated on Fig. 11.

3.8 Web Access to 3DEarDB

The current version of 3DEarDB will be placed at a free of charge disposal of academic and non-profit research people interested in it. An extended description of the 3DEarDB structure, build-in functions, other potentialities, and license agreements will appear on the web site of IICT-BAS very soon.

4 3DEarDB Consistency Experiments

To test the current 3DEarDB functionality, we have experimented using our EGI based approach to ear classification and/or recognition [29]. The EGI representation squeezes appropriately the 3D mesh model data into a sphere, so that it can be visualized and/or used like a 2D (histogram) image, and even like an 1D histogram, by an appropriate re-indexing of facets, e.g. by a spiral, see also [29].

The EGI (Extended Gaussian Image) was initially proposed by B.K.P. Horn, in 1984, [49], see also [50]. Formally, the EGI of a 3D surface represents a histogram of all orientations of the modeled surface on a unit (Gaussian) sphere. Because of surface usual representation by a discrete mesh, every facet from the modeling 3D mesh will be accumulated into the respective point on the Gaussian sphere, according to the unit normal vector and the area of each facet. I.e. the total weight of each EGI point equals the cumulative area of all the mesh facets with the same normal vector direction. In practice, the Gaussian sphere is also discretized by a triangular tessellation, most often based on icosahedron (20 triangular facets). Depending on the level n of the sphere discretization, the number m of 3-angle-facets equals: $m = 4^n 20$, $n = 0, 1, \ldots$

In our experiments, we have chosen the following three levels: $n = 1, 2, 3$ corresponding to $m = 80, 320$, and 1280, see Table 1.

The opportunity of using the simpler EGI representation of 3D ear mesh models (in deviance of their convex/concave ambiguity) was experimentally demonstrated on a small ear DB, containing only 11 ears models, see [29]. The current version of our 3DEarDB consists of more than 100 ear models that by our best knowledge is enough statistically representative. A hundred of these models, obtained at scan resolution of 1 mm, in similar laboratory conditions, and well post-processed as described here, has been experimented (see Table 1), similarly to [29], to believe

Table 1 EGI accuracy results: true recognition rate (TRR)

TRR (%)		0.5 mm (a higher resolution recalculated)			1.0 mm (the original 3D scanning resolution)			1.4 mm (a lower resolution recalculated)		
Noise (mm)		**0.05**	**0.10**	**0.15**	**0.10**	**0.20**	**0.30**	**0.20**	**0.30**	**0.40**
~ % on width		0.16	0.31	0.47	0.31	0.62	0.93	0.62	0.93	1.24
~ % on height		0.10	0.20	0.30	0.20	0.40	0.60	0.40	0.60	0.80
~ % on depth		0.38	0.76	1.14	0.76	1.52	2.27	1.52	2.27	3.03
80 facets	E_2	*100*	*93*	*17*	*100*	*92*	*16*	*100*	*88*	*40*
	E_{BC}	*100*	*100*	*67*	*100*	*100*	*70*	*100*	*100*	*80*
320 facets	E_2	*100*	*84*	*25*	*100*	*83*	*27*	*100*	*80*	*44*
	E_{BC}	*100*	*100*	*80*	*100*	*100*	*66*	*100*	*100*	*80*
1280 facets	E_2	*100*	*63*	*13*	*100*	*52*	*16*	*100*	*56*	*30*
	E_{BC}	*100*	*100*	*62*	*100*	*95*	*48*	*100*	*96*	*78*

one more again in the proposed 3DEarDB plausibility. For evaluation of similarity between EGI histograms, we have considered again the two geometrical scores:

- the Euclidean distance: $E_2 = \sqrt[2]{\sum_{i=1}^{m}(M_i - S_i)^2}$, and

- the Bray Curtis figure of merit [51]: $E_{\mathrm{BC}} = \frac{\sum_{i=1}^{m}|M_i - S_i|}{\sum_{i=1}^{m}(M_i - S_i)}, 0 \leq E_{\mathrm{BC}} \leq 1$;

  where M_i and S_i are both the histogram bins under comparison (of the model and the input objects), $i = 1, 2 \ldots m$; $m = 80$ or 320, or 1280, see Table 1.

4.1   Additional Notes to Table 1

- Nearest-neighbor method has been performed for tests, where each processed 3D ear model is considered a center of a class, i.e. the number of classes now is 100.
- Each 3D ear model in the 3DEarDB has been additively noised before using it for test recognition (retrieving the most similar one from 3DEarDB). Three versions of 3DEarDB, i.e. for 3 scan resolutions have been tested: 1.0 mm that is the original one, and two more, 0.5 and 1.4 mm that are recalculated from the original (see Step 5 in Sect. 3.3).
- The noise is artificially generated randomly in the used intervals of 3D scan, i.e. on average: width = 32.3 mm (on Ox), height = 50.3 mm (on Oy), and depth = 13.2 mm (on Oz). These 3 intervals have been simply averaged using respective eigenvalues at the normalization processing (Step 6 in Sect. 3.3).

- To be comparable with other (or further) experiments, the noise intervals are expressed in percents, respectively towards the averaged width, height and depth.

4.2 Experiment Analysis

The following generalization can be done analyzing the conducted experiments:

- Experiments conducted on the current 3DEarDB (100 ear models) confirm the possibility of using the EGI representation for the unambiguous identification of ears nevertheless of their surface mixture of concavities and convexities. This is confirmed by the evaluated noise limits for each of the three experimented resolutions (0.05, 0.10, 0.20 mm, see leftmost columns of Table 1, where TRR = 100 %) that well overcome 0.05 mm, the declared accuracy of used 3D scanner VIUscan.
- As expected, the Bray-Curtis distance (E_{BC}) is more robust to the corresponding level of noise, than the Euclidean distance (E_2), giving higher TRR.
- A "phenomenon" can be observed for the rest of results of the type TRR < 100 % (at higher level of noise, see middle and rightmost columns), where improvements of either EGI representation (80 $\rightarrow$ 320 $\rightarrow$ 1280) or 3D scanning resolution (0.5 $\leftarrow$ 1.0 $\leftarrow$ 1.4) give an unexpected decrease of TRR at similar levels of noising.
- This "phenomenon" of TRR behavior is considered outside the main positive result for 3DEarDB functionality. Besides of concavities-convexities-mixture of ear surfaces, it can be explained also with combinations of other nonlinearities, like: (i) triangulation irregularities of 3D models, (ii) EGI representation irregularities, (iii) smoothing effect of software manipulation of resolution, etc.
- Because of the opportunities of reducing either the geometric resolution of 3D scanning or the complexity of EGI representation, are always approaching to real time processing, we will keep attention on this phenomenon in our future work.

5 Discussion and Conclusion

The current paper describes and proposes to the ear biometric research community a novel multi-model Ear Database, called 3DEarDB. It is composed from different corresponding sets of ear representations from about 100 subjects of Caucasian race acquired by various capturing devices: 3D Laser Scanner, Kinect Xbox One sensor, and a Digital Photo Camera.

The 3DEarDB distinguishes from the currently known similar DBs for its completeness in ear representations of different formats—3D meshes, 3D depth (range) images, 2D video clips, 2D intensity projections. For this reason, it could be useful for comparative analyses among a large variety of known 2D/3D ear recognition approaches and new ones as well, based on the 3D mesh information itself.

A few extra notes about the 3DEarDB near future:

- The current 3DEarDB consists of more than 100 3D ear models. It will be systematically extended in accordance with the feedback from potential users from biometric community in the country and abroad.
- At present, the 3DEarDB consists of only one 3D ear model per subject. The optimal number of (repeated) models per subject will be evaluated soon on the base of a few model versions for a small number of subjects represented (by their right ear). The same is also intended for the left human ear.
- In order to speed up the model acquisition, besides of Kinect camera, we are planning to experiment also with a 3D scanner of structured light type, perhaps on the price of some precision reduction.

Acknowledgments This research is partly supported by the project AComIn "Advanced Computing for Innovation", grant 316087, funded by the FP7 Capacity Programme "Research Potential of Convergence Regions".

References

1. Day, D.: Biometric applications, overview. In: Li, S.Z., Jain, A.K. (eds.) Encyclopedia of Biometrics, pp. 169–174. Springer, Heidelberg (2015)
2. Hurley, D., Nixon, M., Carter, J.: Ear biometrics by force field convergence. In: Proceedings of the 5th International Conference on Audio- Video- Biometric Person Authentication, pp. 386–394 (2005)
3. Wang, Y., Mu, Z., Zeng, H.: Block-based and multi-resolution methods for ear recognition using wavelet transform and uniform local binary patterns. In: Proceedings of the 19th IEEE International Conference on Pattern Recognition (ICPR), pp. 1–4 (2008)
4. Yan, P., Bowyer, K.: Empirical evaluation of advanced ear biometrics. In: Proceedings of the IEEE Conference on Computer Vision and Pattern Recognition Workshops, pp. 41–42. San Diego, CA, USA, ISBN 0-7695-2372-2 (2005)
5. Bertillon, A.: Signaletic Instructions Including: The Theory and Practice of Anthropometrical Identification (1896)
6. Iannarelli, A.: Ear Identification, Forensic Identification Series. Paramount Publ. Company, Fremont, CA (1989)
7. Cummings, A.H., Nixon, M.S., Carter, J.N.: A novel ray analogy for enrolment of ear biometrics share. In: Proceedings of IEEE Fourth Conference on Biometrics: Theory, Applications and Systems, Washington DC, USA, pp. 1–6 (2010)
8. De Marsico, M., Nappi, M., Riccio, D.: HERO: human ear recognition against occlusions. In: IEEE Conference on Computer Vision and Pattern Recognition Workshops (CVPRW), pp. 178–183. June 13–18 2010

9. Chang, K., Bowyer, K.W., Sarkar, S., Victor, B.: Comparison and combination of ear and face images in appearance-based biometrics. IEEE Trans. Pattern Anal. Mach. Intell. **25**, 1160–1165 (2003)

10. Victor, B., Bowyer, K.W., Sarkar, S.: An evaluation of face and ear biometrics. In: Proceedings of 16th IEEE International Conference on Pattern Recognition (ICPR), pp. 429–432 (2002)

11. Zhang, H., Mu, Z., Qu, W., L Iu, L., Zhang, C.: A novel approach for ear recognition based on ICA and RBF network. In: Proceedings of the 4th IEEE International Conference on Machine Learning and Cybernetics, pp. 4511–4515 (2005)

12. Yuan, L., Mu, Z.: Ear recognition based on 2D images. In: First IEEE International Conference on Biometrics: Theory, Applications, and Systems, pp. 1–5 (2007)

13. Naseem, I., Togneri, R., Bennamoun, M.: Sparse representation for ear biometrics. In: Proceedings of the 4th International Symposium on Advances in Visual Computing (ISVC), Part II, pp. 336–345 (2008)

14. Hurley, D., Nixon, M., Carter, J.: Automatic ear recognition by force field transformations. In: Proceedings of the IEEE Colloquium on Visual Biometrics, pp. 7/1–7/5 (2000)

15. Hurley, D., Nixon, M., Carter, J.: Force field feature extraction for ear biometrics. Comput. Vis. Image Underst. **98**(3), 491–512 (2005)

16. Choras, M., Choras, R.: Geometrical algorithms of ear contour shape representation and feature extraction. In: Proceedings of the 6th IEEE International Conference on Intelligent Systems Design and Applications, pp. 451–456 (2006)

17. Choras, M.: Ear biometrics based on geometrical feature extraction. Electron. Lett. Comput. Vis. Image Anal. **5**(3), 84–95 (2005)

18. Abate, A., Nappi, M., Riccio, D., Ricciardi, S.: Ear recognition by means of a rotation invariant descriptor. In: Proceedings of the 18th IEEE International Conference on Pattern Recognition (ICPR), pp. 437–440 (2006)

19. Hailong, Z., Mu, Z.: Combining wavelet transform and orthogonal centroid algorithm for ear recognition. In: Proceedings of the 2nd IEEE International Conference on Computer Science and Information Technology, pp. 228–231 (2009)

20. Sana, A., Gupta, P.: Ear biometrics: a new approach. In: Proceedings of the 6th International Conference on Advances in Pattern Recognition, 06 Sep. 2006, pp. 1–5 (2007)

21. Nanni, L., Lumini, A.: A multi-matcher for ear authentication. Pattern Recogn. Lett. **28**(16), 2219–2226 (2007)

22. Watabe, D., Sai, H., Sakai, K., Andnakamura, O.: Ear biometrics using jet space similarity. In: Proceedings of the IEEE Canadian Conference on Electrical and Computer Engineering, pp. 1259–1264. Niagara Falls, ON. e-ISBN 978-1-4244-1643-1, May 4–7 2008

23. Dewi, K., Yahagi, T.: Ear photo recognition using scale invariant keypoints. In: Proceedings of the International Computational Intelligence Conference, pp. 253–258 (2006)

24. Kisku, D.R., Mehrotra, H., Gupta, P., Sing, J.K.: SIFT-based ear recognition by fusion of detected key-points from color similarity slice regions. In: Proceedings of the IEEE International Conference on Advances in Computational Tools for Engineering Applications (ACTEA), pp. 380–385 (2009)

25. Chen, H., Bhanu, B.: Human ear detection from side face range images. In: Proceedings of the IEEE International Conference on Pattern Recognition (ICPR), pp. 574–577 (2004)

26. Islam, S., Bennamoun, M., Mian, A., Davies, R.: A fully automatic approach for human recognition from profile images using 2D and 3D ear data. In: Proceedings of the 4th International Symposium on 3D Data Processing, Visualization and Transmission, pp. 131–135. Atlanta, Georgia, USA (2008)

27. Yan, P., Bowyer, K.: Biometric recognition using 3D ear shape. IEEE Trans. Pattern Anal. Mach. Intell. **29**(8), 1297–1308 (2007)

28. Cadavid, S., Abdelmottaleb, M.: 3D ear modeling and recognition from video sequences using shape from shading. IEEE Trans. Inf. Forens. Secur. **3**(4), 709–718 (2008)

29. Cantoni, V., Dimov, D.T., Nikolov, A.: 3D ear analysis by an EGI representation. In: Cantoni, V., Dimov, D.T., Tistarelli, M. (eds.) Proceedings of the 1st International Workshop on Biometrics, BIOMET June 23–24, 2014, Sofia, Bulgaria. Biometric Authentication, LNCS, vol. 8897, pp. 136–150. Springer, Heidelberg (2014)
30. Dimov, D.T., Cantoni, V.: Appearance-based 3D object approach to human ears recognition. In: Cantoni, V., Dimov, D.T., Tistarelli, M. (eds.) Proceedings of the 1st International Workshop on Biometrics, BIOMET June 23–24, 2014, Sofia, Bulgaria. Biometric Authentication, LNCS, vol. 8897, pp. 121–135. Springer, Heidelberg (2014)
31. Barra, S., De Marsico, M., Nappi, M., Riccio, D.: Unconstrained Ear processing: what is possible and what must be done. In: Scharcanski, J., Proença, H., Du, E. (eds.) Signal and Image Proceeding for Biometrics, LNEE, vol. 292, pp. 129–190. Springer, Berlin (2014)
32. Pflug, A.: Ear recognition: biometric identification using 2- and 3-dimensional images of human ears. ISBN: 978-82-8340-007-6, Ph.D. thesis, 205p., Gjøvik Univ. College, 2-2015
33. Prakash, S., Gupta, P.: Ear biometrics in 2D and 3D—localization and recognition. In: Hammoud, R.I., Wolff, L.B. (eds.) Augm. Vision & Reality, vol. 10. Springer, Singapore (2015)
34. Phillips, P.J., Wechsler, H., Huang, J., Rauss, P.J.: The FERET database and evaluation procedure for face recognition algorithms. Image Vis. Comput. **16**(5), 295–306 (1998)
35. Gao, W., Cao, B., Shan, S., Zhou, D., Zhang, X., Zhao, D.: CAS-PEAL database (2004). http://www.jdl.ac.cn/peal/
36. UMIST database (1998). http://www.shef.ac.uk/eee/research/iel/research/face.html
37. MID. NIST mugshot identification database (1994). http://www.nist.gov/srd/nistsd18.cfm
38. XM2VTSDB database (1999). http://www.ee.surrey.ac.uk/CVSSP/xm2vtsdb/
39. AMI Ear Database. http://www.ctim.es/research_works/ami_ear_database/
40. Raposo, R., Hoyle, E., Peixinho, A., Proença, H.: UBEAR: a dataset of ear images captured on-the-move in uncontrolled conditions. In: IEEE Workshop on Computational Intelligence in Biometrics and Identity Management (CIBIM), pp. 84–89. Paris, France (2011)
41. UND Databases. http://www.cse.nd.edu/~cvrl/CVRL/Data_Sets.html
42. USTB Databases. http://www1.ustb.edu.cn/resb/en/index.htm
43. OpenHear Database. http://www2.imm.dtu.dk/projects/OpenHear/
44. SYMARE Database. http://www.ee.usyd.edu.au/carlab/symare.htm
45. HANDY SCAN 3D: The portable 3D scanners for industrial application. http://www.creaform3d.com/sites/default/files/assets/brochures/files/handyscan/Handyscan3D_Brochure_EN_HQ_22052012.pdf
46. Cignoni, P., Callieri, M., Corsini, M., Dellepiane, M., Ganovelli, F., Ranzuglia, G.: MeshLab: an open-source mesh processing tool. In: Proceedings of Eurographics Italian Chapter Conference, pp. 129–136 (2008)
47. Vollmer, J., Mencl, R., Müller, H.: Improved laplacian smoothing of noisy surface meshes. Int. Conf. Eurographics **18**(3), 131–138 (1999)
48. Boyé, S., Guennebaud, G., Schlick, C.: Least squares subdivision surfaces. Comput. Graph. Forum. **29**(7), 2021–2028 (2010)
49. Horn, B.K.P.: Extended Gaussian images. Proc. IEEE. **72**, 1671–1686 (1984)
50. Kang, S.B., Horn, B.K.P.: Extended gaussian image (EGI). In: Ikeuchi, K. (ed.) Computer Vision—A Reference Guide, pp. 275–278. Springer, New York (2014)
51. Bray, J.R., Curtis, J.T.: An ordination of upland forest communities of southern Wisconsin. Ecol. Monogr. **27**, 325–349 (1957)

Modular Digital Watermarking and Corresponding Steganalysis of Images on Publicly Available Web Portals

Svetozar Ilchev and Zlatoliliya Ilcheva

Abstract The protection of images made available on public web portals on the Internet can be implemented only through technical mechanisms intrinsic to the images themselves. A digital watermarking service can be used in an automated way to provide such a mechanism without human involvement beyond the initial setup of the service and its integration with the web portal software. Images are published only after they pass through the service which supplies them with a proper signature identifying the copyright holder and the terms of use. A complementary steganalysis service identifies the presence of the digital watermarking data and, depending on its usage, retrieves the embedded copyright information.

Keywords Image processing · Data hiding · Digital watermarking · Steganalysis · Security

1 Introduction

In the world of today the words security and protection keep gathering importance. At the same time the amount of multimedia content published on the Internet keeps growing. With regard to these trends, a main topic of interest is the creation of suitable mechanisms to secure and protect the published multimedia. In this paper, we focus on digital images but the same principles apply to audio and video content.

The protection of images made available on public web portals on the Internet can best be implemented through technical mechanisms which are intrinsic to the

S. Ilchev (✉) · Z. Ilcheva
Institute of Information and Communication Technologies,
Bulgarian Academy of Sciences, Sofia, Bulgaria
e-mail: svetozar@ilchev.net

Z. Ilcheva
e-mail: zlat@isdip.bas.bg

© Springer International Publishing Switzerland 2016   189
S. Margenov et al. (eds.), *Innovative Approaches and Solutions in Advanced
Intelligent Systems*, Studies in Computational Intelligence 648,
DOI 10.1007/978-3-319-32207-0_12

images themselves. One of the relevant research fields in this direction is digital watermarking [1–5]. Digital watermarking methods embed metadata into the image content in such a way that the metadata remains invisible but readable. Depending on the specific algorithms in use, the readability of the embedded metadata may or may not be guaranteed after image transformations such as lossy compression, cropping, resizing, rotation, etc. [1].

Classic digital watermarking methods are monolithic solutions designed for a concrete problem [6–14]. When the user requirements change, the method must be replaced by a new one, which provides the necessary features. In earlier papers, we presented a modular concept for the creation of steganographic and digital watermarking methods [15, 16]. This concept provides the possibility for reuse of method features in different combinations. The digital watermarking methods are divided into modules and each module provides a certain set of features, e.g. robustness against jpeg compression and recompression, robustness against image changes, etc. By combining different modules together, the user can create a digital watermarking method, which best answers his/her needs. This flexibility is of advantage for contemporary multi-user Internet portals—news publishing platforms, photo gallery websites, artist forums, etc.

Digital watermarking is often used together with cryptographic methods. It does not replace them but it rather provides an added layer of protection over the cryptographic layer (Fig. 1). Before the actual embedding, the data may be subjected to cryptographic processing for security and authentication purposes or for integrity verification. This processing often has the added benefit of randomizing the data bits which improves the performance of the data embedding in general. Our modular methods employ both encryption and compression to deliver better embedding performance [17].

The encryption methods may employ both symmetric and asymmetric algorithms. The data is then embedded into the images as a bit stream. This concept works with almost all types of modules and provides uniform cryptographic protection regardless of the chosen combination of digital watermarking modules. Furthermore, a cryptographic protection in some form may already be in place for the web portal content. In this case, the digital watermarking protection may often be added on top of it without removing the existing solution. This guarantees that the overall level of security of the web portal will not be reduced. In addition to cryptography, the intrinsic security of digital watermarking is further improved by

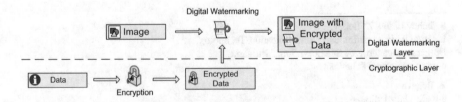

Fig. 1 Digital watermarking as cryptographic enhancement

using pseudo-random number generators to control the distribution of the bit stream across the image blocks during the data embedding and data extraction processes. In this way the image content benefits from both cryptographic and digital watermarking protection mechanisms.

2 Digital Watermarking Service

A digital watermarking service can be used in an automated way to provide a content protection mechanism without human involvement beyond the initial setup of the service and its integration with the web portal software. Images are published only after they pass through the service which supplies them with a proper signature identifying the copyright holder and the terms of use. A complementary steganalysis service identifies the presence of the digital watermarking data and, depending on its usage, retrieves the embedded copyright information. One possible usage of the steganalysis service is to scan portals of interest based on the clients' profiles to identify any images containing digital watermarking data. The scan takes into consideration the specific modular digital watermarking method used by a client of the digital watermarking service and may use correlation-based methods to identify copyrighted images used by competitors as described in [18]. The service then notifies the copyright holder. An alternative use of the steganalysis service is to provide an answer to the needs of creative professions for qualitative images, e.g. for web design purposes, product marketing, presentations, user-interfaces, etc. A professional interested in the fair use of an image can pass it to the service to get information about the terms of use. The major advantage in both use cases is the integral connection between the digital watermarking data and the image which prevents their separation, i.e. the data describing the image remains with the image throughout its use on the Internet. The modularity of the digital watermarking methods allows flexibility and tweaks the use of digital watermarking according to the specific needs of the client.

Figure 2 illustrates the principal integration of the digital watermarking service with the web portal server software. The details depend on the specific software technology used by the web server, e.g. PHP, ASP.NET, Java, etc. After an employee uploads an image, the web portal server automatically forwards it to the digital watermarking service which marks it with the respective copyright information associated with the profile of the web portal server. The service then passes the image back to the web portal server, which makes it available to end users together with the other content.

The digital watermarking service has a profile for each web portal server, which contains the desired copyright and licensing models for different classes of images. The profile also defines the desired level of protection for each image class. This level of protection is interpreted by the service which uses it to select a combination of digital watermarking modules which are then assembled into a customized digital watermarking method. The end goal is to provide a different digital

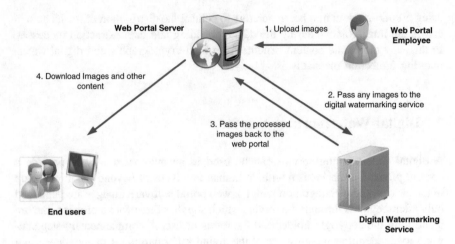

Fig. 2 Automated digital watermarking workflow

watermarking method for each image class with the best possible characteristics for the needs of the web portal.

Another important features associated with the individual web portals are the priority of image processing and the charging model—by number of processed images, by flat-rate subscription or a combination of both.

One specific innovation in this service is the modular structure of the digital watermarking methods. A digital watermarking method consists of modules which can be combined on two levels to form the complete method. For any class of applications, specific modules are created which can be reused later in combinations chosen by the digital watermarking service. This flexibility in the provision of digital watermarking protection is difficult to achieve with classic digital watermarking.

The complementary steganalysis service processes images on the web with the purpose of finding out and extracting digital watermarking data. We will focus on the use case of image producers that seek to maximize the popularity of their images for non-commercial use and minimize the commercial use without paid commercial license. Examples are photographers, artists (incl. computer-generated art). Digital watermarking is ideal for this purpose because it does not impede the distribution of images but instead it provides means for tracking and identification which serves both the copyright holder and any potential users.

Figure 3 shows the workflow of the general use case outlined above. Before publishing the image, the copyright holder passes it to the digital watermarking service which embeds the licensing information. Then, the image is made available on the Internet free of charge for non-commercial distribution. A commercial user downloads the image and decides to use it for business purposes. The embedded license provides the necessary details for the payment.

The payment process is automated so that when the commercial user is registered with the steganalysis service, the image analysis and the money flow are

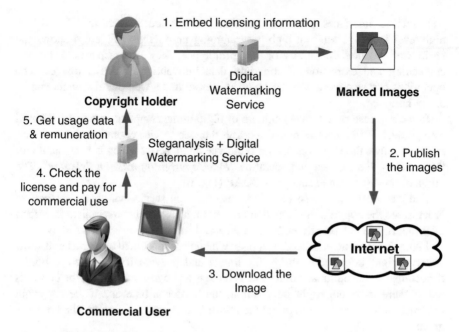

Fig. 3 Digital watermarking and steganalysis

handled in an automated way saving time for both the copyright holder and the business user. The copyright holder receives the payment through the steganalysis service in a previously arranged manner.

The innovation in this case is the use of the digital watermarking and steganalysis services as escrow and gatekeeper services which automate the workflow concerning commercial use of already published images. They also gather statistics about the most popular images of each copyright holder and give expert recommendations on price and marketing venues depending on the profiles of commercial users that have shown interest in the images.

When applied on a large scale—with many customers and images—both services can be employed as business intelligence instruments to conduct large-scale analysis of image preferences, image uses and copyright violations.

3 Technical Implementation of the Digital Watermarking Service Prototype

The technical implementation of our digital watermarking service prototype required the simultaneous use of several technologies. The most time-critical parts of the digital watermarking methods are written in ANSI C++ to obtain good utilization of the cpu and the main memory. The rest of the methods as well as the web service interfaces provided for external use are written in VB.NET—a high-level language offering high programming productivity. Figure 4 shows the basic web service operations of the prototype. They are responsible for the embedding and extraction of copyright data into and from JPEG images. The operations can also work with URLs of images so that a web portal scanner can be easily built.

In the use case of automatic signing of digital images uploaded to a PHP-based web portal, a PHP client script that accesses the web service operations is necessary. Figure 5 shows the PHP code of our prototype client script which is responsible for the access of the web service operation "*hideCopyrightInformationByImage*". The protocol used for data exchange is SOAP (Fig. 6).

End customers must also be able to use the web service via their browsers in order to see the copyright information and act on it. A web browser client is written in JavaScript to extract any embedded data and inform the user accordingly.

Figure 7 shows an enhanced web page which uses this client to extract and show any copyright information inside the image and provide it with a color border depending on the legitimacy of its use. If there is a copyright violation, the border is red, if there is no copyright information, the border is blue and if the copyright information matches the web portal the image is currently residing on, the border is green.

The technical implementation prototype uses one server running the data hiding service itself on IIS, one web server running Apache and a Firefox browser on a client computer. The data hiding service is accessed when new images are uploaded to the Apache web server or when visitors access the web page.

Fig. 4 Digital watermarking web service interface and operations

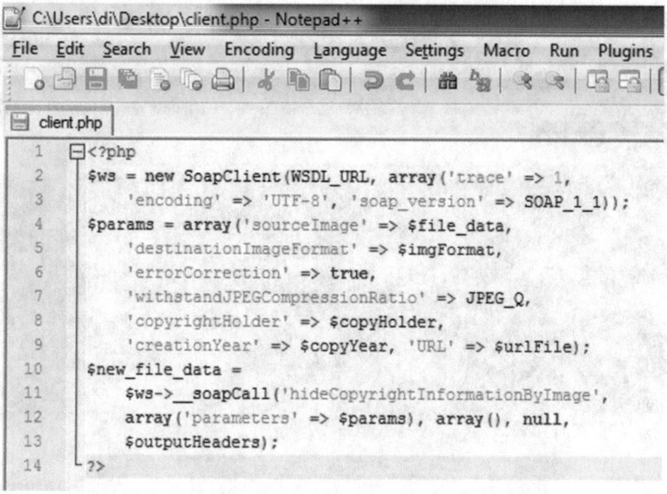
C:\Users\di\Desktop\client.php - Notepad++
File Edit Search View Encoding Language Settings Macro Run Plugins
client.php
<?php
$ws = new SoapClient(WSDL_URL, array('trace' => 1,
'encoding' => 'UTF-8', 'soap_version' => SOAP_1_1));
$params = array('sourceImage' => $file_data,
'destinationImageFormat' => $imgFormat,
'errorCorrection' => true,
'withstandJPEGCompressionRatio' => JPEG_Q,
'copyrightHolder' => $copyHolder,
'creationYear' => $copyYear, 'URL' => $urlFile);
$new_file_data =
$ws->__soapCall('hideCopyrightInformationByImage',
array('parameters' => $params), array(), null,
$outputHeaders);
?>

Fig. 5 PHP client script

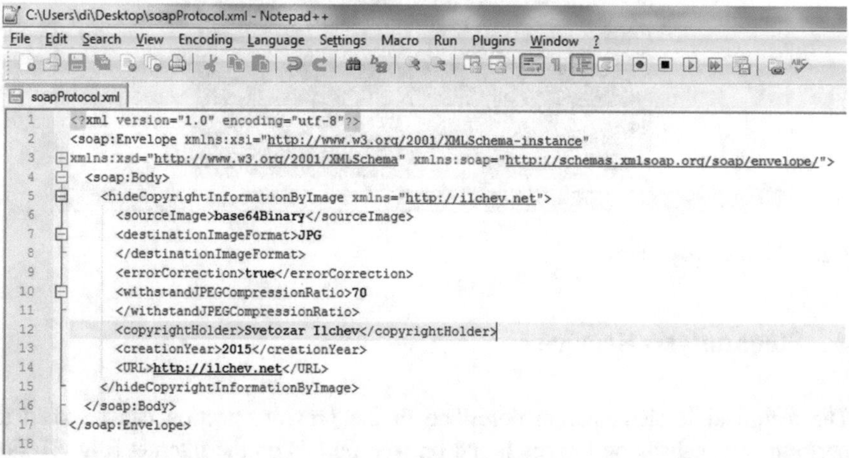
C:\Users\di\Desktop\soapProtocol.xml - Notepad++
File Edit Search View Encoding Language Settings Macro Run Plugins Window ?
soapProtocol.xml
<?xml version="1.0" encoding="utf-8"?>
<soap:Envelope xmlns:xsi="http://www.w3.org/2001/XMLSchema-instance"
xmlns:xsd="http://www.w3.org/2001/XMLSchema" xmlns:soap="http://schemas.xmlsoap.org/soap/envelope/">
<soap:Body>
<hideCopyrightInformationByImage xmlns="http://ilchev.net">
<sourceImage>base64Binary</sourceImage>
<destinationImageFormat>JPG
</destinationImageFormat>
<errorCorrection>true</errorCorrection>
<withstandJPEGCompressionRatio>70
</withstandJPEGCompressionRatio>
<copyrightHolder>Svetozar Ilchev</copyrightHolder>
<creationYear>2015</creationYear>
<URL>http://ilchev.net</URL>
</hideCopyrightInformationByImage>
</soap:Body>
</soap:Envelope>

Fig. 6 SOAP request used by the PHP client

This configuration proved successful when used on a small scale with several client machines. The scaling should not pose any unusual problems because each request is independent of the others and can be processed individually. In addition, the data hiding methods lend themselves to efficient parallelization so that on a multicore machine all cores are fully utilized.

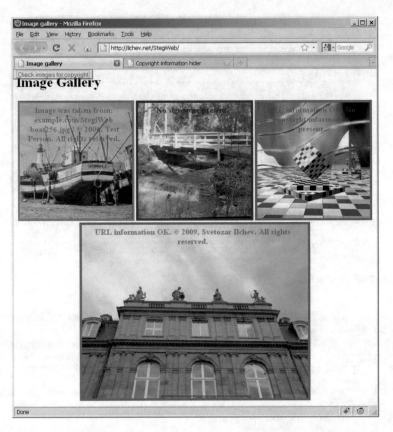

Fig. 7 Web browser client

4 Steganalysis Service

The technical implementation described in the previous section can be used to perform steganalysis on images found on web portals on the Internet (Fig. 8). For the purpose one needs an HTML and JavaScript parser which filters out image URLs. These URLs are then passed to the digital watermarking web service for detection of embedded data.

We run a user script written in JavaScript within the browser in order to use its DOM parsing capabilities to search for images and URLs. When images are found, the user script handles the passing of the images to the service (Fig. 9).

In this way, one can create statistics and make profiles about the use of each individual image, which has been marked by the digital watermarking service before its Internet distribution. These statistics, when coupled with metadata about the web portals themselves such as main customer types, discussion topics, country, language,

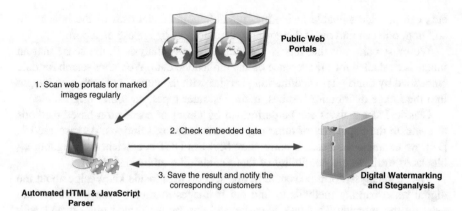

Fig. 8 Steganalysis

```
GM_xmlhttpRequest({

method: 'POST',

url:
        'http://xxx.xxx.xxx.xxx:8888/StegiWeb/StegiWeb.asmx/un
HideCopyrightInformationByURL',

headers: {
        'User-agent': 'Mozilla/4.0 (compatible)
                Greasemonkey/0.3',
        'Accept': 'application/xml,text/xml',
        'Content-type': 'application/x-www-form-urlencoded'
},

data: 'imageURL=' + escape(url),

onerror: function(responseDetails) {
        inf(15, "Service offline");
},

onload: function(responseDetails) {
        var parser = new DOMParser();
        try{
                var dom = parser.parseFromString(
                        responseDetails.responseText,
                        "application/xml");
                checkForParsingErrors(dom);
                checkSignature(dom, url);
        }catch(e){
                inf(12, "No signature present");
        }
}

});
```

Fig. 9 User script

etc., can provide valuable inside into possible commercial uses of the images. In addition, one can find out if there is any unauthorized image use on a portal.

We must point out that this is not a classical steganalysis in the sense that an image is evaluated for the presence of any embedded data. We rather search for data embedded by our digital watermarking service with the purpose of gathering insight into the image distribution venues, main customer types, customer origins, etc.

Classical steganalysis can be performed by means of correlation-based methods to evaluate the popularity of other digital watermarking solutions. As discussed in [18], we examine the Pearson correlation between DCT coefficients of neighboring blocks to evaluate the possibility of them containing embedded data.

The steganalysis form presented in this section uses inside knowledge about the digital watermarking methods to find out if images marked by these methods are used on the Internet. This task is easier and can be performed quicker and with fewer resources than traditional steganalysis. The efficiency of the implementation makes it applicable on a relatively large scale suitable for business intelligence purposes.

5 Conclusion

The modular digital watermarking service coupled with steganalysis, has the potential of alleviating an important problem on the Internet for most of the creative professions: how to gain popularity and distribute the works in a way so that they reach a maximum number of viewers while at the same time preserving the right to obtain license fees for any commercial use. Classic security technologies are not a suitable mechanism because they impede the process of popularization and distribution. Digital watermarking, on the other hand, allows unhindered distribution and it can still provide a degree of control over the images. The modularity and the service orientation of our own methods make possible the automation of these processes for a wide selection of different clients. The digital watermarking service acts as intermediary facilitating the payment flow for commercial uses and can gather and provide statistics about image distribution and popularity.

The technical implementation has proven that the digital watermarking and steganalysis services can be applied successfully at least on a small scale with the potential of covering a much larger client and web portal base.

References

1. Cox, I.J., Miller, M., Bloom, J., Fridrich, J., Kalker, T.: Digital Watermarking and Steganography, 2nd edn. Morgan Kaufmann Publishers (2008)
2. Cole, E.: Hiding in Plain Sight: Steganography and the Art of Covert Communication, 1st edn. Wiley (2003)

3. Lin, E., Delp, J.: A review of data hiding in digital images. In: Proceedings of the Image Processing, Image Quality, Image Capture Systems Conference (PICS '99), Savannah, Georgia, pp. 274–278 (1999)
4. Lu, C.: Multimedia Security: Steganography and Digital Watermarking Techniques for Protection of Intellectual Property, 1st edn. Idea Group Publishing (2005)
5. Curran, K., Bailey, K.: An evaluation of image based steganography methods. Int. J. Digital Evid. **2**(2) (2003)
6. Izadinia, H., Sadeghi, F., Rahmati, M.: A new steganographic method using quantization index modulation. In: International Conference on Computer and Automation Engineering (ICCAE), pp. 181–185 (2009)
7. Zhao, R.-M., Lian, H., Pang, H.-W., Hu, B.-N.: A watermarking algorithm by modifying AC coefficies in DCT domain. In: International Symposium on Information Science and Engieering (ISISE), vol. 2, Shanghai, China, pp. 159–162 (2008)
8. Zhang, X.-P., Li, K., Wang, X.: A novel look-up table design method for data hiding with reduced distortion. IEEE Trans. Circuits Syst. Video Technol. **18**(6), 769–776 (2008)
9. Friedrich, J., Goljan, M., Chen, Q., Pathak, V.: Lossless data embedding with file size preservation. In: Proceedings EI SPIE, San Jose, CA (2004)
10. Chang, C.-C., Chen, T.-S., Chung, L.-Z.: A steganographic method based upon JPEG and quantization table modification. Inf. Sci. Inf. Comput. Sci. **141**(1–2), 123–138 (2002)
11. Fridrich, J.: Image watermarking for tamper detection. In: IEEE International Conference on Image Processing (ICIP), Chicago (1998)
12. Fridrich, J., Goljan, M., Hogea, D.: Steganalysis of JPEG images: breaking the F5 algorithm. In: 5th Information Hiding Workshop, Noordwijkerhout, The Netherlands, pp. 310–323 (2002)
13. Westfeld, A.: F5—a steganographic algorithm. In: Proceedings of the 4th International Workshop on Information Hiding, Lecture Notes In Computer Science, vol. 2137, pp. 289–302 (2001)
14. Lin, C.-Y., Chang, S.-F.: Semi-fragile watermarking for authenticating JPEG visual content. In: SPIE International Conference on Security and Watermarking of Multimedia Contents II, vol. 3971, San Jose, California, USA (2000)
15. Ilchev, S., Ilcheva, Z.: Modular data hiding for digital image authentication. In: Proceedings of the IADIS European Conference on Data Mining, Freiburg, Germany, pp. 122–127. ISBN 978-972-8939-23-6 (2010)
16. Ilchev, S., Ilcheva, Z.: Modular data hiding approach for web based applications. In: Proceedings of the International Conference "Automatics and Informatics'10", Sofia, Bulgaria, pp. I253–I256. ISSN 1313-1850 (2010)
17. Ilchev, S.: Accurate data embedding in JPEG images for image authentication. In: Comptes rendus de l'Acad´emie bulgare des Sciences, vol. 66, no. 9, pp. 1247–1254. ISSN 1310-1331 (2013)
18. Ilchev, S., Ilcheva, Z.: Correlation-based steganalysis of modular data hiding methods. In: 15th International Conference on Computer Systems and Technologies (CompSysTech '14), Ruse, Bulgaria, pp. 108–115. ISBN: 978-1-4503-2753-4. doi:10.1145/2659532.2659611 (2014)

Deblurring Poissonian Images via Multi-constraint Optimization

Stanislav Harizanov

Abstract This paper deals with the restoration of images corrupted by a non-invertible or ill-conditioned linear transform and Poisson noise. The paper is experimental and can be seen as a continuation of "as reported by Harizanov et al. (Epigraphical Projection for Solving Least Squares Anscombe Transformed Constrained Optimization Problems 2013)". The constraint set in the minimization problem, considered there, was too large and the results tend to oversmooth the initial image. Here, we consider various techniques for restricting this set in order to improve the image quality of the result, and numerically investigate them. They are based on image domain decomposition and give rise to multi-constraint optimization problems.

1 Introduction

Industrial computed tomography (CT) scanning uses irradiation (usually with x-rays) to produce three-dimensional representations of the scanned object both externally and internally. The latter is derived from a large series of two-dimensional radiographic images taken around a single axis of rotation. To create each of the planar images, a heterogeneous beam of X-rays is produced and projected toward the object. A certain amount of X-ray is absorbed by the object, while the rest is captured behind by a detector (either photographic film or a digital detector). The local magnitudes of the detected X-ray amount determine the corresponding gray-scale pixel values of the radiographic image. In such processes, where images are obtained by counting particles, Poisson noise occurs. Being interested in improving the quality of the 3D CT reconstruction, we are motivated to investigate 2D Poisson denoising techniques, and to apply them to each radiographic image.

S. Harizanov (✉)
Institute of Information and Communication Technologies,
Bulgarian Academy of Sciences, Sofia, Bulgaria
e-mail: sharizanov@parallel.bas.bg

© Springer International Publishing Switzerland 2016
S. Margenov et al. (eds.), *Innovative Approaches and Solutions in Advanced Intelligent Systems*, Studies in Computational Intelligence 648,
DOI 10.1007/978-3-319-32207-0_13

In mathematical terms, one wants to solve the ill-posed inverse problem of recovering the original 2D image $\bar{u} \in [0, v]^{M \times N}$ from observations

$$f = \mathcal{P}(H\bar{u}),$$

where v is the gray-scale intensity, $\mathcal{P}$ denotes an independent Poisson noise corruption process, and $H \in [0, +\infty)^{n \times n}$ is a blur operator, corresponding to a convolution with a Gaussian kernel. Here $n = MN$, because it is beneficiary to column-wise reshape the image into a long 1D vector. Note that blurring appears naturally in practice (e.g., when the industrial CT scan is not well calibrated) and needs to be incorporated in the problem.

Poisson denoising is a hot and active research field, so it is impossible to list all the related publications. We mention only few of them [1–10] in chronological order, as illustration. All the approaches are based on minimizing a regularization term $\Psi(u)$, where a data fidelity term $F(u, f)$ is either incorporated in the cost function as penalization

$$\operatorname*{argmin}_{u} \Psi(u) + \lambda F(u, f), \quad \lambda \geq 0, \tag{1}$$

or considered as constraint

$$\operatorname*{argmin}_{u} \Psi(u) \quad \text{subject to} \quad F(u, f) \leq \tau, \quad \tau \geq 0. \tag{2}$$

The problems (1) and (2) are closely related and, under some mild assumptions on Ψ and F, there is a one-to-one correspondence $\lambda \leftrightarrow \tau$, such that their solutions coincide (see [11]). In general, (1) is easier to solve, but the optimal parameter λ cannot be well approximated, while (2) is both mathematically and computationally more complex, but the optimal parameter τ is statistically estimated.

There are two main-stream directions for the choice of the data fidelity F. In the first one, the mean/variance dependence of the Poisson distribution can be reduced by using variance-stabilizing transformations (VST), such as the *Anscombe transform* [12]

$$T : [0, +\infty)^n \to (0, +\infty)^n : v = (v_i)_{1 \leq i \leq n} \mapsto 2\left(\sqrt{v_i + \frac{3}{8}}\right)_{1 \leq i \leq n}.$$

It transforms Poisson noise to approximately Gaussian noise with zero-mean and unit variance (if the variance of the Poisson noise is large enough), for which Least Squares estimates are maximum-likelihood ones. The second approach is closely related to a direct Maximum A Posteriori (MAP) estimate, where the neg-log-likelihood of the Poisson noise, i.e., the *I*-divergence (generalized Kullback-Leibler divergence)

$$u \mapsto D(f, Hu) := \begin{cases} \langle \mathbf{1}_n, f \log \frac{f}{Hu} - f + Hu \rangle & \text{if } Hu > 0, \\ +\infty & \text{otherwise,} \end{cases}$$

is used. Here $\langle \cdot, \cdot \rangle$ denotes the standard Euclidean inner product and $\mathbf{1}_n$ denotes the vector consisting of n entries equal to 1.

This paper is a continuation of our previous work [8], thus we deal with the Total Variation (TV) [13] constraint optimization problems

$$\operatorname*{argmin}_{u \in [0, +\infty)^n} \|\nabla u\|_{2,1} \quad \text{subject to} \quad \|T(Hu) - T(f)\|_2^2 \le \tau_A, \tag{3}$$

$$\operatorname*{argmin}_{u \in [0, +\infty)^n} \|\nabla u\|_{2,1} \quad \text{subject to} \quad D(f, Hu) \le \tau_I, \tag{4}$$

where $\nabla \in \mathbb{R}^{2n \times n}$ is the discrete gradient operator (forward differences and Neumann boundary conditions are used), and $\|\cdot\|_{2,1}$ denotes the $\ell_{2,1}$ norm.

To both problems, we apply the **p**rimal- **d**ual **h**ybrid **g**radient algorithm [14, 15] with an extrapolation (**m**odification) of the dual variable (PDHGMp). At each iteration step, we compute n epigraphical projections [16] w.r.t. a 1D convex function related to T [8], respectively solve an I-divergence constrained least squares problem [6, 9]. The algorithms' description can be found in [8, Sect. 4]. We keep the notation Algorithms 1–2 from it.

The paper is organized as follows: In Sect. 2, we experiment with different choices of τ_A, τ_I and measure their effect on the output image quality. In Sect. 3, we propose various domain decompositions in order to improve that quality. Those decompositions give rise to multi-constraint optimization problems, for which Algorithms 1–2 are still applicable. Numerical experiments are conducted and the results are discussed. Conclusions are drawn in Sect. 4.

2 Single-Constraint Optimization with Optimal τ

In this paper, we test the same initial images $\bar{u}$ 'cameraman' (256×256), its central part (130×130), and 'brain' (184×140) as in [8] (Fig. 1), and we work with the same polluted images f. This allows us to compare numerical results. Again, we denote them by B1$_v$, B1part$_v$, and B2$_v$, where v stands for the gray-scale intensity. We recall that the peak signal to noise ratio (PSNR) and the mean absolute error (MAE) are computed via

$$\text{PSNR} = 10 \log_{10} \frac{|\max \bar{u} - \min \bar{u}|^2}{\frac{1}{n}\|u - \bar{u}\|_2^2}, \quad \text{MAE} = \frac{1}{nv}\|\bar{u} - u\|_1.$$

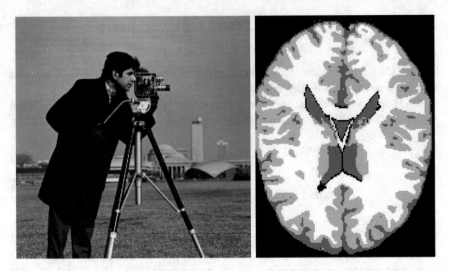

Fig. 1 Original images 'cameraman' (*left*) and phantom of a brain image (*right*)

The statistically motivated choice for the constraint parameter $\tau_A = n$ in (3), resp. $\tau_I = n/2$ in (4), places the true image $\bar{u}$ with high probability very close to the boundary of the constraint sets (see [8, Table 1] for experimental verification).

$$C_A = \{u : ||T(Hu) - T(f)||_2^2 \leq \tau_A\}, \quad C_I = \{u : D(f, Hu) \leq \tau_I\}, \tag{5}$$

thus guaranteeing those sets are non-empty and minimizers u_A, resp. u_I, exist. Moreover, since the TV functional is a semi-norm and every semi-norm is positively-homogeneous, $u_A \in \partial C_A$, resp. $u_I \in \partial C_I$, are unique, unless the constraint sets (5) contain constant images (the global minimizers of the TV functional), i.e., whenever

$$\min_{c \in \mathbb{R}} ||T(f) - c||_2^2 > \tau_A, \quad \text{resp.} \quad \min_{c \geq 0} D(f, c) > \tau_I.$$

We used that H reproduces constants ($H\mathbf{1}_n = \mathbf{1}_n$), which is true for convolution-based blur operators. Since $E(f_i) = (H\bar{u})_i$ and $T'(\bar{u}_i) \geq T'(v) = 1/\sqrt{v + 3/8}$ $\forall i = 1, \ldots, n$,

$$\min_{c \in \mathbb{R}} ||T(f) - c||_2^2 \geq \frac{1}{v + 3/8} \min_{c \in \mathbb{R}} ||H\bar{u} - c||_2^2.$$

Therefore, problems (3) and (4) admit unique solution if

$$\min_{c \in \mathbb{R}} \|H\bar{u} - c\|_2^2 > v\tau_A, \quad \min_{c \geq 0} D(H\bar{u}, c) > \tau_I, \quad v \gg 0. \tag{6}$$

In particular, when $\tau_A = n$, $\tau_I = n/2$, (6) holds true for all *nontrivial* (e.g., edge-containing, not close-to-constant) initial images $\bar{u}$ and moderate blur operators H (such that $H\bar{u}$ remains nontrivial). To conclude, for $\tau_A = n$, $\tau_I = n/2$, and under some natural assumptions on $\bar{u}$, H, and v, problems (3) and (4) are well-posed, with $\bar{u}$ placed around the boundary of their constraint sets (5), thus being an admissible candidate for the unique solution.

Being an admissible candidate is not enough for $\bar{u}$ to be "close" to the actual solution! If the constraint set is too large, then the two images might still differ a lot. As Fig. 2 illustrates, this is indeed the case for our problem (3). The minimizer is significantly "oversmoothened". Hence, in order to let u_A better approximate $\bar{u}$, we need to restrict C_A.

The easiest way to do so is to simply decrease τ_A. We tried it on B1part$_{3000}$ and the results can be viewed on Fig. 3. Since the minimizer of (3) tends to "stretch out" when τ_A decreases, we worked with box constraints $u \in [0, 3000]^n$ in C_A. Due to the $\lambda \leftrightarrow \tau$ relations between (1) and (2), $\tau_A(\lambda_A)$ is monotonically decreasing, thus invertible. We numerically solved the penalized version of (3) for various λ_A, which, as discussed in the introduction, is computationally much more efficient, and cheaply derived the solution u_A of (3) for the corresponding τ_A. We observe that the optimal values for τ_A w.r.t. both PSNR and MAE are smaller than n, which confirms the above conclusion that C_A is too large. Maximal PSNR is obtained for $\tau_A = 0.791n$, while minimal MAE is obtained for $\tau_A = 0.894n$. On the other hand, $\lambda_A = 10^4$ gives rise to $\tau_A = 0.5742n$, while $\lambda_A = 10^5$ gives rise to $\tau_A = 0.5452n$, thus we expect C_A to be empty for $\tau_A = 0.5$.

Fig. 2 The ratio $TV(\bar{u})/TV(u_A)$ as a function of v for the both test images

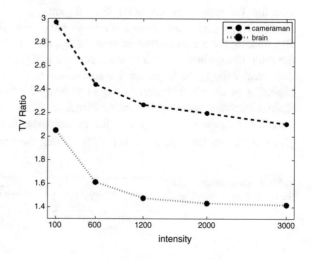

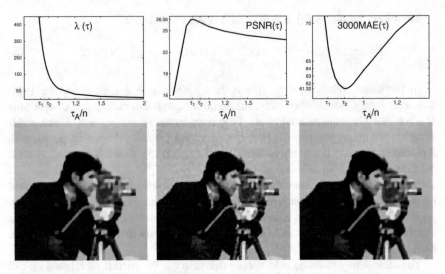

Fig. 3 *Left* λ as a function of τ_A and u_A for $\tau_A = n$. *Center* PSNR (u_A) as a function of τ_A and its minimizer u_A ($\tau_A = \tau_1 n$). *Right* νMAE (u_A) as a function of τ_A and its minimizer u_A ($\tau_A = \tau_2 n$)

From the plots of PSNR (u_A) and 3000MAE (u_A) $(\nu = 3000)$ we see that the former is much more stable to the change of τ_A than the latter, e.g., when $\tau_A \in [0.72, 1]$, PSNR $(u_A) \in [25.5932, 26.5614]$, while ν MAE $(u_A) \in [61.3239, 90.2654]$. Moreover, the optimal minimizer w.r.t. PSNR possesses quite large MAE (namely ≈ 67) and artifacts in the smooth regions of $\bar{u}$, while the optimal minimizer w.r.t. MAE has quite satisfactory PSNR (namely 26.16) and good visual properties (see Table 1). This is due to the different norms involved in the two functions, namely the 2-norm and the 1-norm. The initial image $\bar{u}$ was firstly blurred by a Gaussian kernel with $\sigma = 1.3$, thus smoothed around its edges (the TV semi-norm of the blurred image is 1.3146e+06 which is more than 2 times smaller than the TV semi-norm 2.9390e+06 of B1part$_{3000}$). In particular, the most problematic image part is the camera where we have very "thin" details. Since, we are also looking for the smoothest solution in C_A, the minimizers of (3) fail to recover the jump discontinuities of $\bar{u}$ in these regions and $\|u_A - \bar{u}\|_\infty$ is quite large there. The smaller the τ_A the larger the TV semi-norm of the minimizer and the better the capturing of the singularities. In the same time, the Poisson noise in the smooth regions becomes more and more problematic, since we further deviate from the statistically optimal value $\tau_A = n$ for its complete removal. PSNR is based on $\frac{1}{n}\|\bar{u} - u_A\|_2^2$ and the impact of $\|u_A - \bar{u}\|_\infty$ along the camera edges is very strong.

Table 1 Comparison among the optimal τ_A's from Fig. 2		$\tau_A = 0.791n$	$\tau_A = 0.894n$	$\tau_A = n$
	PSNR	26.5614	26.1581	25.5934
	νMAE	67.0804	61.3239	63.6238

Hence the denoising failure in the smooth regions is undermined as long as the size of the artifacts there is not comparable to the one of the jump discontinuities at the edges. On the other hand, $3000MAE(u_A) = \frac{1}{n}||\bar{u} - u_A||_1$, which is a "sparser" norm and it is small when many entries of $u_A - \bar{u}$ are close to zero. Hence, it captures better the presence of denoising artifacts.

In conclusion, for the constraint optimization problem (3) it seems that, in order to obtain good visual results, it is better to minimize the MAE of u_A than to maximize its PSNR.

3 Multi-constraint Optimization

Even though decreasing τ_A may improve the image qualities of the solution of (3), it doesn't seem like a good strategy. We deviate from its statistical estimation, thus we need to guess the right value of τ_A, which is computationally expensive. Moreover, we lose the nice properties of $\bar{u}$ being close to the boundary of C_A, thus the guaranteed well-posedness of (3) as well as the admissibility of $\bar{u}$ to be the minimizer. In this section we follow a different approach for restricting C_A that is based on image domain decomposition and the use of independent constraints for the regions. In such a setup, Eq. (3) is reformulated into

$$\operatorname*{argmin}_{u\in[0,+\infty)^n} ||u||_{2,1} \quad \text{subject to} \quad ||T(Hu) - T(f)||_{A_i} \le \tau_i, \quad i = 1, \dots, K. \quad (7)$$

where $\{A_i\}_{i=1}^K$ is a tessellation of the image domain $(int(A_i) \cap int(A_j) = \emptyset)$, and $||\cdot||_{A_i}$ is a short notation for the squared 2-norm, restricted to the region A_i. Analogously, the multi-constraint version of (4) is

$$\operatorname*{argmin}_{u\in[0,+\infty)^n} ||\nabla u||_{2,1} \quad \text{subject to} \quad D_{A_i}(f, Hu) \le \frac{1}{2}\tau_i, \quad i = 1, \dots, K. \quad (8)$$

Both Algorithms 1–2 can be straightforwardly modified to such multi-constraint setting, since no correlation among pixel data appear in $||\cdot||_2^2$ and $D(\cdot, \cdot)$, allowing for direct and complete component-wise splitting. Moreover, following the notation in [9], $\lambda_i^{(k+1)}$ is the solution of $D_{A_i}(f, g(p_1^{(k)} + Hu^{(k+1)}, \lambda/\sigma)) = \tau_i/2$ and for $\lambda_i := \lim_{k\to\infty}\lambda_i^{(k)}$ we have that the the minimizer of (8) also minimizes

$$\operatorname*{argmin}_{u\in[0,+\infty)_n} ||\nabla u||_{2,1} + \sum_{i=1}^K \lambda_i D_{A_i}(f, Hu).$$

3.1 Block Subdivision

The first thing we try is a simple block subdivision of the spatial domain as
illustrated on Fig. 4. More precisely, given a level l, we split the image into 4^l
blocks $A_{i,j}^{(l)}$, $i,j = 0,\ldots,2^l - 1$ of "equal" size (if the number of pixels in height or
width is not divisible by 2^l, we of course need to round off and the blocks cannot be
absolutely identical) and take $\tau_{i,j}^{(l)} = \mathrm{card}(A_{i,j}^{(l)}) \approx n/4^l$. Due to triangle inequality, it
is straightforward that the corresponding constraint set $C_A^{(l)}$ in (7) is a proper subset
of C_A. Moreover, if we denote $C_A^{(0)} := C_A$, we have $C_A^{(l)} \subset C_A^{(l-1)}, \forall l \in \mathbb{N}$.

The results are summarized in Table 2. We observe that, as expected, increasing
the level l we restrict the global constraint sets and increase the TV semi-norm of
the solutions of (7) and (8). Up to a certain moment (in the particular case for
B1part$_{3000}$ this is $l = 3$) both PSNR and MAE values of the outputs improve. Then
the MAE value has a significant jump and the visual quality of the output image

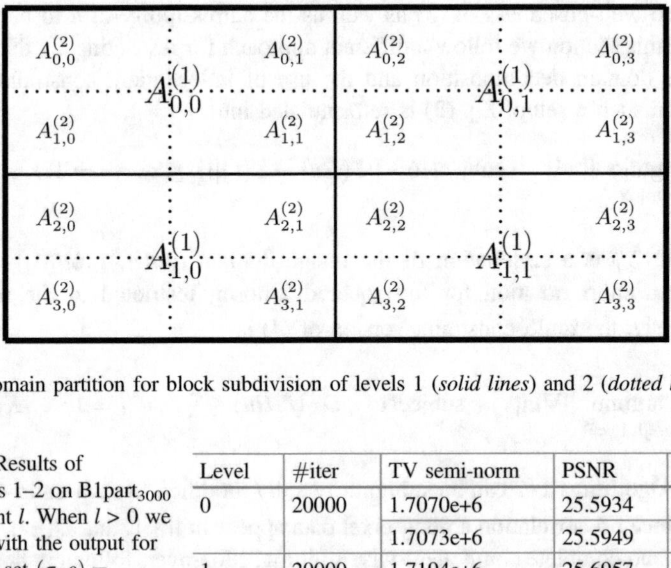

Fig. 4 Domain partition for block subdivision of levels 1 (*solid lines*) and 2 (*dotted lines*)

Table 2 Results of
Algorithms 1–2 on B1part$_{3000}$
for different l. When $l > 0$ we
initialize with the output for
$l - 1$, and set $(\sigma, \rho) = (0.4, 0.3)$ in Algorithm 1

Level	#iter	TV semi-norm	PSNR	MAE·v
0	20000	1.7070e+6	25.5934	63.6238
		1.7073e+6	25.5949	63.6054
1	20000	1.7194e+6	25.6957	62.8892
		1.7197e+6	25.6975	62.8620
2	20000	1.7418e+6	25.8372	62.4555
		1.7423e+6	25.8385	62.4421
3	50000	1.7930e+6	25.9966	62.1405
	20000	1.7934e+6	25.9975	62.1315
4	50000	1.9490e+6	26.0686	64.6669
	20000	1.9649e+6	26.0298	65.1237

Fig. 5 Block subdivision. From *left* to *right*: Algorithm 1 outputs for $l = 3$ and $l = 4$

drops down (see Fig. 5). The reason is that, we are still using the statistically estimated bounds $\tau_{i,j}^{(l)}$ for each of the blocks. However, when l is large the size of the blocks $A_{i,j}^{(l)}$ is small, the law of large numbers that backs up the theoretical arguments in [12, 17] fails, and we cannot guarantee that the value of $\tau_{i,j}^{(l)}$ is adequate or even that the constraint problem (7) is block-wise well-posed.

Indeed, already at the third subdivision level we have trivial regions (constant images are admissible for some of the blocks). The uniqueness of the global minimizer is not affected, because those trivial blocks interfere with the others and, since we minimize the overall TV semi-norm, the constant intensity value on them is uniquely determined by its nontrivial neighbors. On the other hand, existence of the global minimizer becomes problematic at the fourth subdivision level. We have smaller blocks (their size is $\approx 8 \times 8$) and on some of them the original image B1part$_{3000}$ is close-to-constant. When such a region $A_{i,j}^{(4)}$ is of small intensity (i.e., close to black) the Poisson noise is insignificant there and the oscillations of the neighboring pixel values in $f|_{A_{i,j}^{(4)}}$ are negligible. On top of that, the Anscombe transform is not reliable in the sense that $T(f|_{A_{i,j}^{(4)}})$ is not guaranteed to be normally distributed, so the choice $\tau_{i,j}^{(4)} = \text{card}(A_{i,j}^{(4)})$ has no theoretical justification. To summarize,

$$||T(H\bar{u}) - T(f)||_{A_{i,j}^{(4)}} \sim 0 = \text{card}(A_{i,j}^{(4)}) = \tau_{i,j}^{(4)},$$

and $\bar{u}|_{A_{i,j}^{(4)}}$ is well inside the interior of $C_{A_{i,j}^{(4)}}$. In the single-constraint case we have already verified that $\bar{u}$ is with high probability close to ∂C_A, thus the above relation implies for another region $A_{i,j}^{(4)}$, $||T(H\bar{u}) - T(f)||_{A_{i,j}^{(4)}} \gg \tau_{i,j}^{(4)}$, so $C_{A_{i,j}^{(4)}}$ could be empty.

Hence, Algorithm 1 may not converge on $A_{i,j}^{(l)}$. The following example of one such empty-constraint block at level 5 illustrates the problem

$$\bar{u}\big|_{A_{4,19}^{(5)}} = \begin{pmatrix} 2250.0 & 2262.3 & 2262.3 & 2176.2 \\ 2274.6 & 2262.3 & 2299.2 & 2286.9 \\ 2262.3 & 2311.5 & 2311.5 & 2299.2 \\ 2299.2 & 2323.8 & 2299.2 & 2348.4 \end{pmatrix}$$

$$f\big|_{A_{4,19}^{(5)}} = \begin{pmatrix} 2348 & 2243 & 2360 & 2183 \\ 2244 & 2295 & 2205 & 2234 \\ 2190 & 2364 & 2213 & 2393 \\ 2313 & 2269 & 2326 & 2264 \end{pmatrix}.$$

$\bar{u}\big|_{A_{4,19}^{(5)}}$ is almost constant and of high intensity, while the Poisson noise contributes significantly and visibly alters the entries of f.

As a result, the output image of Algorithm 1 for $l = 4$ possesses certain artifacts in some high, close-to-constant intense regions. As discussed in Sect. 2, those artifacts are captured by the MAE value of the output, which immediately jumps up, but not by its PSNR value, which still improves (see Table 2).

The block subdivision might be different. For example, it may be data-dependent and based on the output image at zeroth level $u^{(0)}$. We tested a 2-block subdivision, where A_0 is a 130×75 block that deviates most from the statistical expectation (i.e., it maximizes the quantity $\big| \,\|T(Hu^{(0)} - T(f)\|_A - 130 \cdot 75\big|$), and A_1 is its complement, as well as a 4-block subdivision, where A_0 is a 75×75 block that maximizes the analogous expression, A_1 and A_2 are the "horizontal" and "vertical" complements of A_0, and A_3 is the complement of their union $A_0 \cup A_1 \cup A_2$. For different original images and intensity levels the benefit of such data-dependency is different, but the quality of the output is comparable to that of the "standard" block subdivision presented above, possibly on a higher level. The close similarity between the solutions of (3) and (4), numerically observed in [8], holds true also for their multi-constraint versions (7) and (8). Only for $l = 4$ the outputs of Algorithms 1–2 differ visibly, but this is due to the ill-posedness of the optimization problems and the slow (or even lack of) convergence of the algorithms. This phenomenon appears for all test images and all constraint choices, we considered, therefore from now on we deal only with (7) and Algorithm 1.

3.2 Intensity Tessellation

The statistical choice $\tau_A = n$ in (3) is based not only on the law of large numbers, but also on the Central Limit Theorem. Therefore, it theoretically holds for independent and *identically distributed* Poisson random variables, which in our setting

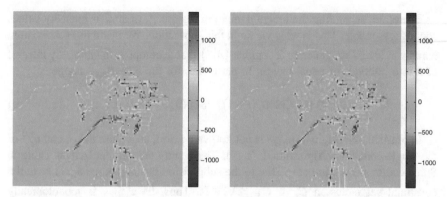

Fig. 6 Noise redistribution in block subdivision. The difference images between the output of Algorithm 1 at level 0 $u_A^{(0)}$ (*left*), respectively at level 3 $u_A^{(3)}$ (*right*) and the original image B1part$_{3000}$

is equivalent to a trivial blurry image $H\bar{u} = const$. This is by far not true for the examples we consider, and even though in [8] $\|T(H\bar{u}) - T(f)\|_2^2/n \approx 1$ is numerically verified, the solution u_A of (3) is oversmoothened due to a "redistribution" of the noise among the pixels. Indeed, as illustrated on Fig. 6, around the edges (jump discontinuities) of the original image, where most of the TV semi-norm of $\bar{u}$ is concentrated, the big positive displacements $H\bar{u}_{i,j} - f_{i,j} \gg 0$ of the high-intensity pixels (i,j) related to the Poisson distribution are wrongly accumulated in the minimizer by the neighboring low-intensity pixels (i',j'), making $Hu_{i,j} \approx f_{i,j}$ and $Hu_{i',j'} \approx f_{i',j'} + (H\bar{u}_{i,j} - f_{i,j})$, while still $u \in C_A^{(l)}$. In other words, if a high-intensity edge pixel value of $\bar{u}$ is decreased by the noise, it is not properly denoised in $u_A^{(l)}$ but rather its neighboring low-intensity edge pixel increases its intensity with a reciprocal amount. Thus f is no longer a realization of independent Poisson random variables over $Hu_A^{(l)}$, meaning that $u_A^{(l)}$ does not approximate well $\bar{u}$ along the edges. On the other hand, away from the edges $u_A^{(l)}$ is a quite good approximation of $\bar{u}$. We back this up with a simple experiment. We replace all pixel intensities of $u_A^{(3)}$ from Table 2, that are more than d away from the corresponding values of the original image $\bar{u}$ with the true intensities and recompute the PSNR and the MAE values of such a "hybrid" image. The difference image $u_A^{(3)} - \bar{u}$ is the right one from Fig. 6. For $d = v/6 = 500$, only 397 pixels (≈ 2.35 % of all the pixels) are modified, while PSNR and vMAE improve to 30.1569 and 45.1862, respectively. For $d = v/15 = 200$, we modify 1334 (≈ 7.9 %) pixels and derive PSNR = 36.5034 and vMAE = 27.1241. Finally, $d = v/30 = 100$ gives rise to 2211 (≈ 13 %) modified pixels with PSNR = 40.1747 and vMAE = 19.7688. Finally if we take the opposite hybrid image for $d = 500$ (i.e., we use the $\bar{u}$ data for all the pixels but those 397 mentioned above), we use ≈ 97.65 % of the original pixels intensities, but PSNR = 28.0989 (worse than the counterpart test!) while vMAE = 16.9543 keeps improving.

Therefore, separating the image domain in regions A_i of similar intensity values sounds reasonable. This is what we do in this subsection. We again use subdivision, namely at level l we generate 2^l regions $\{A_i^{(l)}\}$ that decompose the intensity interval of $\bar{u}$ into intervals of equal length. In particular, for the B1part$_{3000}$ image

$$A_i^{(l)} := \{j | \bar{u}_j \in (i2^{-l}3000, (i+1)2^{-l}3000)\}, \quad \forall i = 0, \ldots, 2^l - 1. \tag{9}$$

In practice, the original image is not known a priori, so we use the output $u_A^{(0)}$ of the single-constraint Algorithm 1 in (9). The results are summarized in Table 3. Some comments are in order. Due to the subdivision technique, we again have the constraint set inclusion $C_A^{(l)} \subset C_A^{(l-1)}, \forall l \in \mathbb{N}$. Thus, $||\nabla u_A^{(l)}||_{2,1}$ is monotonically increasing with respect to l. We observe that PSNR $(u_A^{(l)})$ is also monotonically increasing with $l \geq 1$, while MAE $(u_A^{(l)})$ is not monotone at $l = 5$. In Sect. 3.1 we tessellated the image domain into 4^l regions, while here we used only 2^l. Therefore, it is reasonable to compare the quality of the l-level output images from Table 2 with the quality of the $2l$-level output images from Table 3. We see that, apart from the PSNR value for the image at level 3 in Table 2, respectively 6 in Table 3, both the PSNR and MAE values improve with intensity tessellation. Especially the MAE value which goes below its optimal value 61.3239 for the single-constraint optimization problem (3) (see Fig. 3). As before, there are indications that high l ($l = 7$ and $l = 8$) may lead to empty constraint sets $C_A^{(l)}$, thus the problem may be ill-posed and the algorithm may not converge. However, no visual artifacts appear (see Fig. 7) and MAE $(u_A^{(l)})$ continues to decrease.

The algorithm depends on the initial choice of image u in (9), and different u give rise to different outputs. We have always used $u_A^{(0)}$ in the experiments above, but we have also tested some of the block-subdivision outputs for higher levels, as well as some of the intensity-tessellation outputs for lower levels. The results are more or less comparable, with $u_A^{(0)}$ seeming to be the best option in general. Last but not least, tuning the regularization parameters σ and ρ was the key for the efficient performance

Table 3 Results of Algorithm 1 on B1part$_{3000}$ for different levels of intensity tessellation. For all levels we set $(\sigma, \rho) = (0.4, 0.3)$, and $\tau_i^{(l)} = \mathrm{card}(A_i^{(l)})$	Level	#iter	TV semi-norm	PSNR	MAE·v
	1	20000	1.7080e+6	25.5810	63.6927
	2	20000	1.7228e+6	25.7251	62.7592
	3	20000	1.7439e+6	25.8739	61.4654
	4	50000	1.7535e+6	25.8895	61.2834
	5	50000	1.7695e+6	25.9421	61.5316
	6	50000	1.7775e+6	25.9923	61.1921
	7	50000	1.8046e+6	26.0401	61.0400
	8	50000	1.8491e+6	26.1230	60.7815

Fig. 7 The output images $u_A^{(4)}$ (*left*), $u_A^{(7)}$ (*center*), $u_A^{(8)}$ (*right*) from Table 2

of the single-constraint algorithm, while in both Sects. 3.1 and 3.2 the algorithms' convergence rate seems to be slow and independent of that choice, thus we always use $\sigma = 0.4$, $\rho = 0.3$. The same parameters also work for the other images $B1$ and $B2$ on all the considered intensity levels $v = 100, 600, 1200, 2000, 3000$.

The main drawback of the intensity tessellation algorithm is that we have no control on the size of the tessellated regions. It may happen that even at low levels, some of the regions consist of only few points (some of them might be even empty, but this is not a problem). Thus, the law of large numbers may be violated and $\tau_i^{(l)} = \text{card}(A_i^{(l)})$ may be a bad choice that leads to an ill-posed optimization problem or to a minimizer that is quite different from the initial image. A possible remedy is to adaptively split the intensity interval in order to guarantee $\text{card}(A_i^{(l)}) \approx n/2^l$, $\forall i = 0, \ldots, 2^l - 1$. This is left for future work.

3.3   2-Step Combined Tessellation

So far we saw that both block subdivision and intensity tessellation improve the image quality of the output of Algorithm 1. On the other hand, the former technique violates the Central Limit Theorem but allows for the application of the law of large numbers, while the latter one violates the law of large numbers but allows for the application of the Central Limit Theorem. Thus for both of them, the choice $\tau_i^{(l)} = \text{card}(A_i^{(l)})$ for constraint parameters is not theoretically justified and may lead to diverging algorithms or meaningless results. In this subsection, we try to combine the two approaches in a beneficial way. We use 1-level block subdivision together with 3-region-intensity tessellation of the image domain. The idea is the following: the block subdivision performs very well away from the edges, while along them it leads to noise redistribution (Fig. 6). If we assume that the redistribution always involves 2 neighboring pixels, one of law intensity and one of high intensity, and it simply exchanges their noisy values as discussed in Sect. 3.2, then we need to find all such pixel pairs, separate them into 2 regions—one $\bar{A}_0$ of low and one $\bar{A}_1$ of high intensity,

Fig. 8 Edge detection via difference images. *Left* $u_A^{(1)} - f$. *Center* $\bar{u} - f$. *Right* $u_A^{(1)} - u_A^{(0)}$. The images $u_A^{(i)}$, $i = 0, 1$ are taken from Table 2

and introduce a second constrained $\|T(Hu) - T(f)\|_{\bar{A}_i} \leq \text{card}(\bar{A}_i)$, $i = 0, 1, 2$. (For the sake of symmetry, we take $\bar{A}_2$ to be the complement of $\bar{A}_0 \cup \bar{A}_1$.)

The operator H "smoothens" the edges of $\bar{u}$, but as long as the blur is not very strong (i.e., corresponds to a Gaussian kernel with close-to-one standard deviation) the edges "survive" it. The Poisson noise more or less preserves what is left from them. The block subdivision also smoothens the edges, but not as much as the blur operator. Thus, both the difference images $\bar{u} - f$ and $u_A^{(l)} - f$ contain enough edge information (see Fig. 8). However, the smoothing effect of the blur operator dominates and we cannot say from the second image where noise redistribution appear. Hence, we prefer to work with the difference image $\partial u := u_A^{(1)} - u_A^{(0)}$. Indeed, the subdivision of the image domain into 4 regions alleviates the noise redistribution effect from $u_A^{(0)}$ and part of the edge information is again visualized (the right image in Fig. 8). More importantly, all the pixels with significantly different $u_A^{(1)}$ and $u_A^{(0)}$ values indeed belong to the edges of $\bar{u}$, making edge-detection a plausible application of such multi-constraint optimization.

We compute $M := \max_i \partial u_i > 0$, $m := \min_i \partial u_i < 0$, fix a number $c > 0$, and set $\bar{A}_0 := \{i | \partial u_i > M/c\}$, $\bar{A}_1 := \{i | \partial u_i < m/c\}$. Then we solve

$$\underset{u \in [0, +\infty)^n}{\text{argmin}} \|\nabla u\|_{2,1} \quad s.t. \quad \begin{array}{ll} \|T(Hu) - T(f)\|_{A_{i,j}^{(1)}} \leq \text{card}\left(A_{i,j}^{(1)}\right), & i, j = 0, 1; \\ \|T(Hu) - T(f)\|_{\bar{A}_k} \leq \text{card}(\bar{A}_k), & k = 0, 1, 2. \end{array}$$

We apply a straightforward modification of Algorithm 1, which decouples the two tessellations $\{A_{i,j}^{(1)}\}$ and $\{\bar{A}_k\}$. Results are summarized in Table 4.

We observe that the higher the intensity (thus the sharper the edges) the bigger the improvement in the quality of the result with respect to the corresponding images $u_A^{(1)}$ and $u_A^{(0)}$. For very low intensity v the combined technique may even worsen the MAE of the output (see B2$_{100}$). The PSNR always improves.

Table 4 Results of the 2-steps combined tessellation for different images and intensity levels. For all examples we set $(\sigma, \rho) = (0.4, 0.3)$, and $c = 9$

Image	#iter	TV semi-norm	PSNR	MAE-v	$TV(u^{(1)})$	$PSNR(u^{(1)})/PSNR(u^{(0)})$	$MAE(u^{(1)})/MAE(u^{(0)})$
B1part$_{3000}$	100000	1.7573e+6	26.0000	60.9231	1.7194e+6	25.6957/25.5934	62.8892/63.6238
B1$_{100}$	20000	1.2602e+5	24.4036	3.0603	1.2312e+5	24.3061/24.2844	3.0659/3.0761
B1$_{600}$	20000	9.0592e+5	25.6788	15.2529	8.9419e+5	25.5992/25.5738	15.3439/15.4200
B1$_{1200}$	20000	1.9653e+6	26.2329	28.6136	1.9339e+6	26.1012/26.0742	28.8040/29.0003
B1$_{2000}$	20000	3.3570e+6	26.4457	46.1726	3.3308e+6	26.3738/26.3476	46.3732/46.6200
B1$_{3000}$	20000	5.2662e+6	26.7679	66.7055	5.2092e+6	26.6491/26.6338	67.1725/67.5230
B2$_{100}$	20000	9.9370e+4	19.9442	6.1998	9.3734e+4	19.8992/19.8949	5.9867/5.9804
B2$_{600}$	20000	7.5496e+5	21.9439	23.7810	7.4231e+5	21.8235/21.8230	23.8747/23.8499
B2$_{1200}$	20000	1.5802e+6	22.4779	42.5783	1.5561e+6	22.3595/22.3457	43.0012/43.0531
B2$_{2000}$	20000	2.7236e+6	22.9163	64.9695	2.6747e+6	22.7625/22.7535	66.1393/66.1365
B2$_{3000}$	20000	4.1120e+6	23.1652	93.2223	4.0658e+6	23.0049/22.9832	94.8958/94.9391

4   Conclusions

The constraint sets (5) for the optimization problems (3) and (4) are too large, thus their minimizers tend to oversmooth the image. We experimented with various restriction techniques on C_A, C_I.

Simply decreasing τ_A, τ_I does improve the image quality of the output at the beginning, but we deviate from their statistical estimations and need to guess their optimal values, which is computationally very expensive. Moreover, those optimal values depend on the quality measures we consider, differ significantly from minimizing MAE to maximizing PSNR, and does not necessary lead to an output with good visual properties.

Another option, suggested in [8] as a future work direction, is to consider multi-constraint optimization. We investigated such approach, within the framework (7) and (8). We considered spatial, intensity, and mixed domain decompositions of the image, and summarized the numerical results in Tables 2, 3 and 4.

In all the setups, we observed that the image quality of the output improved up to a certain level. After that, the optimization problems became ill-posed, the algorithms' convergence was unclear, and artifacts appeared. This effect was caught by the MAE output values, but not by the PSNR ones, which still increased. Multiple constraints slowed down Algorithm 1, and its convergence rate was poor, independent on the choice of the accelerators σ, ρ. Therefore, parallel implementation of Algorithm 1 is practically important and is an object of ongoing work. While the problems (7) and (8) were well-posed, the close similarity between their solutions u_A, u_I, numerically observed in [8], hold true. Hence, their difference image $u_A - u_I$ can be used as a criterion for well-posedness.

Acknowledgments This research is supported by the project AComIn "Advanced Computing for Innovation", grant 316087, funded by the FP7 Capacity Program.

References

1. Dupé, F.X., Fadili, J., Starck, J.L.: A proximal iteration for deconvolving Poisson noisy images using sparse representations. IEEE Trans. Image Process. **18**(2), 310–321 (2009)
2. Zanella, R., Boccacci, P., Zanni, L., Bertero, M.: Efficient gradient projection methods for edge-preserving removal of Poisson noise. Inverse Prob. **25**(4), 045010 (2009)
3. Figueiredo, M.A.T., Bioucas-Dias, J.M.: Restoration of Poissonian images using alternating direction optimization. IEEE Trans. Image Process. **19**(12), 3133–3145 (2010)
4. Setzer, S., Steidl, G., Teuber, T.: Deblurring Poissonian images by split Bregman techniques. J. Vis. Commun. Image Represent. **21**(3), 193–199 (2010)
5. Mäkitalo, M., Foi, A.: Optimal inversion of the Anscombe transformation in low-count Poisson image denoising. IEEE Trans. Image Process. **20**(1), 99–109 (2011)
6. Carlavan, M., Blanc-Féraud, L.: Sparse Poisson noisy image deblurring. IEEE Trans. Image Process. **21**(4), 1834–1846 (2012)

7. Jezierska, A., Chouzenoux, E., Pesquet, J.-C., Talbot, H.: A primal-dual proximal splitting approach for restoring data corrupted with Poisson-Gaussian noise. In: IEEE International Conference on Acoustics, Speech, and Signal Processing (ICASSP 2012), pp. 1085–1088, Kyoto, Japan (2012)
8. Harizanov, S., Pesquet, J.C., Steidl, G.: Epigraphical projection for solving least squares Anscombe transformed constrained optimization problems. In: Kuijper et al., A. (eds.) Scale-Space and Variational Methods in Computer Vision (SSVM 2013), LNCS 7893, pp. 125–136, Springer, Berlin (2013)
9. Teuber, T., Steidl, G., Chan, R.H.: Minimization and parameter estimation for seminorm regularization models with I-divergence constraints. Inverse Prob. **29**, 1–28 (2013)
10. Li, J., Shen, Z., Yin, R., Zhang, X.: A reweighted ℓ_2 method for image restoration with Poisson and mixed Poisson-Gaussian noise. Inverse Prob. Imaging **9**(3), 875–894 (2015)
11. Ciak, R., Shafei, B., Steidl, G.: Homogeneous penalizers and constraints in convex image restoration. J. Math. Imaging Vis. **47**(3), 210–230 (2013)
12. Anscombe, F.J.: The transformation of Poisson, binomial and negative-binomial data. Biometrika **35**, 246–254 (1948)
13. Rudin, L.I., Osher, S., Fatemi, E.: Nonlinear total variation based noise removal algorithms. Physica D **60**, 259–268 (1992)
14. Chambolle, A., Pock, T.: A first-order primal-dual algorithm for convex problems with applications to imaging. J. Math. Imaging Vis. **40**(1), 120–145 (2011)
15. Pock, T., Chambolle, A., Cremers, D., Bischof, H.: A convex relaxation approach for computing minimal partitions. In: IEEE Conference on Computer Vision and Pattern Recognition, pp. 810–817 (2009)
16. Chierchia, G., Pustelnik, N., Pesquet, J.-C., Pesquet-Popescu, B.: Epigraphical projection and proximal tools for solving constrained convex optimization problems. SIViP **9**(8), 1737–1749 (2015)
17. Bardsley, J.M., Goldes, J.: Regularization parameter selection methods for ill-posed Poisson maximum likelihood estimation. Inverse Prob. **25**(9), 095005 (2009)

Innovative Graphical Braille Screen for Visually Impaired People

Dimitar Karastoyanov, Ivan Yatchev and Iosko Balabozov

Abstract The graphical interfaces based on visual representation and direct manipulation of objects made the adequate use of computers quite difficult for people with reduced sight. Within the European Union, the problem with the access of blind people to computer resources is quite pressing. A new type graphical Braille screen is developed. The Braille screen is a matrix with linear electromagnetic micro drives and non magnetic needles, passing trough the axes of the electromagnets. Over the electromagnets is mounted a grid with holes. The needles go through the holes and move up pimples. The visually impaired peoples feel them tactile way and can adopt symbols and graphics. Permanent magnet linear actuator intended for driving a needle in Braille screen has been proposed and optimized. Finite element analysis, response surface methodology and design of experiments have been employed for the optimization. The influence of different parameters of the construction of the developed permanent magnet linear electromagnetic actuator for driving a needle in a Braille screen is also studied. Static force characteristics of the actuator for driving a needle in Braille screen are presented. Using the graphical Braille screen, we can present symbols and simple graphics—Windows icons for example. The results of experiments for high speed moving of Braille needles using linear electromagnetic micro drive are presented. The experiments were conducted in the laboratory "Smart lab" with using of high speed camera NAC Memrecam. The international for Bulgarian patent application is presented as follows: Braille Screen—WIPO Patent Application, No P C T/B G 2014/000038, October 24, 2014; Braille Screen—Bulgarian Patent Application, No 111638, November 29, 2013

D. Karastoyanov (✉)
Institute of Information and Communication Technologies – BAS, bl.2,
ac. G. Bonchev street, 1113 Sofia, Bulgaria
e-mail: dkarast@iinf.bas.bg

I. Yatchev · I. Balabozov
Technical University – Sofia, 8 Kl. Ochridski bul., 1756 Sofia, Bulgaria
e-mail: yatchev@tu-sofia.bg

I. Balabozov
e-mail: i.balabozov@tu-sofia.bg

© Springer International Publishing Switzerland 2016
S. Margenov et al. (eds.), *Innovative Approaches and Solutions in Advanced Intelligent Systems*, Studies in Computational Intelligence 648,
DOI 10.1007/978-3-319-32207-0_14

219

Keywords Braille screen · Visually impaired · Linear electromagnetic microdrive

1 Introduction

Within the European Union, the problem with the access of blind people to computer resources is quite pressing. Studies on European and world scale are carried out in many directions:

- A basic direction is the attempt for social integration of the visually impaired, [1]. Within the Centers for social adaptation of blind peoples optimal conditions for assisting and integration of such peoples were provided
- Development of Braille terminals and printers and adaptation to computer systems. The impediment here is the fact that there is no unified system for representation of graphical and mathematical elements (e.g. integral, square root, etc.), [2].
- Since the communication man-computer was quite simple (mainly based on text instructions), solution of the problem was sought on the basis of voice synthesis or other forms of feedback, [3]
- Development of haptic interfaces based on electrically addressable and deforming polymer layer. Practically, the efforts are aimed at the manufacturing of a haptic dynamic input-output device allowing visually impaired people to obtain video information in other form, [4].

2 Voice Interface Modelling

So far as the modeling of human speech is concerned, it could be formed by separate components combined in a common system, [5]. For this purpose, it is necessary to model a vocal tract which will be the basis for the design of the voice synthesizer, [6]. Formal modeling could be realized through a model of the oral cavity from the larynx to the lips. To realize comparatively adequate model, certain number of parameter must be introduced to form articulate vector and define the personal characteristics of each individual. The model of human vocal tract basically consists of three components:

Oral cavity;
Glottal functional apparatus;
Acoustic impedance at the lips.

Generally, the oral cavity is modeled as an acoustic tube with slowly changing (in time and space) cross-section $A(x)$ where the acoustic waves propagate unidirectionally. Under these conditions, the following equations are suggested to calculate the pressure p(x, t) and volume velocity u(x, t):

$$-\frac{\partial p}{\partial x} = \frac{\rho}{A(x,t)} \frac{\partial u}{\partial t} \qquad (1)$$

$$-\frac{\partial u}{\partial x} = \frac{A(x,t)}{\rho c^2} \frac{\partial p}{\partial t} \qquad (2)$$

Differentiating Eqs. (1) and (2) with respect to time and space and eliminating the mixed partials, the equation of Webster is obtained:

$$\frac{\partial^2 p}{\partial x^2} + \frac{1}{A(x,t)} \frac{\partial p}{\partial x} \frac{\partial A}{\partial x} = \frac{1}{c^2} \frac{\partial^2 p}{\partial t^2} \qquad (3)$$

The eigenvalues of Eq. (3) are the frequencies of the formants. Solving Eq. (3), it is possible to find a stable sinusoidal transfer function for the acoustic tube, including thermal and viscous effects like losses along tube walls. For this purpose, substituting p(x, t) = P(x, ω)·e and u(x, t) = U(x, ω)·e, where ω is the angular frequency and j—imaginary unit, and introducing the terms acoustic impedance Z (x, ω) and acoustic conductivity Y(x, ω) to provide possibilities to calculate the losses in the acoustic tube, Eqs. (1) and (2) can be transformed to obtain:

$$\frac{d^2 U}{dx^2} = \frac{1}{Y(x,\omega)} \frac{dU}{dx} \frac{dY}{dx} - Y(x,\omega)Z(x,\omega)U(x,\omega) \qquad (4)$$

The sinusoidal transfer function of the vocal tract can be calculated by discretization of Eq. (4) in space and finding an approximated solution of the differential Eq. (4). Let us assume the denotation Uik for U(iΔx, kΔω), and allowing spatial discretization Δx = L/n, at i = 0 at the glottis and i = n at the lips, as shown in Fig. 1. Similar to these considerations, Δω = Ω/N and let k be c 0 < k < N. Based on

Fig. 1 Scheme of the vocal tract

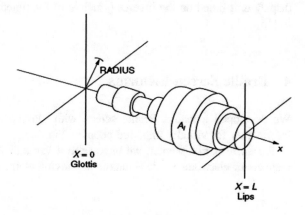

these initial assumptions and denotations, Eq. (4) is transformed into differential equation which, after slight mathematical transformations can be written as:

$$U_{i+1}^k = U_i^k \left(3 + (\Delta \Delta x^2 Z_i^k Y_i^k - \frac{Y_{i-1}^k}{Y_i^k}\right) + U_{i-1}^k \left(\frac{Y_{i-1}^k}{Y_i^k} - 2\right) \tag{5}$$

Figure 1 shows the general scheme of the vocal tract model.

The operation system of the autonomous device should guarantee its performance rate which implies modification of some of its kernel functions. According to the discussion on the job of the autonomous device, its general structure should be built in modules using as much as possible standard interfaces for communication with computer systems.

3 Permanent Magnets

There has been intensive use of permanent magnets in wide area of applications for different actuators in recent years. An incentive is the possibility for development of energy efficient actuators. New constructions of permanent magnet actuators are employed for different purposes. One such purpose is the facilitation of perception of images by visually impaired people using the so called Braille screens. Recently, different approaches have been utilized for the actuators used to move Braille dots [7–15]. A linear magnetic actuator designed for a portable Braille display application is presented in [7]. Actuators based on piezoelectric linear motors are given in [8, 9]. A phase-change microactuator is presented in [10] for use in a dynamic Braille display. Similar principle is employed in [12], where actuation mechanism using metal with a low melting point is proposed. In [13], Braille code display device with a polydimethylsiloxane membrane and thermopneumatic actuator is presented. Braille sheet display is presented in [14] and has been successfully manufactured on a plastic film by integrating a plastic sheet actuator array with a high-quality organic transistor active matrix. A new mechanism of the Braille display unit based on the inverse principle of the tuned mass damper is presented in [15].

4 Braille Screen Elements

We developed a matrix Braille screen with moving elements, giving graphical information for visually impaired people—Fig. 2.

For each matrix element we have several variants. All of them use permanent magnets or electromagnetic actuators as moving elements.

Fig. 2 Graphical Braille
screen

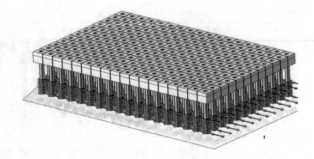

Variant 1: Electromagnetic driven balls (shots), [16].
The magnetic field pull up the balls
(with a spring for neutral position)—Fig. 3
(b) The magnetic field beat off the balls—Fig. 4
Variant 2: Rotating balls—each ball has North and South poles—Fig. 5, [17].
Variant 3: Actuator with lifters and springs—Fig. 6, [18].

For computer modelling and simulation of the behavior of the driving elements, numerical approach employing magnetic field analysis using the finite element method has been implemented. The problem has been solved as axisymmetric, nonlinear and steady-state. FEMM program was used for the finite element

Fig. 3 Actuators with springs

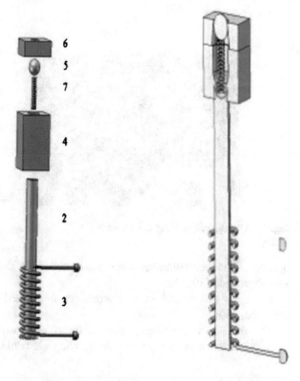

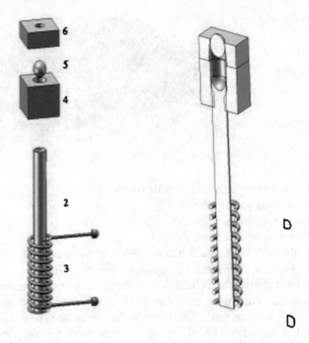

Fig. 4 The field beat off the balls

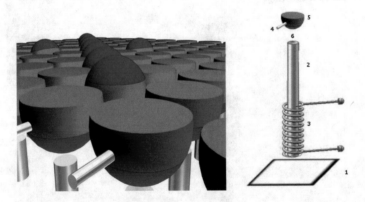

Fig. 5 The magnetic field rotates the balls

analysis. The computations were automated with the help of the Lua Script®
language code [19].

For cost, power and force reasons we developed a combined actuator construction, [20].

The developed permanent magnet linear actuator is energy efficient, as the
energy is used only for driving the moving elements up and down and not for

Fig. 6 The pixel lifted and
held by the spring

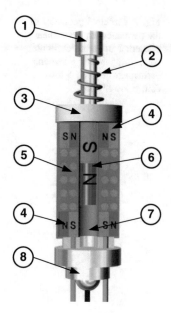

keeping them in upper or lower position. At the lower and at the upper position, the
moving elements are kept fixed due to the permanent magnet force [21].

For the analysis of the magnetic field of the actuator, again the finite element
method and its implementation in FEMM program have been used together with
command code written in Lua Script® language. The problem has been solved as an
axisymmetric one, employing weighted stress tensor approach for obtaining the
force on the moving permanent magnet [21].

Computations of series of characteristics for the static force have been carried
out. For obtaining these characteristics, different actuator parameters have been
varied. The varied parameters were the clearance (air gap) between the two parts of
the magnetic circuit—lower and upper core, the permanent magnet length and the
height of the coils [21].

5 Actuator Construction

The geometry of the actuator is outlined in Fig. 7. The mover is a permanent
magnet. Its magnetization direction is along the axis of rotational symmetry.

The upper and lower coil are connected in series. This connection is realized so
that the flux created by each of them is in opposite directions in the mover zone.
Thus by choosing proper power supply polarity, the motion of the mover will be in
desired direction. For example, in order to have motion of the mover in upper
direction, the upper coil has to be supplied in a way to create air gap magnetic flux,
which is in the same direction as the one of the flux created by the permanent

Fig. 7 Principal geometry of
the permanent magnet linear
actuator *1* upper core; *2* outer
core; *3* upper coil; *4* moving
permanent magnet; *5* lower
coil; *6* lower core

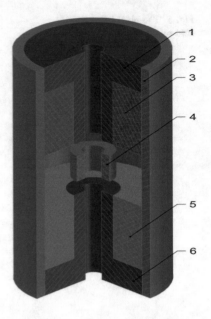

magnet. The lower coil in this case will create magnetic flux which is in opposite
direction to the one of the magnetic flux created by the permanent magnet. In this
case motion up will be observed. In order to have motion down, the lower coil
should be supplied in a way so that its flux is in the same direction as the flux by the
permanent magnet. The upper coil then will create magnetic flux in opposite
direction. In order to fix the moving part to the Braille dot, non-magnetic shaft, not
shown in Fig. 7 is used. Additional construction variants of the actuator have also
been considered, in which two small ferromagnetic discs (not shown in Fig. 7) are
placed on both sides—upper and lower—of the moving permanent magnet.

This actuator is also energy efficient, as energy is used only for changing the
position of the moving part from lower to upper and vice versa. Both at lower and at
upper position, no energy is used. At these positions the mover is kept fixed due the
force ensured by the permanent magnet.

6   Finite Element Modelling

Magnetic field of the construction variant of the permanent magnet linear actuator
with two ferromagnetic discs on both sides of the permanent magnet is analysed
with the help of the finite element method. The program FEMM [22] has been used
and additional codes written Lua Script® language are developed for faster computation. The magnetic field is analysed as an axisymmetric one due to the rotational symmetry of the actuator. The weighted stress tensor approach has been
utilized for evaluating the electromagnetic force on mover.

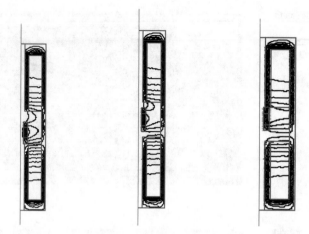

Fig. 8 Typical flux lines distribution for three different mover positions

As an illustration, the flux lines distribution for three positions of the permanent magnet is shown in Fig. 8. The force for the three positions of the mover is in upward direction.

7 Static Force Characteristics

The static force characteristics are obtained for different construction parameters of the actuator. The outer diameter of the core is 7 mm. The air gap between the upper and lower core, the length of the permanent magnet and the coils height have been varied.

In Figs. 9, 10, 11, 12, 13, 14, 15 and 16, the force-displacement characteristics are given for varied values of the permanent magnet height hm, coil height hw, magnetomotive force Iw and apparent current density in the coils J. With c1 and c2,

Fig. 9 Force-displacement characteristics for hm = 2 mm, δ = 3 mm, hw = 5 mm, Iw = 180 A, J = 20 A/mm^2

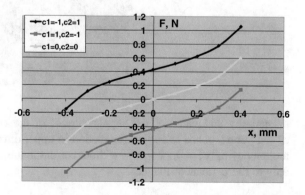

Fig. 10 Force-displacement characteristics, for $hm = 2$ mm, $\delta = 4$ mm, $hw = 5$ mm, $Iw = 180$ A, $J = 20$ A/mm^2

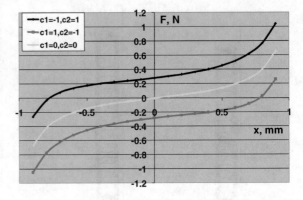

Fig. 11 Force-displacement characteristics for $hm = 2$ mm, $\delta = 5$ mm, $hw = 5$ mm, $Iw = 180$ A, $J = 20$ A/mm^2

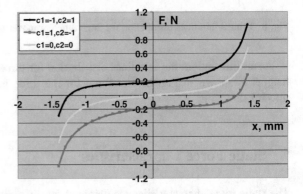

Fig. 12 Force-displacement characteristics for $hm = 3$ mm, $\delta = 4$ mm, $hw = 5$ mm, $Iw = 180$ A, $J = 20$ A/mm^2

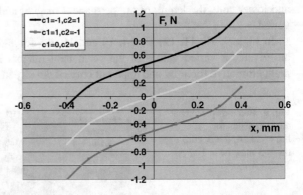

the polarity and value of the supply current of the coils is denoted. "$c_1 = -1$, $c_2 = 1$" corresponds to supply for motion in upper direction; "$c_1 = 1$; $c_2 = -1$"—motion down, "$c_1 = 0$, $c_2 = 0$"—without current in the coil, i.e. this the force is due only to the permanent magnet.

Fig. 13 Force-displacement characteristics for
hm = 3 mm, δ = 5 mm,
hw = 5 mm, Iw = 180 A,
J = 20 A/mm^2

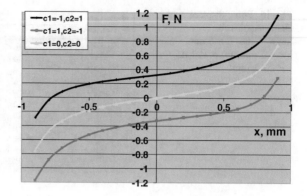

Fig. 14 Force-displacement characteristics for
hm = 3 mm, δ = 6 mm,
hw = 5 mm, Iw = 180 A,
J = 20 A/mm^2

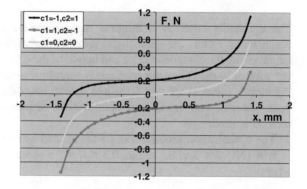

Fig. 15 Force-displacement characteristics for
hm = 4 mm, δ = 5 mm,
hw = 5 mm, Iw = 180 A,
J = 20 A/mm^2

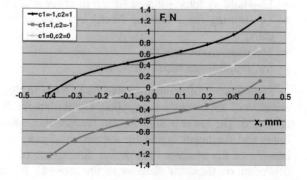

Fig. 16 Force-displacement
characteristics for
hm = 4 mm, δ = 6 mm,
hw = 5 mm, Iw = 180 A,
J = 20 A/mm²

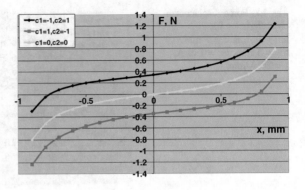

8 Optimization

The optimality criterion is minimal magnetomotive force NI of the coils. The
optimization factors are geometrical parameters (height of the permanent magnet,
height of the ferromagnetic discs and height of the coils. The optimization is carried
out subject to the following constraints—minimal electromagnetic force acting on
the mover, minimal starting force and overall outer diameter of the actuator have
been set. For performing the optimization, sequential quadratic programming
method is used.

The optimization problem can be formulated as follows:

min {NI} 5 ≤ hw, 0.5 ≤ hm, 0.3 ≤ 0 ≤ J ≤ A/mm², Fh ≥ 0.3 N, Fs ≥ 0.05 N,
where:

- NI—magnetomotive force—minimizing energy consumption with satisfied
  force requirements;
- Fh—holding force—mover (shaft) in upper position, no current in the coils;
- Fs—starting force—mover (shaft) in upper or lower position and energized
  coils;
- J—coils current density;
- hw, hm, hd—geometric dimensions.

Minimization of magnetomotive force NI is in direct correspondence to minimization of the energy consumption. Constraints for Fs and Fh have already been
discussed. The lower bounds for the dimensions are imposed by the manufacturing
limits and the upper bound for the current density is determined by the thermal
balance of the actuator. The radial dimensions of the construction are directly
dependent on the outer diameter of the core—D. Its value is fixed, the reason for
which has been discussed earlier. The influence of those parameters on the behavior
of the construction have been studied in previous work [23] from where a conclusion can be drawn that there is no necessity radial dimensions to be included in
the list of optimization factors. The results of the optimization are as follows [24]:

$NI_{opt} = 79.28$ A, $hw_{opt} = 5$ mm, $hm_{opt} = 2.51$ mm, $hd_{opt} = 1.44$ mm, $J_{opt} = 19.8$ A

The optimal parameters have been set as input values to the FEM model. The force-displacement characteristics of the optimal actuator are shown in Figs. 17 and 18.

In Figs. 19 and 20, the magnetic field distribution of the optimal actuator is given for two cases. The force constraints for Fs and Fh are active which can be expected when minimum energy consumption is required. The active constraint for hw is also expected because longer upper and lower cores size which respectively means longer coils will increase the leakage coil flux and corrupted coil efficiency.

Fig. 17 Force-displacement characteristic of the optimal actuator. The force is created by the permanent magnet only (no current in the coils)

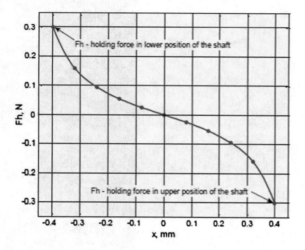

Fig. 18 Force-displacement characteristic of the optimal actuator. Coils are energized. The mover is displaced from upper to lower position

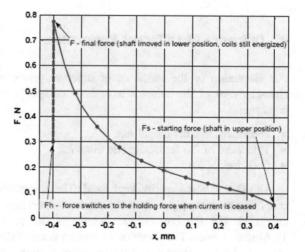

Fig. 19 Magnetic field of the optimal actuator with mover in upper position and coils energized

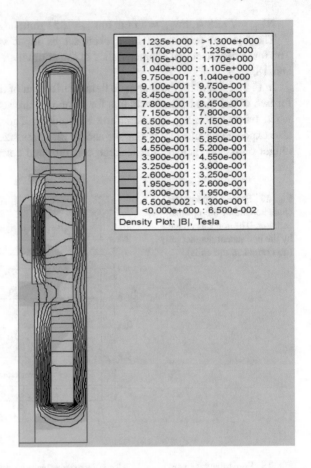

1.235e+000 : >1.300e+000	
1.170e+000 : 1.235e+000	
1.105e+000 : 1.170e+000	
1.040e+000 : 1.105e+000	
9.750e-001 : 1.040e+000	
9.100e-001 : 9.750e-001	
8.450e-001 : 9.100e-001	
7.800e-001 : 8.450e-001	
7.150e-001 : 7.800e-001	
6.500e-001 : 7.150e-001	
5.850e-001 : 6.500e-001	
5.200e-001 : 5.850e-001	
4.550e-001 : 5.200e-001	
3.900e-001 : 4.550e-001	
3.250e-001 : 3.900e-001	
2.600e-001 : 3.250e-001	
1.950e-001 : 2.600e-001	
1.300e-001 : 1.950e-001	
6.500e-002 : 1.300e-001	
<0.000e+000 : 6.500e-002	

Density Plot: |B|, Tesla

9 Influence of Different Parameters

For estimation of the influence of different parameters, the force-displacement characteristics of the actuator have been computed when varying the following parameters:

- Permanent magnet height hm;
- Thickness of the ferromagnetic discs hd;
- Coils height hw.

All other dimensions are kept fixed. The computations are performed for two cases of power supply—when the coils are energized and when the coils are not energized. For the coil current density in the case of energizing a value of 15 A/mm^2 is used. The direction of motion is taken downwards. All computations are carried out for three values of the maximal stroke, namely 0.8, 1.8 and 2.8 mm.

Fig. 20 Magnetic field of the optimal actuator with no current in the coils

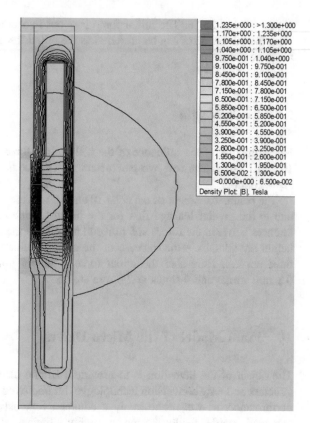

1.235e+000 : >1.300e+000
1.170e+000 : 1.235e+000
1.105e+000 : 1.170e+000
1.040e+000 : 1.105e+000
9.750e-001 : 1.040e+000
9.100e-001 : 9.750e-001
8.450e-001 : 9.100e-001
7.800e-001 : 8.450e-001
7.150e-001 : 7.800e-001
6.500e-001 : 7.150e-001
5.850e-001 : 6.500e-001
5.200e-001 : 5.850e-001
4.550e-001 : 5.200e-001
3.900e-001 : 4.550e-001
3.250e-001 : 3.900e-001
2.600e-001 : 3.250e-001
1.950e-001 : 2.600e-001
1.300e-001 : 1.950e-001
6.500e-002 : 1.300e-001
<0.000e+000 : 6.500e-002

Density Plot: |B|, Tesla

9.1 Permanent Magnet Height

The values of the permanent magnet height used for estimating its influence are 2, 3 and 4 mm. This study is performed while keeping other parameters constant: $hd = 1$ mm; $hw = 5$ mm.

The results show that the maximal variation of the force for the different permanent magnet heights is about 10–12 % and this is valid for all studied strokes. There is a difference, though, in the values of the initial force—it is much higher at 0.8 mm stroke—more than 50 % than the initial force at 1.8 and 2.8 mm.

9.2 Thickness of the Ferromagnetic Discs

Two values of the thickness of the ferromagnetic discs are—1 mm and 2 mm. The values of the other two parameters are kept constant: $hm = 2$ mm; $hw = 5$ mm.

The absolute differences for the different strokes are close to each other. The relative difference for 0.8 mm stroke is about 10 %, while for 2.8 mm stroke it is higher—up to 25 %. The initial force is about two times higher for 2 mm discs. Thus the thicker discs can be preferred as the initial force is an important parameter for these actuators.

9.3 Coils Height

For estimation of the influence of the coils height three values have been used –5, 10 and 15 mm. The rest two parameters are kept constant at values: $hm = 2$ mm; $hd = 1$ mm.

The results for height of the coils 10 and 15 mm are practically the same. This is due to the greater leakage flux for the highest value of the coil height. The differences between the forces at 5 mm coil height and the other two values of the coil height are valuable—the increase of the average force is from 23 to 27 % for the three strokes. Here also the initial force is highly influenced, being 2 times at 0.8 mm stroke and 4 times at 2.8 mm stroke.

10 Final Model of the Micro Drives

The object of the invention is to provide a Braille display (Fig. 21), with simple structure and easy conversion technology with improved static, dynamic and energy performance, as well as to apply a common link between all moving parts, an extended tactile feedback and a highly efficient start up with low energy consumption.

This object is solved by a Braille display [25, 26], representing a matrix comprised of a base with fixed electromagnets, arranged thereon, including an outer cylindrical magnetic core, in which a winding magnetic core locking up the

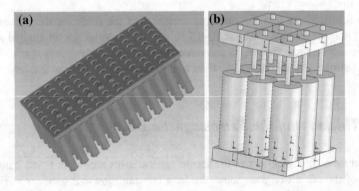

Fig. 21 Braille display. **a** Braille screen matrix, **b** one Braille cell—symbol or graphic

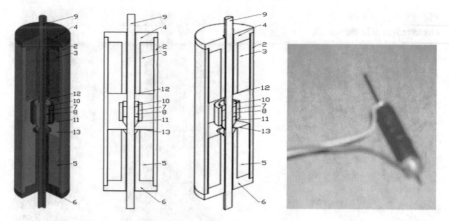

Fig. 22 Linear electromagnetic micro drive (model and device), where: *1* base *2*, *4*, *6* magnetic cores *3*, *5* coils *7* non magnetic cylindrical body *8* permanent magnet *9* non magnetic needle *10*, *11* ferromagnetic washers *12*, *13* magnetic poles *14* grid

cylindrical magnetic core at the top side and a winding magnetic core locking up the cylindrical magnetic core at the bottom side are placed, where the magnetic cores are with axial holes, and into the space between the windings a movable non-magnetic cylindrical body is placed, carrying a cylindrical permanent magnet axially magnetized and a non-magnetic needle, passing axially through the permanent magnet and the axial holes of the magnetic cores, and the top side of the permanent magnet a ferromagnetic disc is arranged having an axial hole, and on its underside a ferromagnetic disc is arranged having an axial hole wherein, the upper disc and the upper magnetic core have cylindrical poles and the lower magnetic core and the lower disc have conical poles, and above the electromagnets a lattice is placed with openings through which the needles pass—Fig. 22.

The electromagnets can be placed in one line in the matrix as well as in two, three or more lines, side by side at an offset along two axes (x and y) and a different length of the movable needles along the third axis (z), such that they overlap and occupy less space in the matrix, and the tips of the needles are in one plane—the plane of the lattice with openings of the Braille display—Fig. 23.

An advantage of the invention is that the retention of the needles in their final positions does not need an external power supply, as it is provided by the permanent magnet.

Another advantage is the lack of additional springs which allows for very precise control of the feed force for realizing of the tactile feedback.

Advantage is the intermittent power supply voltage with which an extremely low power consumption is achieved (only for moving the needle from one end to the other), as well as more efficient use of materials and reducing of the size of the matrix.

Fig. 23 Place of the
electromagnets in the matrix

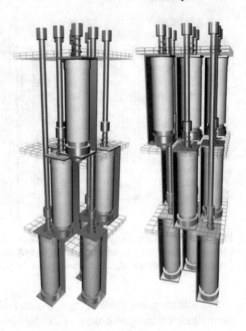

An advantage of the present invention are the extremely low electric and mechanical time constants, resulting in very good velocity and dynamic characteristics.

Furthermore, an advantage is the considerably wider range of the realizable tactile feedback due to the fact that all electromagnets can be operated simultaneously and synchronously, which gives the operator a more realistic and closer to reality tactile perception.

Another advantage is that besides letters and numbers, graphic symbols (roots, integrals), images of icons, paintings in the case of larger matrix and others can be written on the tactile (Braille) matrix.

Similarly, an advantage is the extremely high positioning accuracy of the needles and the stability of the retaining forces for them into their final positions.

Another advantage is that the chosen propulsion method is distinguished by its exceptional reliability and trouble-free operation and requires no additional settings and service.

The technology, convenient for the realization of the matrix is also an advantage.

11   Study of the Micro Drive Using High Speed Camera

In order to determine experimentally the movement speed of the movable part of the studied electromagnet (Fig. 24), a high-speed camera is employed—Fig. 25. The results can used to verify and prove the results of the dynamic characteristics

Fig. 24 Linear
electromagnetic Microdrive

Fig. 25 High speed camera

derived by computer simulation. The results can also be compared with results obtained using the accelerometer (acceleration sensor) obtained in the previous experimental studies of the electromagnetic module.

To achieve this purpose an experimental arrangement is made (Fig. 26), involving a research prototype, an electromagnet, a power and a control device, a high-speed camera and a computer to analyze the results from the camera—displacements, speed and acceleration. There have been two tests (one when moving from bottom to top position and one back) with 10 prototype of the experimental electromagnet. The carried out experimental studies with a high-speed camera confirm the previous results obtained from computer models made in Comsol, and also the reliability of the results obtained from the accelerometer in a previous study of prototypes of the electromagnet.

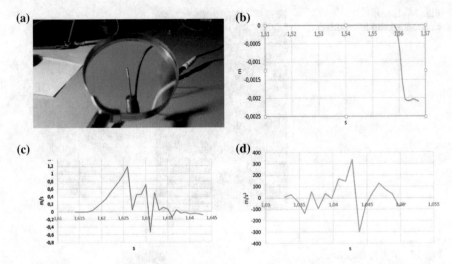

Fig. 26 The experiments. **a** Testing stand, **b** displacement, **c** speed, **d** acceleration

12 Conclusion

Based on the results obtained the following conclusions can be drawn:

- the developed actuator has static force characteristics which are suitable for Braille screen application;
- increasing the height of the coil has important influence on the force-displacement characteristics and the holding force. Above a certain value, thought, further increase does not lead to significant change;
- the maximal stroke influences more significantly the initial force than the holding one and its minimal value could be recommended;
- higher outer diameter of the actuator leads to significant increase of both holding and initial force;
- current density of 15 A/mm^2 could ensure enough initial force at lower starting position of the mover.

Acknowledgments The research work reported in the paper is partly supported by the project AComIn "Advanced Computing for Innovation", grant 316087, funded by the FP7 Capacity Programme (Research Potential of Convergence Regions).

References

1. Simeonov, S.: Modeling the interaction of groups of interfaces designed for unsighted people. Int. J. Pure Appl. Math. **104**(3), pp. 375–388, 0.379, November 2015, SJR: 0,322 (2015). ISSN: 1314-3395
2. Kamel, H.M., Roth, P., Sinha, R.R.: Graphics and user's exploration via simple sonics: providing interrelational representation of objects in a non-visual environment. In: Proceedings of the International Conference on Auditory Display, Espoo, Finland, pp. 261–266 (2006)
3. Kennel, A.R.: Audiograf: a diagram reader for the blind. In: Proceedings ACM Conference on Assistive Technologies, pp. 51–56 (1996). ISBN:0-89791-776-6
4. Simeonov, S., Simeonova, N.: Graphical interface for visually impaired people based on Bi-stable solenoids. Int. J. Soft Comput. Softw. Eng. **3**(3), 844–847 (2014). e-ISSN:2251-7545
5. Pitt, I.J., Edwards, A.D.N.: Improving the usability of speech-based interfaces for blind users. In: Proceedings ACM Conference on Assistive Technologies (ASSETS), pp. 124–130 (1996). ISBN:0-89791-776-6
6. Raman, T.V.: Emacspeak—Direct speech access. In: Proceedings ACM Conference on Assistive Technologies (ASSETS), pp. 32–36 (1996). ISBN:0-89791-776-6
7. Nobels T., Allemeersch, F., Hameyer, K.: Design of a high power density electromagnetic actuator for a portable Braille display. In: International Conference EPE-PEMC 2002, Dubrovnik & Cavtat, pp. p1–p5 (2002)
8. Cho, H.-C., Kim, B.-S., Park, J.-J., Song, J.-B.: Development of a Braille display using piezoelectric linear motors. In: International Joint Conference SICE-ICASE, pp. 1917–1921, Busan, Korea (2006)
9. Hernandez, H., Preza, E., Velazquez, R.: Characterization of a piezoelectric ultrasonic linear motor for Braille displays. In: Electronics, Robotics and Automotive Mechanics Conference CERMA, pp. 402–407, Cuernavaca, Mexico (2009)
10. Green S., Gregory, B., Gupta, N.: Dynamic Braille display utilizing phase-change microactuators. In: 5th IEEE Conference on Sensors, pp. 307–310, Daegu, Korea (2006)
11. Nakashige M., Hirota, K., Hirose, M.: Linear actuator for high-resolution tactile display. In: 13th IEEE International Workshop on Robot and Human Interactive Communication ROMAN 2004, pp. 587–590 (2004)
12. Chaves, D., Peixoto, I., Lima, A.C.O., Vieira, M., de Araujo, C.J.: Microactuators of SMA for Braille display system. In: IEEE International Workshop on Medical Measurements and Applications. MeMeA 2009, pp. 64–68 (2009)
13. Kwon, H.-J., Lee, S.W., Lee, S.S.: Braille code display device with a PDMS membrane and thermopneumatic actuator. In: 21st IEEE International Conference on Micro Electro Mechanical Systems MEMS 2008, pp. 527–530, Tucson, AZ, USA
14. Kato, Y., et al.: A flexible, lightweight Braille sheet display with plastic actuators driven by an organic field-effect transistor active matrix. In: IEEE International Electron Devices Meeting (2005). http://xplorebcpaz.ieee.org/xpl/mostRecentIssue.jsp Punumber = 10701
15. Kawaguchi, Y., Ioi, K., Ohtsubo, Y.: Design of new Braille display using inverse principle of tuned mass damper. In: SICE Annual Conference, Taiwan, pp. 379–383, 2010
16. Karastoyanov, D.: Braille screen. Bulgarian Patent announce No 110731, 10.08.2010 г
17. Karastoyanov, D., Simeonov, S., Dimitrov, A.: Braille display. Bulgarian Patent announce No 110795, 11.11.2010 г
18. Karastoyanov, D., Simeonov, S.: Braille display. Bulgarian Patent announce No 110794, 11.11.2010 г
19. Yatchev, I., Hinov, K., Balabozov, I., Gueorgiev, V., Karastoyanov, D.: Finite element modeling of electromagnets for Braille screen. In: 10 International Conference on Applied Electromagnetics PES 2011, pp. O8.1–O8.4, Nis, Serbia, 25–29 Sept 2011
20. Karastoyanov, D., Yatchev, I., Hinov, K., Rachev, T.: Braille screen. Bulgarian Patent announce No 111055, 13.10.2011

21. Yatchev, I. Hinov, K., Gueorgiev, V., Karastoyanov, D., Balabozov, I.: Force characteristics of an electromagnetic actuator for Braille screen. In: International Conference ELMA 2011, pp. 338–341, Varna, Bulgaria, 21–22 Oct 2011
22. Meeker, D.: Finite element method magnetics version 3.4. User's Manual (2005)
23. Karastoyanov, D.: Energy efficient control of linear micro drives for Braille screen. In: International Conference on Human and Computer Engineering ICHCE 2013, pp. 860–864, Osaka, Japan, 14–15 Oct 2013. pISSN 2010-376x, eISSN 2010-3778
24. Yatchev, I., Balabozov, I., Hinov, K., Gueorgiev, V., Karastoyanov, D.: Optimization of permanent magnet linear actuator for Braille screen. In: International Symposium IGTE 2012, pp. 59–63, Graz, Austria, 16–18 Sept 2012
25. Karastoyanov, D., Yatchev, I., Hinov, K., Balabozov, I.: Braiile Screen—Bulgarian Patent Application, No 111638, 29.11.2013
26. Karastoyanov, D., Yatchev, I., Hinov, K., Balabozov, I.: Braiile Screen—WIPO Patent Application, No P C T/B G 2014/000038, 24 Oct 2014

Smart Feature Extraction from Acoustic Camera Multi-sensor Measurements

**Petia Koprinkova-Hristova, Volodymyr Kudriashov, Kiril Alexiev,
Iurii Chyrka, Vladislav Ivanov and Petko Nedyalkov**

Abstract The paper applies recently developed smart approach for feature extraction from multi-dimensional data sets using Echo state networks (ESN) to the focalized spectra obtained from the acoustic camera multi-sensor measurements. The aim of the study is development of distance diagnostic system for prediction of wearing out of bearings. The procedure for initial features selection and features extraction from the focalized spectra was developed. Then the k-means clustering algorithm and Support vector machine (SVM) classifiers were applied to differentiate the tested bearings into two classes with respect to their condition ("Good" or "Bad"). The results using different dimensions of the extracted features space were compared.

Keywords Smart signal processing · Multi-sensor system · Feature extraction · Classification · Echo state network · IP tuning · K-means clustering · Support vector machine

P. Koprinkova-Hristova (✉) · V. Kudriashov · K. Alexiev · I. Chyrka
Institute of Information and Communication Technologies, Bulgarian Academy of Sciences,
Acad. G. Bonchev street, bl.25A, Sofia, Bulgaria
e-mail: pkoprinkova@bas.bg

V. Kudriashov
e-mail: kudriashovvladimir@gmail.com

K. Alexiev
e-mail: alexiev@bas.bg

I. Chyrka
e-mail: yurasyk88@gmail.com

V. Ivanov · P. Nedyalkov
Technical University of Sofia, 8 Kl. Ohridski Blvd., Sofia, Bulgaria
e-mail: vvi@tu-sofia.bg

P. Nedyalkov
e-mail: nedpetko@tu-sofia.bg

© Springer International Publishing Switzerland 2016
S. Margenov et al. (eds.), *Innovative Approaches and Solutions in Advanced
Intelligent Systems*, Studies in Computational Intelligence 648,
DOI 10.1007/978-3-319-32207-0_15

241

1 Introduction

Nowadays high-dimensional data analyses are needed for a wide range of applications in many fields, such as genomics and signal processing [5]. Particularly, the clustering of such data sets faces two major problems—scalability and capacity for dealing with multidimensionality [1]. This led to numerous developments towards dimensionality reduction (DR), i.e. mapping of the original high-dimensional data into low-dimensional spaces preserving at the same time the important structural data properties. The vast majority of DR methods work in a unsupervised manner, i.e. they process data features without taking into account information about class labels etc. [4]. The main aims of DR are two: visualization needed for inspection of the data and "defeating" the "curse of dimensionality" in order to be able to solve computationally demanding tasks like regression or classification.

A recently developed smart DR approach [13] proposes two steps algorithm: mapping of high dimensional data onto a new space with different dimension (lower or higher) using a kind of recurrent neural network called Echo state network (ESN) and then projection to a two-dimensional space using only two out of numerous extracted features from this new space. These two dimensional projections are considered as different "views" of the original data set that are used for clustering or classification purposes. By far this new approach was applied to variety of data sets in combination with different clustering and/or classification methods and the results were summarized in [2, 12].

The main aim of the present work is to apply our smart features extraction approach to another task that produces a multi-dimensional data set—classification of the focalized spectra of noise measurements obtained using acoustic camera. The practical task to be solved was to create a distance diagnostic system able to predict the failure of rolling bearings measuring the noise they generate in real working conditions. For this purpose a data base of noise measurements from undamaged and worn out bearings was created using the Brüel & Kjær acoustic camera composed by 18 microphones. For the aims of DR and classification of sample bearings the mentioned above DR approach [13] was applied to the collected data set. Then two types of classifiers (unsupervised k-means clustering and supervised Support Vector Machines) were created.

The rest of the paper is organized as follows: next section presents known by far approaches to bearings diagnostics; then experimental set-up for data set collection is described; section four represents raw data preprocessing; next steps of expert features extraction and DR approach was described; finally the results from clustering and classification of our data collection are presented and analyzed; conclusions including directions of future works finish the paper.

2   Problem Formulation

Rolling bearings are key elements supporting the rotating machine parts installed in about 95 % of the machines with moving parts. Phenomena related to the internal dynamics of rolling bearings cause wear and damage limitation and/or shorten the life of the bearings. On the other hand the same phenomena and processes related to the internal dynamics of the bearings are sources of vibro-acoustic emissions, which are the basis for the development of external diagnostic systems and methods for evaluation of diagnostic performance.

The main issues related to the vibro-acoustic diagnosis of rolling bearings are based on the analysis of the internal bearing frequencies (Frequencies Inner Bearing), which are associated with the bearing parts. The analysis of internal bearing narrowband frequencies is due to the fact that the individual elements emit at fixed frequencies and some of their harmonics, while maintaining a fixed speed of the bearing shaft.

The analysis of remaining emissions is hampered by complex interactions between multiple bearing elements and emission mixing. Modern diagnostic methods develop towards the analysis and separation of sources of emission frequency, intensity, band, emission direction and other determining factors.

Diagnostic methods for the control of bearing units are classified into two main groups. The first of these groups includes the systems with continuous operation, which are installed on machines and thus working and reporting in real working conditions. Advance of these systems over the second group (described below) is the general possibility to directly compare the data in different time periods or different kinematic and power arrangement of a given machine. As a result of the advanced techniques and instruments for analysis in systems with continuous learning, algorithms (procedures) are introduced in the analyzing software, which enables modeling and a limited (partial) prediction of the life of the bearing unit while maintaining operational factors affecting on the bearings [3, 7, 8, 10, 15].

The second group includes systems with discontinuous operation. In this case it is possible to separate the diagnostic system from the tested machine, which is typical for the majority of diagnostic test rigs. Such systems allow besides the recognition of Inner Bearing Frequencies, recognition of abrasion, presence of pitting and spalling defects, presence of pollutants, etc. Thus the base problem is the recognition of particular defects and processes because of their mixture like a source of vibro-acoustic signals.

Processing of vibration signal of rolling-element bearing enables its non-destructive testing and health monitoring. The procedure may require dismantling of the bearing for its further analysis using standalone equipment for vibration estimation. Assembling and disassembling of parts of production lines causes unwanted costs and may require production stops.

From the other hand, since each vibration generates also sound waves, their measurement in real working conditions can be used as another mean for distance diagnostic without stopping of machinery and dismounting of its bearings. Localization of sound sources is a task with numerous practical applications and variety of special purpose devices were created for this aim. Among these acoustic cameras consisting of multiple microphones are the most sophisticated multi-sensor systems that allow precise location and differentiation of different noise sources. Besides their primary application for position localization of the sound sources in the observed area, the distance diagnostics of machinery can be another useful for industry practical application. Estimates of acoustic emission enable to provide the condition estimation remotely. The latter may withdraw the noted dismantling.

3 Experimental Set-Up

In present study a collection of 10 SKF ball bearings with the same geometry (nine rolling elements (balls) with diameter ~7.938 mm) and with known quality (in good condition or damaged) was investigated. Fundamental noise frequencies generated during work of a bearing are a function of its geometry and relative speed between its inner and outer rings [8]. Hence every destructive change in balls or rings will result in respective changes in the noise frequencies of a bearing. The experiment carried out aims classification of the bearings into two classes (healthy and worn-out) based on the distance measurement of the noise generated by them in working conditions. For this purpose data set consisting of the noise generated by all bearings in the investigated collection during their work on a laboratory test-bed was collected.

The measurements were done by the Brüel & Kjær acoustic camera shown on Fig. 1. It consists of 18 microphones, input module, and acquisition Pulse LabShop software. The microphones frequency band pass is from 10 Hz to 20 kHz. The

Fig. 1 Brüel & Kjær acoustic camera

microphones are placed in 2D "slice wheel" array. Diameter of the array is about 0.32 m. The arrangement of the microphones in the array is optimized by the manufacturer to ensure high enough angular resolution with respect to the allowed maximum side lobe level in field of view (about ±30° away from center axis), for wide frequencies range. Currently, the camera maintains beam forming and acoustic holography methods for the acoustic noise source identification. The microphone array beam width is up to about 3° at frequencies about 20 kHz, using beam forming. As well, the camera enables export of data for user-defined post processing.

The camera was placed in front of bearings vibration diagnosis equipment shown on Fig. 2. The belt drive passes the electric motor rotation to the spindle. The spindle has mounting place for bearing inner ring. Bearing outer ring is loaded with the pressing screw.

The experiments were carried out in a laboratory room, which is not optimized for acoustic measurements. The interference sources are the voltage transformer, the equipment motor with gears, people activities and urban noises from outdoor. Their emissions impact is comparable to some of the target incoming emissions. Distance from the microphone array to bearing seat on the spindle is about 0.4 m (Fig. 3). Cross-range separation between the bearing seat and center of the motor is about 0.12 m. Their separation in range is about 0.2 m. Plexiglas panel is between them. Pitch diameter is ~39.04 mm. Contact angle is zero. The spindle rotational speed slightly varies from 1820 to 1830 rpm that corresponds to shaft frequency from 30.3 to 30.5 Hz.

Fig. 2 Picture of the experimental setup

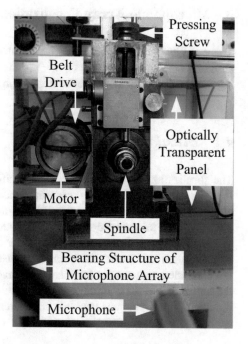

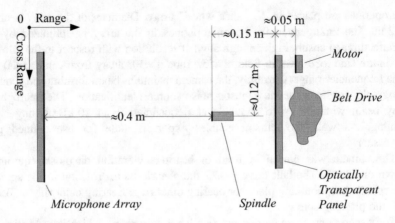

Fig. 3 Scheme of the experimental setup (*top view*)

In order to assess the frequency of the noise generated by the bearing mainly, the multi-sensor data of 18 microphones in the acoustic camera were "fused" to produce focalized spectra characterizing noise at the position of interest (the tested bearing) accounting for the distance and position of the bearing relative to the "observed" by the acoustic camera area. The spectra of incoming signals had been focalized based on the approach described in next section.

4 Microphone Array Focalization

There are two basic approaches for focalization spectra estimates (FSE) at the chosen focalization point $\mathbf{r} = (x, y, z)$ in the three-dimensional space—beamforming for far distances and acoustic holography for close distances.

4.1 Beamforming

The focalization requires introducing specific delays to the microphones' signals [6]. The delays are defined considering incoming wave propagation model, spatial coordinates of position under focalization and coordinates of microphones in the array. Further, the delayed signals may be added squared or processed in other way. We assume that stationary signals are under consideration. As well, we assume that unwanted intrinsic noises of the microphones and signal acquisition hardware are not correlated between the hardware's channels, unlike the incoming signals. Thus, the unwanted noises contribute in autocorrelation terms of cross-spectral matrix of the microphones' signals [6], at infinite averaging time. In [6], it was proposed to exclude the autocorrelation terms by nullifying of main diagonal of the cross-spectral matrix, to enhance quality of the focalization. The focalization enables suppression of the unwanted signals as well as maintaining signals from the defined position.

The mentioned above focalization approach may be used for obtainment of spectra estimates from particular spatial volume around the target point $\mathbf{r} = (x, y, z)$. Elements $s(f, \mathbf{r})$ of FSE in Pa^2 are obtained as:

$$s(f, \mathbf{r}) = \left| X F X^H \right| \tag{1}$$

where X is row vector:

$$X = \left[\exp\left(-j \ \frac{f d_m}{c} \right) \right]_{m=1 \div M} \tag{2}$$

Here m denotes microphone number, $m = 1 \div M$; f is frequency; d_m is the Euclidean distance from the focalization point to the microphone m; c is sound speed in the air; superscript symbol H denotes conjugate transpose; F is a modified cross-spectral matrix in which the main diagonal is replaced by zeros [6]:

$$F = \left[\frac{S^H S}{M^2} \right] \tag{3}$$

where S is row vector:

$$S = [s_m]_{m=1 \div M} \tag{4}$$

and $s_m(f, \mathbf{r})$ denotes estimated using Fourier transform complex amplitude of the signal measured by the microphone m for frequency f. Equation (3) was used for center axis direction.

Finally the FSE are converted into dB/Pa as follows:

$$s_{dB/Pa}(f, \mathbf{r}) = 10 \log_{10} s(f, \mathbf{r}) \tag{5}$$

4.2 Acoustic Holography

In the lower frequency range focalization can be implemented by using statistically optimized nearfield acoustic holography (SONAH) [9]. In this case elements $s(f, \mathbf{r})$ of FSE at the focalization point $\mathbf{r} = (x, y, z)$ are obtained as sound pressure values calculated by weighting of sound pressures $p(\mathbf{r}_{h,m})$ measured at all M microphone positions:

$$s(f,\mathbf{r}) = \sum_{m=1}^{M} \mathbf{c}(f,\mathbf{r})p(\mathbf{r}_{h,m}) = \mathbf{p}_h^{\mathrm{T}}\mathbf{c}(f,\mathbf{r}), \qquad (6)$$

where a weighting matrix $\mathbf{c}(\mathbf{r})$ is calculated for each frequency f as a least squares solution that provide optimal projection of elementary waves from the hologram plane to the prediction plane (source surface). After that it also is converted by (5).

Despite comparatively big computational complexity due to calculation of approximate values of integrals in the matrix $\mathbf{c}(\mathbf{r})$, SONAH is more preferable in comparison to similar (by effectiveness) method of holography called equivalent source method (ESM) [16], because it allows calculation of a single pressure value in the single point in the space instead of reconstruction some piece of the acoustic field in front of the source object.

Nevertheless, using of acoustic holography has two main disadvantages: it is limited by upper frequency (usually not more than 2–2.5 kHz) and it requires close placement of the measurement microphone array to the object under investigation that means "hard" focalization by itself.

Since in our experimental set-up (see Fig. 3) the acoustic camera was at about 40 cm distance from the sound source (tested bearing), beamforming approach was used to obtain FSEs. It was implemented in a frequency domain with resolution of 4 Hz that is limited by the Pulse LabShop software. The received acoustic wave front is not flat, at the 0.4 m range, which is comparable with the microphone array diameter. The acoustic wave propagation velocity is assumed equal for all frequencies and for all microphones. The wave attenuation during propagation to center and border of the microphone array is assumed equal. The introduced delays account exact distances between microphones of the array and position of the bearing seat.

The obtained in this way FSE in dB/Pa of all bearings in our collection were considered as "features" describing their condition. Since they are multidimensional data sets, application of any kind of classification procedure needs data dimensionality reduction or extraction of a low dimensional vector of features. In present study we used recently developed intelligent feature extraction approach based on Echo state networks (ESN), described in the next section.

5 Feature Extraction Approach

The feature extraction approach developed recently in [13] is shown on Fig. 4.

Here ESN is a special kind of recurrent neural network [14] consisting of randomly generated dynamic reservoir of non-linear neurons which current state depends on the current network input as well as on their previous state as follows:

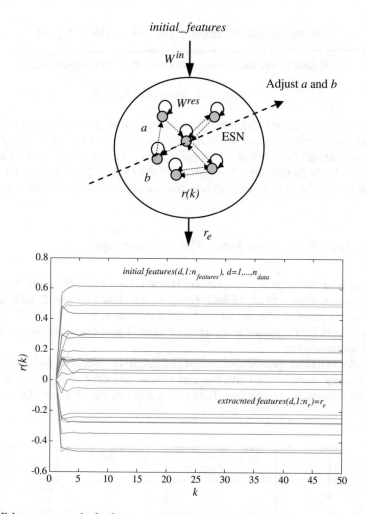

Fig. 4 Echo state networks for features extraction

$$r(k) = \tanh\left(diag(a)W^{in}initial\_features + diag(a)W^{res}r(k-1) + b\right) \quad (7)$$

Here W^{in} and W^{res} are $n_r \times n_{features}$ and $n_r \times n_r$ matrices that are randomly generated and are not trainable; n_r is number of neurons in the reservoir; $n_{features}$ is the dimension of the input vector (in this case it is vector of the initial features of data set); $r(k)$ is current reservoir state; the vectors a and b contain the tunable parameters of the ESN that are adjusted to the data structure using IP training algorithm [17].

The reservoir equilibrium r_e is iteratively calculated using Eq. (7) until the reservoir state become steady as follows:

$$r_e = \tanh\left(diag(a)\,W^{in}\,initial\_features + diag(a)\,W^{res}r_e + b\right) \qquad (8)$$

This equilibrium forms the extracted features from the original once:

$$extracted\_features = r_e \qquad (9)$$

In [13] it was shown that generation of variety of 2D projections (using all possible combinations of two extracted features) allows us to "see" the multidimensional data set from different viewpoints. Thus it is possible to find a view that discovers clearly its structure. The features in the chosen 2D projection are further used to train a classifier of or to cluster the original multi-dimensional data set.

6 Feature Extraction from FSE of Bearings

Obtained after raw signals preprocessing features, i.e. FSE extracted from acoustic camera measurements of all 10 SKF bearings, are shown on Fig. 5. Although we were able to generate FSE for wider frequency domain, it was observed that FSE differ manly in the frequency range up to 12 kHz. Hence, this range was considered further in our work.

In order to extract specific characteristics from these spectra we decided to divide the frequency range into 50 intervals and then to select the maxima and its corresponding frequency for each one. Thus totally 100 "specific features" for each FSE

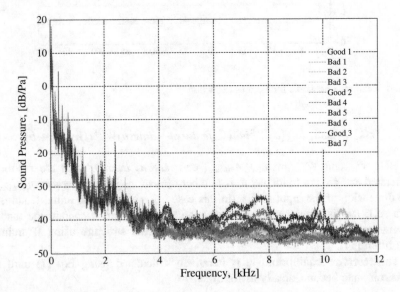

Fig. 5 Focalized spectra of the 10 SKF bearings in investigated collection

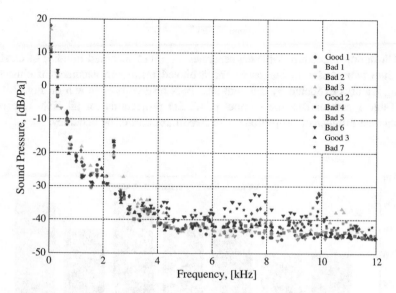

Fig. 6 Features obtained from focalized spectra

were defined. They are shown on Fig. 6. Each "dot" represents a couple of features (maximum, frequency) and different colors denote different bearings as in previous Fig. 5. The extracted in this way "specific features" of FSE compose multidimensional data set of initial features used further in the described above procedure for features extraction and classification.

Healthy bearings are denoted as "Good" while the worn-out are denoted as "Bad" type respectively on both figures.

We tuned several ESNs with different reservoir sizes varying from 10 up to 500 neurons. Thus, the original features were projected onto smaller as well as onto bigger feature spaces.

Two types of classifiers were tested—unsupervised k-means clustering algorithm and supervised trained Support vector machines (SVM) with radial basis kernels. In both cases we divided the data into two classes—"Good" and "Bad". The results of clustering and classification are summarized further.

7 Results and Discussion

The tuned ESN reservoirs with sizes varying from 10 up to 500 neurons produce the corresponding to the reservoir size number of extracted features. Thus, the number of possible 2D projections depends on the reservoir size and increases rapidly with it, i.e.:

$$n_{proj} = n_r(n_r - 1)/2 \tag{10}$$

Clustering procedure k-means separates data into specified number of clusters, i.e. into two classes in our case. The achieved maximum accuracy of clustering using features extracted by different size ESN reservoirs is shown on Fig. 7.

Table 1 summarizes the number of all 2D projections for all ESN reservoirs versus the number of projections with maximum clustering accuracy.

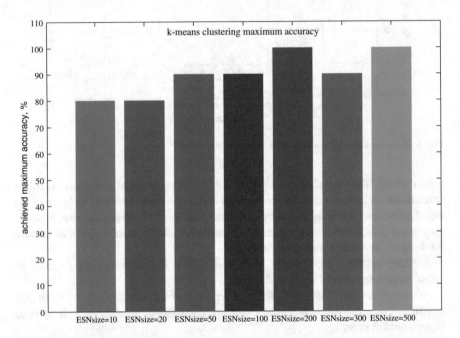

Fig. 7 Maximum achieved accuracy by k-means clustering procedure

Table 1 Number of all 2D projections versus number of projections yielding maximum accuracy

ESN size n_r	Total number of 2D projections	Number of best projections
10	45	2
20	190	3
50	1225	29
100	4950	6
200	19,900	23
300	44,850	211
500	124,750	9

Table 2 Accuracy of k-means clustering using initial features and all ESNextracted features	ESN size	Accuracy (%) using all ESN extracted features	Accuracy (%) using initial features
	10	60	30
	20	50	30
	50	40	30
	100	40	30
	200	40	30
	300	50	30
	500	40	30

From Fig. 7, it becomes clear that increasing of the total number of projections increases clustering accuracy. The Table 1 shows that the number of the best projections is random however—a fact that can be explained by the random generation of the initial ESN reservoir connections.

The second observation from the results shows that projection to a bigger feature space than the original one allows better data clustering in comparison to the clustering results using original features set. Table 2 compares clustering accuracy achieved by k-means using initial features and all features extracted by ESN reservoirs.

Comparison with Fig. 7 shows that the achieved accuracy in both cases is much lower than accuracy achieved using 2D features sub-spaces. Besides, the accuracy of clustering using initial features is the worse one.

Next we used extracted features to train SVM classifiers in supervised manner. Since we have only 10 samples, 3 from class one ("Good") and 7 from class two ("Bad"), we apply leave-one-out cross validation to test accuracy of the trained classifiers. The maximum achieved mean correct rate in percents from all SVM classifiers and all reservoir sizes are compared on Fig. 8.

We observe that the reservoir that produces extracted features vector with the same size as the initial features vector yielded the best mean accuracy.

The achieved mean classification accuracy of trained SVMs using initial features as well as all ESN extracted features was about 70 %—a strange fact that initially puzzled us. Investigation of the properties of the trained SVMs showed that their support vectors match the data class with prevailing number of samples. Since our data base consisted of only 10 samples, only 3 of which were classified by experts as "Bad" and the rest of the bearings (70 %) was classified as "Good", all trained SVM classifiers failed to predict correctly and always classified the test sample into this prevailing class of "Good" bearings. In contrast, the trained SVMs using only two out of all extracted features allowed finding a 2-dimensional projection that yields increased classification accuracy up to 90 % (case of ESN with 100 neurons in the reservoir, shown on Fig. 8). This fact revealed the advantage of our feature extraction procedure that maps the data into a new feature space thus overcoming imbalanced number of samples from different classes in the collection of data set.

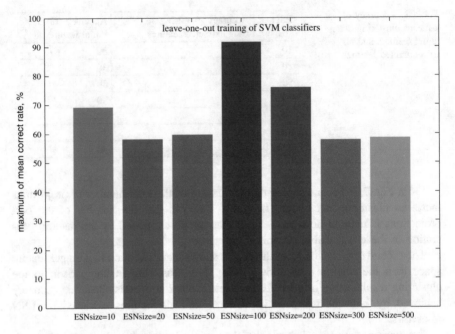

Fig. 8 Maximum achieved mean accuracy of SVM trained using "leave-one-out" procedure

8 Conclusions

The presented here application of our newly developed approach for smart feature extraction demonstrated once again its ability to map the initial features of high-dimensional data sets onto a new features space that reveals better the complicated data structure thus increasing the quality of data clustering and classification.

A new theoretical outcome of the present study is experimental proof that our approach can also help to overcome imbalanced number of samples from different classes in collection data set as it happened in the case of trained SVM classifiers. Extension of the classification task to higher dimensions of features spaces (as in [2]) will be our next step towards further refinement and testing of our DR approach.

A practical outcome of the presented work is the developed approach for a particular task—feature extraction from the FSE of acoustic camera measurements aimed at distance diagnostics of bearings. The directions for future work will be towards improvement of the trained classifiers accuracy by accumulation of bigger collection of data as well as towards training of classifiers able to distinguish intermediate classes of bearings between "Bad" and "Good" that will support early prediction of bearings quality deterioration during their exploitation life.

Acknowledgments This work was partly supported by the project AComIn, grant 316087, funded by the FP7 Capacity Program (Research Potential of Convergence Regions).

References

1. Assent, I.: Clustering high dimensional data. Wiley Interdisc. Rev. Data Min. Knowl. Discovery **2**(4), 340–350 (2012)
2. Bozhkov, L., Koprinkova-Hristova, P., Georgieva, P.: Learning to Decode Human Emotions with Echo State Networks, Neural Networks, In Press, Corrected Proof. doi:10.1016/j.neunet. 2015.07.005. Accessed 7 Sept 2015
. Buscarello, R.: Practical Solutions to Machinery and Maintenance Vibration Problems. http://www.update-intl.com/VibrationBook8g.htm (2002)
4. Bunte, K., Lee, J.A.: Unsupervised dimensionality reduction: the challenges of big data visualization. In: ESANN 2015 proceedings, European Symposium on Artificial Neural Networks, Computational Intelligence and Machine Learning. Bruges (Belgium), 22–24 April 2015, i6doc.com publ. ISBN 978-287587014-8. http://www.i6doc.com/en/
5. Cai, T., Shen, X. (eds.): High-Dimensional Data Analysis, Frontiers of Statistics, vol. 2. World Scientific Publishing, Singapore (2010)
6. Christensen, J.J., Hald, J.: Beamforming. Brüel Kjær Tech. Rev. **1**, 1–48 (2004)
7. Dempsey, P.J., Kreider, G., Fichter, T.: Tapered Roller Bearing Damage Detection Using Decision Fusion Analysis, NASA/TM—2006-214380, NASA (2006)
8. Graney, B.P., Starry, K.: Rolling element bearing analysis. Mater. Eval. **70**(1), 78–85 (The American Society for Nondestructive Testing Inc.) (2011)
9. Hald, J.: Patch near-field acoustical holography using a new statistically optimal method. Brüel Kjær Tech. Rev. **1**, 40–50 (2005)
10. Halme J., Andersson, P.: Rolling contact fatigue and wear fundamentals for rolling bearing diagnostics—state of the art. In: JET656 Proceedings IMechE, Part J: Journal Engineering Tribology, vol. 224 (2009)
11. James, C.R., Berry, J.E.: Description of PEAKVUE® and illustration of its wide array of applications in fault detection and problem severity assessment. In: Emerson Process Management Reliability Conference (2001)
12. Koprinkova-Hristova, P.: Multi-dimensional data clustering and visualization via Echo state networks. In: Kountchev, R., Nakamatsu, K. (eds.) Intelligent Systems Reference Library, New Approaches in Intelligent Control and Image Analysis: Techniques, Methodologies and Applications. Springer (2016)
13. Koprinkova-Hristova, P., Tontchev, N.: Echo state networks for multi-dimensional data clustering. Lecture Notes in Computer Science (including subseries Lecture Notes in Arti-ficial Intelligence and Lecture Notes in Bioinformatics), vol. 7552 (PART 1), pp. 571–578 (2012)
14. Lukosevicius, M., Jaeger, H.: Reservoir computing approaches to recurrent neural network training. Comput. Sci. Rev. **3**, 127–149 (2009)
15. Michael J.R., Kacpryznski, G.J., Orsagh, R.F.: Advanced Vibration Analysis to Support Prognosis of Rotating Machinery Components. Impact Technologies, New York
16. Sarkissian, A.: Method of superposition applied to patch near-field acoustic holography. J. Acoust. Soc. Am. **118**(2), 671–678 (2005)
17. Schrauwen, B., Wandermann, M., Verstraeten, D., Steil, J.J., Stroobandt, D.: Improving reservoirs using intrinsic plasticity. Neurocomputing **71**, 1159–1171 (2008)

Multistatic Reception of Non-stationary Random Wiener Acoustic Signal

Volodymyr Kudriashov

Abstract The chapter presents a detection rule for multistatic reception of the non-stationary random Wiener acoustic signal. The rule is obtained using the maximum likelihood approach. In the chapter, the localization is carried out using time difference of arrival estimates from multistatic receiving positions. It enables simultaneous processing of the frequency bandwidth, which is close to human hearing range. The approach is promising in detection and localization enhancement. The chapter shows an experimental result on the localization of a source with a fractional bandwidth of emission equal to 2. The localization is provided in range—cross range—elevation coordinates. The proposed technique is suitable in acoustic emitters' detection and localization for non-destructive testing. Particular applications concern testing of an aircraft's landing regime and health monitoring of its engines at landing/take off.

Keywords Acoustic signal processing · Ultra wideband antenna · Detection · Localization

1 Introduction

Systems for acoustic noise source identification are suitable to localize sound sources and to characterize them. Such systems are used to estimate acoustic emissions in a wide range of applications from occupational health to machine diagnostics and acoustic emission of aircrafts, and cars. Modern acoustic cameras employ microphone arrays to generate sound pressure images [1–3]. The most frequently, the images are generated in azimuth—elevation coordinates, for the

V. Kudriashov (✉)
Mathematical Methods for Sensor Information Processing Department,
Institute of Information and Communication Technologies,
Bulgarian Academy of Sciences, Acad. G. Bonchev St.,
Block 25A, 1113 Sofia, Bulgaria
e-mail: KudriashovVladimir@Gmail.com

© Springer International Publishing Switzerland 2016   257
S. Margenov et al. (eds.), *Innovative Approaches and Solutions in Advanced*
Intelligent Systems, Studies in Computational Intelligence 648,
DOI 10.1007/978-3-319-32207-0_16

predefined range [1–3]. The cited works do not contain a mathematical description of detection rule of incoming signals. Aircraft flyover noise occupies few octave bands of frequencies (about 9 kHz), in a frequency region below 11.3 kHz [4]. The maximal bandwidth of a transducer array, which estimates a direction of arrival through a beamforming, is lower one octave [5].

The increasing of frequency bandwidth may enhance signal-to-noise ratio and quality of estimates. Multilateration enables localization of emission source using multiple receiver sites and cross-correlation of their signals. The frequency bandwidth of the signals may include few octave bands. Existing systems employ a spatial diversity of several sensors for acoustic noise source localization through triangulation of the acquired data [6, 7]. The systems do not apply multilateration through time difference of arrival measurements using the cross-correlation of incoming signals.

Detection-measurement of non-stationary random Wiener signal sources requires an appropriate detection rule [8, 9]. The rule enables to define the threshold level and the detector block diagram [10–12]. Additionally, it enables localization of ultra wideband acoustic noise sources in cross range, elevation, and range coordinates.

In this chapter, the rules and threshold levels for the detection of the signal against the non-stationary random Wiener interference via the bistatic and the multistatic acoustic systems are proposed. The 4D acoustic image has been generated indoor using the proposed approach, for fractional bandwidth equal to 2.

2 Detection Rule for Bistatic Acoustic System

An acoustic Camera fuses images from a microphone array and an optic camera to assist its user in a noise source detection [1–3]. A priori known type of incoming signal enables to obtain its detection rule.

Emission of an object is considered as a realization of non-stationary random Wiener signal. The desired signal frequency bandwidth is wide [1–3]. Microphones and microphone signal acquisition channels limit it by their bandwidth B. The microphones are significantly spaced (Fig. 1). Baselines denote distances between

Fig. 1 A basic geometry of the multistatic acoustic system

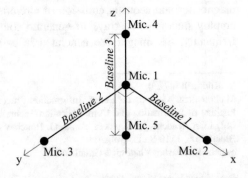

three pairs of microphones, in Fig. 1. Estimated parameter is a time difference of arrival of the signal to microphones. Output signals of the pair of microphone signal acquisition channels are denoted as $y_I(t)$ and $y_{II}(t)$, correspondingly. The signals may contain the incoming signal (condition $A = 1$) or not contain it (condition $A = 0$) [10, 11].

The detection rule is derived for the incoming signal $x(t)$ against mix of non-stationary random Wiener interfering signals $c_I(t)$, $c_{II}(t)$ and intrinsic noises of microphones and microphone signal acquisition channels $n_I(t)$, $n_{II}(t)$. The intrinsic noises' power spectral density is N_0, for B of the equipment. The signal model is denoted as [12, 13]:

$$
\begin{aligned}
y_I(t) &= A x(t - t_x) + n_I(t) + c_I(t - t_c), \\
y_{II}(t) &= A x(t - \tau) + n_{II}(t) + c_{II}(t - \tau), \\
& 0 < t < T,
\end{aligned}
\tag{1}
$$

where $x(t)$, $n_I(t)$, $n_{II}(t)$, $c_I(t)$ and $c_{II}(t)$ are not correlated in pairs; t_x and t_c are time difference of arrival of the incoming signal and interference (industrial noise, multipath propagation of a signal on the scene etc.); τ is an introduced time delay, which compensates the time differences of arrival; and T is the acquisition time. It is assumed that microphones and microphone signal acquisition channels have equal transfer functions and the spectral densities of interfering signals are known. Mean squares of components of (1) are expressed as: $\overline{x^2(t)} = \sigma_s^2$, $\overline{n_I^2(t)} = \overline{n_{II}^2(t)} = \sigma^2$ and $\overline{c_I^2(t)} = \overline{c_{II}^2(t)} = \sigma_c^2$, where superscript character $^-$ denotes statistical averaging.

According to the Wiener process property, the considered $x(t)$, $n_I(t)$, $n_{II}(t)$, $c_I(t)$ and $c_{II}(t)$ have independent increments those obey normal distribution [8, 9]. Sampling of these signals (1) in time is done with a sampling interval $\Delta t \leq (2B)^{-1}$, which is defined by the Nyquist sampling theorem [13]. Variation of increments of a channel output signal is expressed as:

$$
\overline{\left[y_{I,i}(t) - y_{I,(i-1)}(t) \right]^2} = \Delta y_i^2 \Delta t,
\tag{2}
$$

where

$$
\Delta y_i^2 = \left[\sigma_{s,i}^2 - \sigma_{s,(i-1)}^2 \right] + \left[\sigma_i^2 - \sigma_{(i-1)}^2 \right] + \left[\sigma_{c,i}^2 - \sigma_{c,(i-1)}^2 \right] = \Delta\sigma_s^2 + \Delta\sigma^2 + \Delta\sigma_c^2;
$$

underline indexes i and $(i - 1)$ denote sequential time instances of the samples [13]. The exact time interval, which enables to obtain the normal distribution of the increments, may be obtained by further experimental investigations. Variant of obtainment of Δy_i values is given in Fig. 2.

Fig. 2 Block diagram for Δy_i
obtainment

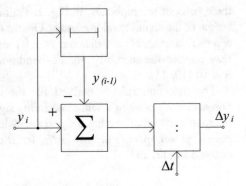

For the case of incoming signal absence ($A = 0$ or $\sigma_s^2 = 0$), variation of the increments equals:

$$\overline{\Delta y_{I,i}^2(t)} = \overline{\Delta y_{II,i}^2(t)} = \Delta\sigma^2 + \Delta\sigma_c^2 = \Delta\sigma^2(1 - \rho_n)^{-1}, \tag{3}$$

where $\rho_n = \Delta\sigma_c^2 \left(\Delta\sigma_c^2 + \Delta\sigma^2\right)^{-1}$ is the correlation coefficient of the input signals. For the case $A = 1$:

$$\overline{\Delta y_{I,i}^2(t)} = \overline{\Delta y_{II,i}^2(t)} = \Delta\sigma_s^2 + \Delta\sigma^2 + \Delta\sigma_c^2 = \Delta\sigma^2 (1 - \rho)^{-1}, \tag{4}$$

where $\rho = \left(\Delta\sigma_s^2 + \Delta\sigma_c^2\right)\left(\Delta\sigma_s^2 + \Delta\sigma_c^2 + \Delta\sigma^2\right)^{-1}$ is the correlation coefficient. Elements of a row $\overrightarrow{Y}$ are the noted above increments $\Delta y_{I,i}$ and $\Delta y_{II,i}$.

Probability densities of the $\overrightarrow{Y}$ have to be obtained for the two cases: $A = 1$ and $A = 0$, in order to obtain likelihood ratio $L\left(\overrightarrow{Y}\right)$ and the detection rule. For case $A = 1$, the incoming signal is correlated, as well as the interference. Joint probability density of corresponding samples $\Delta y_{I,i}$ and $\Delta y_{II,i}$ obeys two-dimensional distribution function of two normally distributed random variables [9]. The corresponding probability density function is obtained using (2)–(4) and equality: $p\left(\overrightarrow{Y}/A = 1\right) = \prod\limits_{i=1}^{k} p\left(\Delta y_{I,i}, \Delta y_{II,i}/A = 1\right)$, where $k = 2BT$. The function is expressed as:

$$p\left(\overrightarrow{Y}/A = 1\right) = \left(\frac{\sqrt{1-\rho}}{2\pi\Delta\sigma^2\sqrt{1+\rho}}\right)^k \exp\left[-\frac{\sum\limits_{i=1}^{k}\left(\Delta y_{I,i}^2 + \Delta y_{II,i}^2 - 2\rho\,\Delta y_{I,i}\,\Delta y_{II,i}\right)}{2\Delta\sigma^2(1+\rho)}\right]. \tag{5}$$

For the case $A = 0$, the probability density function obtainment is similar. At this case, no elements of the $\overrightarrow{Y}$ are correlated, except the interference. An expression for $p\left(\overrightarrow{Y}\big/A = 0\right)$ is similar to (5), in which ρ is replaced by ρ_n. The likelihood ratio $L\left(\overrightarrow{Y}\right)$ is obtained as relation of $p\left(\overrightarrow{Y}\big/A = 1\right)$ to $p\left(\overrightarrow{Y}\big/A = 0\right)$ as:

$$
L\left(\overrightarrow{Y}\right) = \left[\frac{(1-\rho)(1+\rho_n)}{(1-\rho_n)(1+\rho)}\right]^{\frac{k}{2}} \exp\left\{\frac{1}{2\,\Delta\sigma^2}\left[\frac{\sum\limits_{i=1}^{k}\left(\Delta y_{I,i}^2 + \Delta y_{II,i}^2 - 2\,\rho_n\,\Delta y_{I,i}\,\Delta y_{II,i}\right)}{1+\rho_n}\right.\right.
$$

$$
\left.\left. - \frac{\sum\limits_{i=1}^{k}\left(\Delta y_{I,i}^2 + \Delta y_{II,i}^2 - 2\,\rho\,\Delta y_{I,i}\,\Delta y_{II,i}\right)}{1+\rho}\right]\right\}.
$$

$$(6)$$

For technical implementation, the natural logarithm of the obtained $L\left(\overrightarrow{Y}\right)$ is more appropriate. It is assumed, that $\sigma^2 = N_0\,(2\,\Delta t)^{-1}$ and $\Delta t = (2\,B)^{-1}$. Above assumptions and replacement of summation by integration enables to obtain logarithm of (6) as:

$$
\ln L[\Delta y_I(t), \Delta y_{II}(t)] = BT\,\ln\left[\frac{(1-\rho)(1+\rho_n)}{(1-\rho_n)(1+\rho)}\right]
$$

$$
+ N_0^{-1} \int\limits_0^T \left[\frac{\Delta y_I^2(t) + \Delta y_{II}^2(t) - 2\,\rho_n\Delta y_I(t)\Delta y_{II}(t)}{1+\rho_n}\right.
$$

$$
\left. - \frac{\Delta y_I^2(t) + \Delta y_{II}^2(t) - 2\,\rho_n\Delta y_I(t)\Delta y_{II}(t)}{1+\rho}\right]dt.
$$

$$(7)$$

For considered practical conditions, the variances of increments of the noise and the interference are larger than the variance of increments of the signal: $\Delta\sigma_s^2 \ll \Delta\sigma_c^2 + \Delta\sigma^2$. Thus, the first of summands of (7) does not depend on the incoming signal. The summand defines a threshold level $P_{Thr.}$ of the bistatic system. The correlation coefficient use to be $|\rho| \to 0$, thus (7) may be replaced by:

$$
\ln L[\Delta y_I(t), \Delta y_{II}(t)] \approx BT\,\ln\left(\frac{1+\rho_n}{1-\rho_n}\right) - \frac{\rho_n}{N_0(1+\rho_n)}\int\limits_0^T [\Delta y_I(t) + \Delta y_{II}(t)]^2 dt, \quad (8)
$$

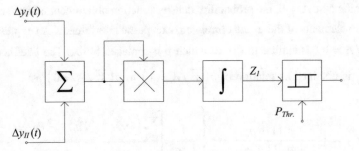

Fig. 3 Block diagram for signal processing in the bistatic system

where $\rho_n \left[N_0 \left(1 + \rho_n \right) \right]^{-1}$ defines the weight of the integration. Thus, optimal signal processing in the bistatic system is defined by the integral Z_1 (Fig. 3) as:

$$Z_1 = \int\limits_0^T \left[\Delta y_I(t) + \Delta y_{II}(t) \right]^2 dt$$

$$= \int\limits_0^T \Delta y_I^2(t)\, dt + \int\limits_0^T \Delta y_{II}^2(t)\, dt + 2 \int\limits_0^T \Delta y_I(t)\, \Delta y_{II}(t) dt. \tag{9}$$

The obtained detection rule estimates autocorrelation functions of increments (Fig. 2) of signals (1) and their cross-correlation function.

Only the last summand of (9) depends on time difference of arrival of the incoming signal. Thus, the detection rule envisages calculating following expression:

$$Z_1 \approx \int\limits_0^T k_I \Delta y_I(t)\, k_{II} \Delta y_{II}(t) dt, \tag{10}$$

where $k_{I,\,II}$ define gain values for channels 1 and 2, correspondingly; $\Delta y_{I,\,II}(t)$ define increments of the signals (1).

The obtained rule (10) concerns detection of the non-stationary random Wiener signal with a bistatic system (Fig. 1, Microphones 1 and 2).

The cross-correlation technique allows processing of signals with wide enough frequency bandwidth. It is useful for localization of sources of ultra-wideband acoustic noise sources, as far as the noise cannot be processed by beamforming-based techniques simultaneously, i.e. without splitting its passband bandwidth to subbands [5]. The resolution on time difference of arrival is in direct ratio to the frequency bandwidth of the acquired signal [12]. Thus, the achieved spatial resolution may be much better than those received by a microphone array, operating in a beamforming regime. Additionally, the spatial resolution is independent of center frequency of the acquired signal. It may be applied for testing of landing regime of an aircraft at far ranges.

3 Detection Rule for Multistatic Acoustic System

Localization of the sound emitter requires estimating of time difference of arrival from multiple bistatic reception systems (Fig. 1). The detection rules for other bistatic reception systems (Fig. 1) may be expressed similarly to (10). The detection rule for the multistatic acoustic system is obtained in a similar way. Output signals of the bistatic reception systems are denoted as $u_1(t)$, $u_2(t)$ and $u_3(t)$, correspondingly. These signals are expressed as:

$$u_{1,2,3}(t) = As_{1,2,3}\left(t - t_{1,2,3}\right) + n_{1,2,3}(t), \tag{11}$$

where $s_{1,2,3}(t)$ and $n_{1,2,3}(t)$ are stationary random processes that denote signals and noises in corresponding bistatic reception systems 1–3 (Fig. 1), correspondingly; $t_{1,2,3}$ denote mutual delays of the signal for these bistatic reception systems. These signals and noises are not correlated in pairs.

Variances of these signals and noises (11) are denoted as:

$$
\begin{aligned}
\overline{s_1^2(t)} &= \sigma_{s1}^2, \\
\overline{s_2^2(t)} &= \sigma_{s2}^2, \\
\overline{s_3^2(t)} &= \sigma_{s3}^2, \\
\overline{n_1^2(t)} &= \overline{n_2^2(t)} = \overline{n_3^2(t)} = \sigma^2.
\end{aligned} \tag{12}
$$

For the case $A = 0$: $\overline{u_1^2(t)} = \overline{u_2^2(t)} = \overline{u_3^2(t)} = \sigma^2$. For the case $A = 1$:

$$
\begin{aligned}
\overline{u_1^2(t)} &= \sigma_{s1}^2 + \sigma^2 = \frac{\sigma^2}{1 - \rho_{12}}, \\
\overline{u_2^2(t)} &= \sigma_{s2}^2 + \sigma^2 = \frac{\sigma^2}{1 - \rho_{23}}, \\
\overline{u_3^2(t)} &= \sigma_{s3}^2 + \sigma^2 = \frac{\sigma^2}{1 - \rho_{31}},
\end{aligned} \tag{13}
$$

where $\rho_{12} = \sigma_{s1}^2\left(\sigma_{s1}^2 + \sigma^2\right)^{-1}$ denotes correlation coefficient for $u_1(t)$ and $u_2(t)$; $\rho_{23} = \sigma_{s2}^2\left(\sigma_{s2}^2 + \sigma^2\right)^{-1}$ denotes correlation coefficient for $u_2(t)$ and $u_3(t)$; $\rho_{31} = \sigma_{s3}^2\left(\sigma_{s3}^2 + \sigma^2\right)^{-1}$ denotes correlation coefficient for $u_3(t)$ and $u_1(t)$. These correlation coefficients are expressed for the case of presence of the desired signal.

Further detection rule obtainment is similar to one in Sect. 2 of the chapter. New row $\overrightarrow{Y}$ consists of samples of: $u_{1,\,i}(t)$, $u_{2,\,i}(t)$ and $u_{3,\,i}(t)$. At the case $A = 1$, the signal and the interference components of the $\overrightarrow{Y}$ are correlated in pairs. Joint probability density of corresponding samples $u_{1,i}(t)$, $u_{2,i}(t)$ and $u_{3,i}(t)$ obeys distribution function of normally distributed random variables [9]. At this case, the corresponding probability density function is obtained using following equality: $p\left(\overrightarrow{Y}\big/A=1\right)=\prod_{i=1}^{k}p\left(u_{1,i},u_{2,i},u_{3,i}\big/A=1\right)$. The probability density function is expressed as:

$$p\left(\overrightarrow{Y}\big/A=1\right)=\left(\frac{\sqrt{(1-\rho_{12})(1-\rho_{23})(1-\rho_{31})}}{(2\pi)^{\frac{3}{2}}\sigma^3\sqrt{1-\rho_{12}^2-\rho_{23}^2-\rho_{31}^2+2\,\rho_{12}\,\rho_{23}\,\rho_{31}}}\right)^{k}$$

$$\times\exp\left\{-\frac{(2\sigma^2)^{-1}}{(1-\rho_{12}^2-\rho_{23}^2-\rho_{31}^2+2\rho_{12}\rho_{23}\rho_{31})}\sum_{i=1}^{k}\left[u_{1,i}^2\left(1-\rho_{23}^2\right)(1-\rho_{12})\right.\right.$$

$$+u_{2,i}^2\left(1-\rho_{31}^2\right)(1-\rho_{23})+u_{3,i}^2\left(1-\rho_{12}^2\right)(1-\rho_{31})$$

$$-2u_{1,i}u_{2,i}(\rho_{12}-\rho_{23}\rho_{31})\sqrt{1-\rho_{12}}\sqrt{1-\rho_{23}}$$

$$-2u_{2,i}u_{3,i}(\rho_{23}-\rho_{31}\rho_{12})\sqrt{1-\rho_{23}}\sqrt{1-\rho_{31}}$$

$$\left.\left.-2u_{3,i}u_{1,i}(\rho_{31}-\rho_{12}\rho_{23})\sqrt{1-\rho_{31}}\sqrt{1-\rho_{12}}\right]\right\}.$$

$$(14)$$

For the case $A=0$, the samples of $\overrightarrow{Y}$ are independent. Variations of these samples are same. Corresponding probability density is expressed as:

$$p\left(\overrightarrow{Y}\big/A=0\right)=\left((2\pi)^{\frac{3}{2}}\sigma^3\right)^{-k}\exp\left[-\left(2\sigma^2\right)^{-1}\sum_{i=1}^{k}\left(u_{1,i}^2+u_{2,i}^2+u_{3,i}^2\right)\right].\quad(15)$$

Relation of the latter probability density functions is new likelihood ratio $L\left(\overrightarrow{Y}\right)$:

$$L\left(\overrightarrow{Y}\right)=\left[(1-\rho_{12})(1-\rho_{23})(1-\rho_{31})\left(1-\rho_{12}^2-\rho_{23}^2-\rho_{31}^2+2\,\rho_{12}\,\rho_{23}\,\rho_{31}\right)^{-1}\right]^{\frac{k}{2}}$$

$$\times\exp\left\{-\frac{(2\sigma^2)^{-1}}{(1-\rho_{12}^2-\rho_{23}^2-\rho_{31}^2+2\,\rho_{12}\,\rho_{23}\,\rho_{31})}\times\sum_{i=1}^{k}\left[u_{1,i}^2\left(1-\rho_{23}^2\right)(1-\rho_{12})\right.\right.$$

$$+u_{2,i}^2\left(1-\rho_{31}^2\right)(1-\rho_{23})+u_{3,i}^2\left(1-\rho_{12}^2\right)(1-\rho_{31})-2u_{1,i}u_{2,i}(\rho_{12}-\rho_{23}\rho_{31})$$

$$\times\sqrt{1-\rho_{12}}\sqrt{1-\rho_{23}}-2\,u_{2,i}u_{3,i}(\rho_{23}-\rho_{31}\rho_{12})\sqrt{1-\rho_{23}}\sqrt{1-\rho_{31}}$$

$$\left.\left.-2u_{3,i}u_{1,i}(\rho_{31}-\rho_{12}\rho_{23})\sqrt{1-\rho_{31}}\sqrt{1-\rho_{12}}\right]+\sum_{i=1}^{k}\left(u_{1,i}^2+u_{2,i}^2+u_{3,i}^2\right)\right\}.$$

$$(16)$$

For low signal-to-noise-plus-interference ratio, the variances of desired signals are weak ($|\rho_{12}|\approx|\rho_{23}|\approx|\rho_{31}|\to 0$), as for the bistatic acoustic system. Logarithm of (16) is expressed as:

$$\ln L\big[u_{1,i}(t),u_{2,i}(t),u_{3,i}(t)\big] = BT \ln\left[\frac{(1-\rho_{12})(1-\rho_{23})(1-\rho_{31})}{1-\rho_{12}^2-\rho_{23}^2-\rho_{31}^2+2\rho_{12}\rho_{23}\rho_{31}}\right]$$

$$-N_0^{-1}\int\limits_0^T \Big\{[u_1(t)-u_2(t)]^2+[u_2(t)-u_3(t)]^2+[u_3(t)-u_1(t)]^2+2u_{1,i}^2u_{2,i}^2u_{3,i}^2\Big\}dt.$$

$$(17)$$

The first summand of (17) defines the threshold level $P_{Thr.}$. The optimal signal processing in these bistatic reception systems is defined by the integral Z_{123} as:

$$Z_{123}=\int\limits_0^T \Big\{[u_1(t)-u_2(t)]^2+[u_2(t)-u_3(t)]^2+[u_3(t)-u_1(t)]^2+2u_{1,\,i}^2u_{2,\,i}^2u_{3,\,i}^2\Big\}dt$$

$$=2\int\limits_0^T \big[u_1^2(t)+u_2^2(t)+u_3^2(t)\big]dt-2\int\limits_0^T [u_1(t)u_2(t)+u_2(t)u_3(t)+u_3(t)u_1(t)]dt$$

$$+2\int\limits_0^T u_1^2(t)u_2^2(t)u_3^2(t)\;dt.$$

$$(18)$$

The first summand of (18) depends on the sum of power estimates in the considered bistatic reception systems (Fig. 1). All possible cross-correlation functions are subtracted from the last summand of (18) by its second summand. The last summand provides multiplication of power estimates of output signals of the bistatic reception systems. It is agreed to detection quality at a limited number of samples [12]. Squaring of the input signals is valuable for small signal-to-noise-plus-interference ratio at outputs of the bistatic reception systems. Spatial localization of the emission source is utilized by the considered multistatic system (Fig. 1) by the latter summand of (18) as:

$$Z_{123}\approx\int\limits_0^T k_1u_1^2(t)\,k_2u_2^2(t)\,k_2u_3^2(t)dt,\qquad(19)$$

where $k_{1,2,3}$ denote gain values of corresponding bistatic systems 1–3.

The obtained requires to estimate time difference of arrival of the signal by each bistatic reception system and to provide further calculation according to (19), for each node of a spatial grid (Fig. 4).

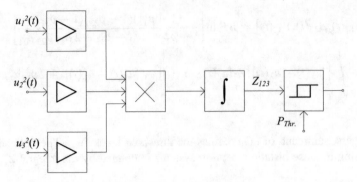

Fig. 4 Block diagram for signal processing in the multistatic system

The non-stationary random Wiener signal detection rule for three bistatic reception systems is obtained according to the maximum likelihood method with respect to the threshold level.

4 Experimental Results

The indoor experiment is focused on localization of acoustic noise source with the multistatic system (Fig. 1). Five microphones (Fig. 1) and acquisition channels of an acoustic camera, manufactured by Brüel & Kjaer are used. The camera uses microphones type 4958, input modules type 3053-B-120 and 3050-B-060 and Pulse LabShop software. The software was used to transfer the acquired signals for further post-processing. The amplitude calibration of the equipment was done in advance.

The localization approach consists of estimates of the cross-correlation functions (10) and their joint processing (19). For each bistatic reception system, the cross-correlation function is obtained along the introduced time delay τ. The known positions of the pair of microphones of the bistatic reception system (Fig. 1) enable assignment each value of the cross-correlation function to its spatial locus, to generate an image. Thus, each spatial node of the image is valued with respect to its slant ranges difference to the pair of microphones. In the experiment, a distance between spatial nodes equals 2 cm, in range, cross range, and elevation. Microphone 1 is placed in the origin of Cartesian coordinate system (Fig. 1). The emitter coordinates are as following: range 1 m, cross range −0.8 m and elevation 0.15 m. The baselines of these bistatic reception systems (Fig. 1) are perpendicular to each other. Each baseline equals to 1 m. Imaging region is from −1.25 to 0.25 m in cross range, from 0 to 1.2 m in range and from −0.15 to 0.45 m in elevation. The center frequency of the incoming signal is about 5 kHz. The passband bandwidth of the signal is 10 kHz.

A propagation of acoustic wave between the emitter and the microphones is assumed non-decaying, through homogeneous media, along rectilinear paths. Meteorological conditions and turbulence-induced coherence losses are not considered, at these short ranges [3].

At the emitter location in experimental setup, the path difference of arrival resolution is in the range 3.4–4.9 cm, for input signal-to-noise ratio about 5 dB to 1 Pa and time-bandwidth product 34 dB (Fig. 5). Along these baselines, the resolution is

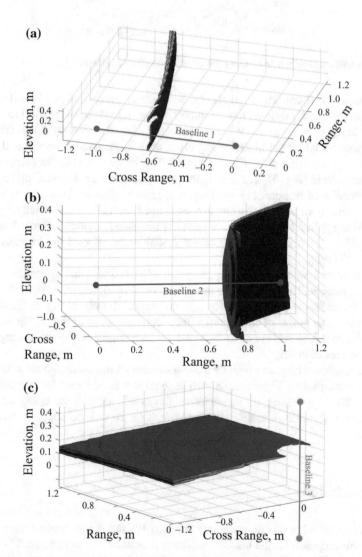

Fig. 5 Acoustic images generated according to the proposed approach using: **a** the first bistatic reception system; **b** the second bistatic reception system; **c** the third bistatic reception system

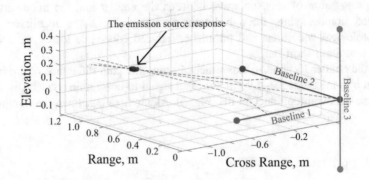

Fig. 6 Acoustic image generated using the using the multistatic system

in the range 1.6–2 cm. The fundamental accuracy of measurement of the time difference of arrival depends on the resolution and on signal-to-noise ratio [13].

The images were generated for three bistatic reception systems (Fig. 1). Microphones 1 and 2 are placed along cross range axis; microphones 4 and 5 are placed along elevation axis (Figs. 1 and 5). The images of (19) are displayed in logarithmic scale (Fig. 5). The multipath propagation inside a typical office room affects equality of responses of the bistatic reception systems. Thus, the generated images were normalized, that is not envisaged in (19). Additionally, −6 dB threshold is applied to the results, to omit the unwanted responses (Figs. 5 and 6). The insufficiently small distance between spatial nodes affected the irregularity of the results (Fig. 5).

The obtained signal processing approach (19) includes multiplication of the generated images. The result of the multiplication (Fig. 6) shows an opportunity to localize the emission source with the obtained approach (19). In Fig. 6, the dashed lines show locus points of maximum values, corresponding to Fig. 5. Dimensions of the obtained response are as follows: 2.1 cm in cross range, 2.9 cm in range and 0.9 cm in elevation (Fig. 6).

False responses originate with increasing number of incoming signals at input of the multistatic system. For example, two false responses occurred for two emission sources. The application of an adaptive threshold, as well as using more of microphones (bistatic reception systems) might be suitable to suppress the false responses.

5 Conclusions

New detection rules of non-stationary random Wiener signal against such interference are proposed for the bistatic and the multistatic acoustic systems. Threshold levels for three bistatic systems are shown. The proposed approach uses time difference of arrival estimates of the incoming signal instead of its phase difference of arrival estimates, to localize its source. It enables increasing of incoming signal passband bandwidth, which is processed simultaneously.

The generated 4D acoustic image enables localization of the test source with the center frequency about 5 kHz and the fractional bandwidth equal to 2. At zero ranges, the mean resolution on path difference of arrival is about 1.8 cm. Maximum dimensions of the obtained response are less than 3 cm, in range—cross range—elevation coordinates.

The approach is suitable for testing of an aircraft landing regime and for monitoring of its engine noise at landing/take off, and for testing of a car engine.

Acknowledgments The research work reported in the chapter was partly supported by the Project AComIn "Advanced Computing for Innovation", grant 316087, funded by the FP7 Capacity Programme (Research Potential of Convergence Regions).

References

1. PULSE Array-based Noise Source Identification Solutions: Beamforming—Type 8608, Acoustic Holography—Type 8607, Spherical Beamforming—Type 8606. In: Brüel & Kjaer Product Data, 12 p. Brüel & Kjær Sound and Vibration Measurement A/S, Naerum (2009)
2. Christensen, J.J., Hald, J.: Beamforming. In: Zaveri, H.K. (ed.) Brüel & Kjær Technical Review, No. 1, pp. 1–48. Brüel & Kjær Sound and Vibration Measurement A/S, Naerum (2004)
3. Hald, J., Ishii, Y., Ishii, T., Oinuma, H., Nagai, K., Yokoyama, Y., Yamamoto, K.: High-resolution Fly-over beamforming using a small practical array. In: Zaveri, H.K. (ed.): Brüel & Kjaer Technical Review, No. 1, pp. 1–28. Brüel & Kjær Sound and Vibration Measurement A/S, Naerum (2012)
4. Khardi, S.: An experimental analysis of frequency emission and noise diagnosis of commercial aircraft on approach. J. Acoust. Emiss. **26**, 290–310 (2008)
5. Li, J., Stoica, P. (eds.) Robust Adaptive Beamforming, 440 pp. Wiley, Hoboken 2006
6. Kondratyev, N.L. (ed.) Zvukometricheskaya stanciya SChZ-6. Rukovodstvo sluzhbi, Moscow: Voennoe Izdat el'stvo Ministerstva oboroni SSR, 151 pp (1958) (in Russian)
7. Fernández Comesaña, D., Jansen, E., Rollán Seco, Y., Elias de Bree, H.: Acoustic multi-mission sensor (AMMS) system for illegal firework localization in an urban environment, 21-st International Congress on Sound and Vibration (ICSV 21), 7 pp (2014)
8. Wentzell, A.D.: Course of the theory of random processes (in Russian). Moscow, Science (1996)
9. Levin, B.R.: Theoretical Basics of Statistical Radio Engineering, vol. 1 (in Russian). Moscow, Soviet Radio (1969)
10. Rozov, A.K.: Detection of Signals in Non-stationary Hydroacoustic Conditions (in Russian). Leningrad, Shipbuilding (1987)
11. Gusev, V.G.: Systems for space-time processing of hydroacoustic information (in Russian). Leningrad, Shipbuilding (1988)
12. Shirman, Y.D., Bagdasaryan, S.T., Malyarenko, A.S., Lekhovytskiy, D.I., Leschenko, S.P., Losev, Y.I., Nikolayev, A.I., Gorshkov, S.A., Moskvitin, S.V., Orlenko, V.M., Shirman, Y.D. (ed.): Radio Electronic Systems: Fundamentals of Theory and Design. Handbook. 2nd edn., revised and expanded (in Russian). Moscow, Radio Engineering (2007)
13. Barton, D.K., Leonov, S.A. (eds.) Radar Technology Encyclopedia (Electronic Edition), 511 pp. Artech House Inc., Boston (1998)

Interpolation of Acoustic Field from Nearby Located Single Source

Iurii Chyrka

Abstract The problem of performance increasing of a small measurement array in the task of single narrowband source localization in the nearfield is considered. Two different approaches for acoustic field interpolation in combination with a conventional acoustic holography are proposed. One of them, based on the autoregressive model of a sinewave, gives better resolution in localization of the source near the center of the array. The second one, based on the estimation of instantaneous phases of signals from each microphone, decreases error of peak localization from sources near the boundaries of the measurement array.

Keywords Nearfield acoustic holography · SONAH · Single source localization · Sound field interpolation · Resolution improvement

1 Introduction

The near-field acoustic holography (NAH) [1] is a technique that deals with a noise source localization problem. Its original idea comes from Fourier acoustics and provide, based on the 2D recording of sound waves, reconstruction of the entire 3D sound field between the source's boundary and the measurement plane [1–4].

Modern NAH approaches [5–9] like the statistically optimized NAH (SONAH), Inverse boundary element method (IBEM) or the equivalent source method (ESM) do not require location of microphones on specific positions and usually works with measurements that are made over a limited surface and by relatively low amount of microphones.

Precision of reconstruction and source localization depends mostly on a microphone aperture (area, microphone number and spacing distance), sound

I. Chyrka (✉)
Mathematical Methods for Sensor Information Processing Department of Institute of Information and Communication Technologies of Bulgarian Academy of Sciences, 25 A, Acad. G. Bonchev str., 1113 Sofia, Bulgaria
e-mail: Yurasyk88@gmail.com

© Springer International Publishing Switzerland 2016
S. Margenov et al. (eds.), *Innovative Approaches and Solutions in Advanced Intelligent Systems*, Studies in Computational Intelligence 648,
DOI 10.1007/978-3-319-32207-0_17

frequency and distance between source and measurement plane. If we talk about technical characteristics of the measurement system, larger aperture or bigger amount of microphones may be disadvantageous from economical point of view. This situation requires search of some alternative ways of numerical extension and data improvement based on the available limited equipment.

Papers [10, 11] are focused on the numerical enlarging the measurement surface tangentially. The work [11] deals with the analytic continuation of a coherent pressure field located over the surface of a vibrator. This analytic continuation is an extrapolation of the measured field into a region outside and tangential to the original finite sheet, and is based on the Green's functions.

When number of available measurement channel is limited, precision of reconstruction and method resolution can be improved by acoustic field interpolation and extrapolation also known as a virtual microphone technique [1]. In contraposition to mentioned works, the present one deals mostly with calculation of the sound field inside the measurement array aperture and adding virtual microphones. In addition, this paper is focused on the specific problem of a single source localization, which allows simplification of an acoustic field model and using proposed approaches for its interpolation.

2 Statistically Optimized Holography

Based on features of the available measurement equipment and a built-in software provided by the manufacturer, research in this work is focused on the SONAH.

SONAH uses the representation of the sound pressure at some point in the space $\mathbf{r} = (x, y, z)$ as a weighted sum of sound pressures $p(\mathbf{r}_{h,n})$ measured at N microphone positions:

$$p(\mathbf{r}) = \sum_{n=1}^{N} \mathbf{c}(\mathbf{r}) p(\mathbf{r}_{h,n}) = \mathbf{p}_h^{\mathrm{T}} \mathbf{c}(\mathbf{r}), \tag{1}$$

where a weighting matrix $\mathbf{c}(\mathbf{r})$ depends only on the point coordinates $\mathbf{r}$. It can be obtained from the requirement that an infinite set of elementary waves

$$\Phi_m(\mathbf{r}) = e^{-j(k_{x,m}x + k_{y,m}y + k_{z,m}z)}, \quad m = \overline{1, M}, \ M \to \infty, \tag{2}$$

where k_x, k_y, k_z are wavenumbers for corresponding directions, must be optimally projected from the hologram plane to the prediction plane. The corresponding least square solution with Tikhonov regularization for transfer matrix $\mathbf{c}(\mathbf{r})$ can be written as

$$\mathbf{c}(\mathbf{r}) = \left(\mathbf{A}^{\mathrm{H}} \mathbf{A} - \theta^2 \mathbf{I} \right)^{-1} \mathbf{A}^{\mathrm{H}} \boldsymbol{\alpha}. \tag{3}$$

$\mathbf{A}^H\mathbf{A}$ is a square matrix that depends on positions of microphones on the hologram plane; $\mathbf{A}^H\boldsymbol{\alpha}$ is a vector matrix that depends on the N positions of microphones and position of the prediction point; $\mathbf{I}$ is an identity matrix; and θ is a regularization parameter. In view of unlimited number M of elementary waves all elements of matrices $\mathbf{A}^H\mathbf{A}$ and $\mathbf{A}^H\boldsymbol{\alpha}$ actually are integrals and must be calculated numerically via Gauss and Gauss-Laguerre quadratures [5].

3 Acoustic Field Interpolation Approaches

Two different approaches of interpolation of acoustic fields, generated by the single point source, are considered in this paper. One of them is based on the sinewave model of the acoustic field and the second one is based on the instantaneous phase measurement of signals from each microphone.

It is well known that any sinewave signal can be represented by an autoregressive model of the second order [12]. After simple trigonometric transformations it can be derived in the next form

$$s_i = \alpha s_{i-1} - s_{i-2}, \quad \alpha = 2\cos(\omega\Delta t). \tag{4}$$

Having signal values at any two points, a time difference between them (usually it is a sampling period) and the signal frequency, we can calculate the signal value at the point inside (interpolation) or outside (extrapolation) the segment of two points. In this way it is possible to obtain an additional signal from a virtual microphone for each pair of real microphones.

The model (4) can be used for signal interpolation only after assumptions about narrowband nature of the signal (that can be achieved by preliminary bandpass filtering of measured signals) and minor difference in the signal power at two real microphones that can be neglected in calculations. The interpolated signal is calculated as

$$s_v = (s_1 + s_2)/\alpha = (s_1 + s_2)/2\cos(\omega\Delta t), \tag{5}$$

where s_1 and s_2 are signal values from the first and second microphones in an arbitrary pair. Time difference and frequency give us a phase growth between two points of signal. The angular frequency is obtained from known signal frequency F as $\omega = 2\pi F$. The time difference Δt for the Eq. (5) can be estimated from known coordinates of real microphones and estimated position of the sound source as

$$\Delta t = \frac{|l_1 - l_2|}{2\lambda F}, \tag{6}$$

where λ is a wavelength, l_1, l_2 are distances from the source to corresponding real microphones. This formula is derived from the next considerations.

Fig. 1 Positioning of the
virtual microphone

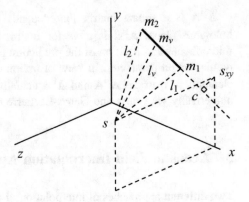

It is necessary to assure the equal phase shift $\omega\Delta t$ of the interpolated signal to two other signals in the Eq. (5). This is achieved by placing the virtual microphone (an interpolation point) in the position where the distance to the signal source is a middle between distances from real microphones to the source. This position is calculated for each pair of microphone. A rough estimation of the source position is taken as an argmax of a reconstructed map from the first stage of processing by a conventional holography. The illustration for calculation of the virtual microphone coordinates is shown in the Fig. 1.

Here m_1, m_2, m_v are positions of the real first, second and virtual microphones respectively; s, s_{xy} is the estimated source point and its projection on the xy (holography) plane; c is the point of cross-section of the line m_1, m_2 with a perpendicular from the point s_{xy}. l_1, l_2, l_v are corresponding distances from the source to microphones.

Usage of the Eq. (5) has one substantial limitation: it requires continuous growth of phase between two real microphones. This condition is fulfilled only if the point c is outside the segment (m_1, m_2). Hence, position of virtual microphones depends on source position and not every microphone pair can be used for interpolation. The example of microphone locations for the boresight source position and distance to object 10 cm is shown in Fig. 2.

The black cross in the center shows the source position. Blue circles shows positions of real microphones and red asterisks represent calculated positions of virtual microphones. One can see that mentioned limitation leads to absence of microphones in the center of the array. In practical cases, this technique can be applied only for sources that are located near the center of the array when generated virtual microphone aperture is almost symmetrical. In other cases it becomes highly asymmetrical and worsens the quality of the reconstructed acoustic field.

The second approach is rather the extrapolation than interpolation, because it allows calculation of the signal value in the any arbitrary point on the measurement plane. In the case of the single source, the acoustic field on some distance around each real microphone can be easily predicted, if position of the source, its frequency, power and instantaneous phase of the signal from microphone is known.

Fig. 2 Positions of virtual microphones for the first interpolation approach

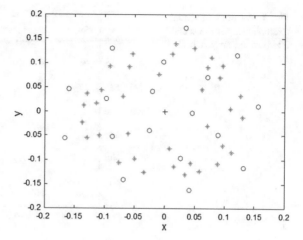

The idea of signal generation at interpolation points is based on the next simplest model of the spherical wave [1]:

$$p(r) = \frac{Ae^{-j(\omega t - kr)}}{r}, \qquad (7)$$

where A is unknown single source amplitude, r is a distance between source and interpolation point, $k = 2\pi/\lambda$ is a wavenumber for the corresponding signal frequency F and wavelength λ in the medium. Because the phase and amplitude of the source are unknown, the interpolated signal should be calculated by the modified equation

$$p(r) = \frac{\hat{A}e^{-j(\hat{\varphi} + \omega \Delta t)}}{r} \qquad (8)$$

where $\hat{A}$, $\hat{\varphi}$ are estimates of signal amplitude and instantaneous phase at the closest microphone, time difference between the microphone and the interpolation point is calculated as $\Delta t = \Delta r/(\lambda F)$, Δr is a difference of distance from source for the microphone and the interpolation point.

As in previous case, it requires rough estimate of source position and additionally estimates of signal amplitude and initial phase at each microphone, that can be obtained from the transformed likelihood equations [13]

$$\begin{cases} \Lambda_x \displaystyle\sum_{i=1}^{N} \sin^2(\omega\tau(i-1)) + \Lambda_y \displaystyle\sum_{i=1}^{N} \sin(\omega\tau(i-1))\cos(\omega\tau(i-1)) = \displaystyle\sum_{i=1}^{N} x_i \sin(\omega\tau(i-1)) \\[2mm] \Lambda_x \displaystyle\sum_{i=1}^{N} \cos(\omega\tau(i-1))\sin(\omega\tau(i-1)) + \Lambda_y \displaystyle\sum_{i=1}^{N} \cos^2(\omega\tau(i-1)) = \displaystyle\sum_{i=1}^{N} x_i \cos(\omega\tau(i-1)) \end{cases},$$

$$\qquad (9)$$

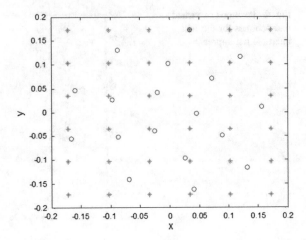

Fig. 3 Positions of virtual microphones for the second interpolation approach

and calculated from the solution as

$$\hat{A} = \sqrt{\Lambda_x^2 + \Lambda_y^2}, \quad \hat{\varphi} = \arctan(\Lambda_y/\Lambda_x), \tag{10}$$

here τ is a sampling period, N is a size of the sample in the time domain.

Thus, such approach allows to rearrange an existing real aperture into an arbitrary shaped virtual one. In this work the rearrangement into the regular virtual array with microphone lattice distance equal half wavelength is considered for a single source localization. In the Fig. 3 the example of virtual microphones positioning for the signal frequency 2 kHz is shown.

4   Simulation Results

The developed software enables simulation of the acoustic pressure data from one or several point sources and obtaining of sound pressure maps, calculated as root-mean-squares (RMS) of pressure at any desired distance and signal frequency. It uses a matrix-form implementation of the SONAH algorithm with proposed modifications.

Signals were simulated for a Brüel & Kjær (Sound and Vibration Measurement A/S) acoustic camera that will be described more detailed in the next section.

The signal measured by each microphone was simulated using (7) and additionally polluted by an additive white Gaussian noise with SNR = 20 dB.

The carried out research has shown that this interpolation technique allows increasing of SONAH localization resolution of sources with frequency bigger than 1.5 kHz. Examples of acoustic maps for frequency 2.3 kHz are shown in the Fig. 4.

It is well visible, that the proposed approach allows up to two times improvement in the width of the peak in comparison to conventional SONAH. Research

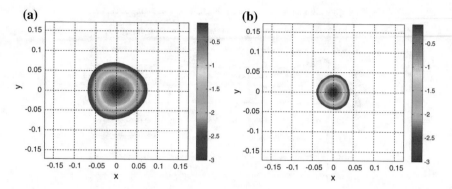

Fig. 4 The reconstructed acoustic map of the point source of a single-tone signal on the distance 10 cm: center frequency 2.3 kHz without interpolation (**a**) and with interpolation (**b**)

results obtained by extrapolation in the similar manner have not shown additional gain in resolution due to bigger distances between real and virtual microphones.

For the second type of interpolation the preliminary research has shown that the most preferable and optimal by computational load is a regular geometry of virtual microphones with spacing distance between them equals half of a wavelength.

The research results have shown that this interpolation technique increases precision of source localization by the SONAH. Examples of reconstructed acoustic maps for frequency 2.5 kHz are shown in the Fig. 5.

As one can see, ordinary SONAH can give a significant shift of the pressure peak, when source is located far from the center of the array. The proposed technique gives much better precision of the source localization, even with a coarse initial estimate of the source position.

Additionally there where done statistical simulations by the Monte-Carlo method for finding resolution improvement on different frequencies for the first type of

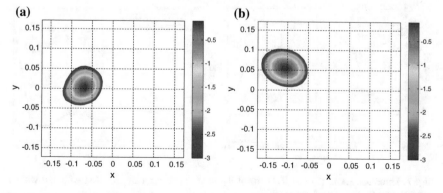

Fig. 5 The reconstructed acoustic map of the point source of the 2.5 kHz signal located in the point (−0.1, 0.05) on the distance 10 cm: without interpolation (**a**) and with interpolation (**b**)

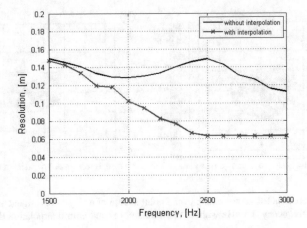

Fig. 6 Plots of estimated resolution values for different frequencies

interpolation. In the Fig. 6 the plots of resolution values after 1000 identical simulations for a single source measured as a width of a main peak on the −3 dB level and distance to source 10 cm.

The resolution is improved up to two times at higher frequencies. The gain of using the interpolation is the biggest at 2.5 kHz and decreases when approach 1.5 kHz.

Similar simulation where done for the second type of interpolation but for a fixed frequency 2 kHz for different frequencies when source was located randomly on the radius 10 cm from the center of the array an on the distance 10 cm to it. In the Fig. 7 plots of error mean before and after using interpolation are shown.

The improvement in error mean is achieved after 1.4 kHz. As a disadvantage of the proposed approach, one can consider a slight increasing of the error variation.

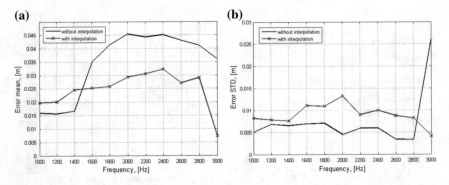

Fig. 7 Estimated mean (**a**) and STD (**b**) of the localization error by SONAH with and without interpolation by the second approach

5 Experimental Results

The Brüel & Kjær acoustic camera has technical characteristics, described in the specification [14]. It consists of three main blocks: a microphone array with an optic camera, a front-end with input modules and a laptop with acquisition and processing software. A non-uniform slice wheel microphone array type WA-1558-W-021 includes 18 microphones type 4958 and has physical extent with radius about 17 cm.

Simulation results were verified in indoor conditions with presence of some reflections from variety of sound reflecting objects like room floor, ceiling, walls, office furniture and equipment. All experiments were done with a small loudspeaker. The narrowband signal central frequencies has been changed from 2 to 3 kHz. The captured signal was additionally pre-filtered by the bandpass filter with bandwidth equals 10 % of the central frequency. The input signal power is chosen large enough in comparison with ambient noise power (SNR is approximately 20 dB). The acquisition time in all situations was 250 ms, but only a short sample of data (equal three periods of the signal) was processed. Experimental results were obtained for the plane located on the actual distance to the source; the reconstruction area is limited to ±0.17 m from the center of the array in horizontal and vertical directions. Signal parameters and source position has been set as during simulations.

Examples of the reconstructed fields for the first interpolation approach are presented in Fig. 8. One can see, that the pressure peak is slightly unsymmetrical due to impact of external factors, but generally it shows similar behavior as in simulations.

The results for the second approach are presented at Fig. 9.

It is well visible that improvement in a real situation with noise and interferences is not as big as in a simulation. Experimental results for other frequencies show similar behavior and performance of the algorithms.

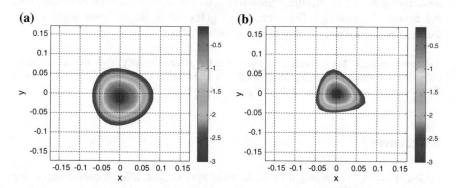

Fig. 8 The reconstructed acoustic map of the point source of a single-tone signal on the distance 10 cm: center frequency 2.5 kHz without interpolation (**a**) and with interpolation (**b**)

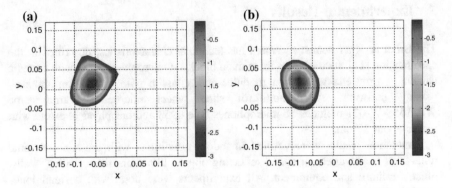

Fig. 9 The reconstructed acoustic map of the point source of a single-tone signal on the distance 10 cm in position (−0.097, 0.026): center frequency 2.5 kHz without interpolation (**a**) and with interpolation (**b**)

Taking into account above presented results, one can conclude that both mentioned methods can be used in specific cases for improvement of SONAH at high frequencies, but they can not increase the top limit for holography. They can be applied automatically after some detection and localization steps that show whether conditions are suitable or not for particular method. Such processing procedure can be used, as example, for suppression of one or several weaker sources near the main one, but this opportunity must be additionally investigated.

6 Conclusions

Two resolution improvement algorithms for SONAH were proposed. It was proved that sound field interpolation algorithm provides improvement in localization precision in the high frequency range for frequencies upper than 1.5 kHz. The first algorithm gives up to two times improvement of resolution for sources located near the center of the array. The second one gives up to 30 % lower mean of the localization error with slightly bigger variance of it.

Acknowledgments The research work reported in the paper was partly supported by the Project AComIn "Advanced Computing for Innovation", grant 316087, funded by the FP7 Capacity Programme (Research Potential of Convergence Regions).

References

1. Bai, R.M., Ih, J.G., Benesty, J.: Acoustic Array Systems. John Wiley & Sons Singapore Pte. Ltd., Singapore (2013)
2. Graham, T.S.: Long-wavelength acoustic holography. J. Acoust. Soc. Amer. **46**(1A), 116 (1969)

3. Maynard, J.D., Williams, E.G., Lee, Y.: Nearfield acoustic holography: I. Theory of generalized holography and the development of NAH. J. Acoust. Soc. Amer. **78**(4), 1395–1413 (1985)
4. Williams, E.G.: Fourier Acoustics. Academic Press, 1999
5. Wu, S.F.: Methods for reconstructing acoustic quantities based on acoustic pressure measurements. J. Acoust. Soc. Amer. **124**(5), 2680–2697 (2008)
6. Hald, J.: Patch near-field acoustical holography using a new statistically optimal method. Brüel & Kjær Tech. Rev. (1), 40–50 (2005)
7. Wang, Z., Wu, S.F.: Helmholtz equation–least-squares method for reconstruction of the acoustic pressure field. J. Acoust. Soc. Amer. **102**(4), 2020–2032 (1997)
8. Sarkissian, A.: Method of superposition applied to patch near-field acoustic holography. J. Acoust. Soc. Amer. **118**(2), 671–678 (2005)
9. Bi, C., Chen, X., Chen, J., Zhou, R.: Near-field acoustic holography based on the equivalent source method. Sci. China Ser. E Eng. Mater. Sci. **48**(3), 338–353 (2005)
10. Sarkissian, A.: Extension of measurement surface in near-field acoustic holography. J. Acoust. Soc. Amer. **115**(4), 1593–1596 (2004)
11. Williams, E.G.: Continuation of acoustic near-fields. J. Acoust. Soc. Amer. **113**(3), 1273–1281 (2003)
12. Kuc, R., Li, H.: Reduced-order autoregressive modeling for center-frequency estimation. Ultrason. Imaging (7), 244–251 (1985)
13. Prokopenko, I.G., Omelchuk, I.P., Chyrka, Y.D.: Radar signal parameters estimation in the MTD tasks. Int. J. Electron. Telecommun. (JET) **58**(2), 159–164 (2012)
14. Web-based information. http://www.bksv.com/

A New Method for Real-Time Lattice Rescoring in Speech Recognition

Petar Mitankin and Stoyan Mihov

Abstract We introduce a novel efficient method, which improves the performance of speech recognition systems by providing the option to partially compile the word lattice into a deterministic finite-state automaton, making it suitable for the rescoring step in the speech recognition process. In contrast to the widely used n-best method our method permits the consideration of significantly larger number of alternatives within the same time-constraint and thus provides better recognition results. In this paper we present a description of the new method and empirical evaluation of its performance in comparison with the n-best method. The achieved WER reduction is up to 3.77 % at a p-value below 3 %. An important advantage of our method is its applicability for real-time applications.

1 Introduction

Most speech recognition systems [1] use a rescoring method. This consists of first using a simple acoustic and grammar model to produce a word lattice, and then to reevaluate these alternative hypotheses with a more sophisticated model on a subset of the lattice. The consideration of hypotheses other than one corresponding to the best path increases the chances of finding the correct transcription. Often, some information source or model not used for building the lattice, such as a more precise grammar or language model, is used to rescore the best hypotheses and determine the best candidate.

P. Mitankin (✉) · S. Mihov
Institute of Information and Communication Technologies,
Bulgarian Academy of Sciences, Sofia, Bulgaria
e-mail: petar@lml.bas.bg

S. Mihov
e-mail: stoyang@lml.bas.bg

P. Mitankin
Faculty of Mathematics and Informatics, Sofia University, Sofia, Bulgaria

© Springer International Publishing Switzerland 2016
S. Margenov et al. (eds.), *Innovative Approaches and Solutions in Advanced
Intelligent Systems*, Studies in Computational Intelligence 648,
DOI 10.1007/978-3-319-32207-0_18

Currently, most systems select the n-best paths from the word lattice [2], which are then subject to rescoring using the more sophisticated models. Since the lattice contains significantly more alternative paths, there is the chance that the correct transcription will not be included in the n-best selection. Thus, if we could increase the number of hypotheses without sacrificing processing time we eventually improve the recognition performance. An alternative approach, on-the-fly rescoring, is to rescore the lattice while it is being constructed, [3, 4].

Here we present a method, which could be considered as a natural development of the method presented in [2] for efficiently generating the n-best hypotheses. We apply essentially a similar procedure, but instead of outputting the n-best sentences, we produce within the same time limit a deterministic finite-state automaton, which represents together with the n-best candidates also a significantly larger number of hypotheses. Since the result is in the form of a deterministic automaton, the rescoring of the result by another n-gram language model will not require additional time for finding the best candidate after rescoring.

In the framework of the AComIn project[1] we have implemented our new method for a Large vocabulary continuous speech recognition system (LVCSR), for Bulgarian [5]. The evaluation of the method has been performed over a newly compiled Bulgarian speech corpus using the AComIn equipment.

In the next section we present the formal preliminaries. Afterwards in Sect. 3 we proceed with the method description. Section 4 presents the empirical evaluations. In the final section some concluding remarks are given.

2   Formal Preliminaries

A *string-to-weight finite-state transducer* is a 6-tuple $\mathcal{A} = \langle \Sigma, Q, E, i, F, \rho \rangle$ where Σ is a finite label set, Q is a finite set of states, $E \subseteq Q \times \Sigma \times \mathbb{R}_+ \times Q$ is a set of transitions, i is an initial state, F is a set of final states and $\rho : F \to \mathbb{R}_+$ is a final weight function. A *path* in $\mathcal{A}$ from q_0 to q_n is a sequence of transitions $\pi = q_0 \overset{\langle x_1, w_1 \rangle}{\to} q_1 \overset{\langle x_2, w_2 \rangle}{\to} q_2 \ldots \overset{\langle x_n, w_n \rangle}{\to} q_n$. The path π may be the empty sequence of transitions if $q_0 = q_n$. The *label of the path* π is defined as $l(\pi) = x_1, x_2 \ldots x_n$ and the *weight* of the path π is defined as $w(\pi) = \sum_{i=1}^{n} w_i$. The set of all paths from q' to q'' is denoted as $P(q', q'')$. The *function represented by* $\mathcal{A}$, $F_{\mathcal{A}} : \Sigma^* \to \mathbb{R}_+$, is defined as follows:

$$F_{\mathcal{A}}(v) = \min\{w(\pi) \mid \exists f \in F : \pi \in P(i, f), l(\pi) = v\}$$

as min $\emptyset$ is undefined. The transducer $\mathcal{A}$ is *deterministic* if for each $q \in Q$ and each $x \in \Sigma$ we have that $|\{p \in Q \mid \exists w \in \mathbb{R}_+ : \langle q, x, w, p \rangle \in E\}| \leq 1$. An algorithm for

[1]AComIn "Advanced Computing for Innovation", grant 316087, funded by the FP7 Capacity Programme (Research Potential of Convergence Regions).

determinization of string-to-weight transducers is presented in [6]. As shown in [6], given an arbitrary acyclic transducer $\mathcal{A}$ we can construct a deterministic transducer $\mathcal{A}' = \langle \Sigma, Q', E', i', F', \rho' \rangle$ which represents F_A as follows. The initial state is $i' = \{\langle i, 0 \rangle\}$. Let S be the set of all sets of pairs from $Q \times \mathbb{R}_+$, i.e. $S = 2^{Q \times \mathbb{R}_+}$. The set of states Q' consists of nonempty finite sets of pairs from $Q \times \mathbb{R}_+$. To define Q' we use two auxiliary partial functions $\lambda : S \times \Sigma \to \mathbb{R}_+$ and $\delta : S \times \Sigma \to S$ where

$$\lambda(\{\langle q_1, c_1 \rangle, \ldots, \langle q_n, c_n \rangle\}, x) = \min\{c_i + w | 1 \leq i \leq n, w \in \mathbb{R}_+, \exists p \in Q : \langle q_i, x, w, p \rangle \in E\},$$

$$\delta(\{\langle q_1, c_1 \rangle, \ldots, \langle q_n, c_n \rangle\}, x) = \begin{cases} \text{not defined if } \lambda(\{\langle q_1, c_1 \rangle, \ldots, \langle q_n, c_n \rangle\}, x) \text{ is not defined} \\ \{\langle p, c \rangle \, | \, \exists i : 1 \leq i \leq n, \exists w \in \mathbb{R}_+ : \langle q_i, x, w, p \rangle \in E, \\ c = \min\{c_j + w' - \lambda(\{\langle q_1, c_1 \rangle, \ldots, \langle q_n, c_n \rangle\}, x) \, | \, 1 \leq j \leq n, \\ w' \in \mathbb{R}_+, \langle q_j, x, w', p \rangle \in E\}\} \text{ otherwise} \end{cases}$$

The set of states Q' is the smallest set $Q' \subseteq S$ such that $i' \in Q'$ and $(q' \in Q' \wedge \delta(q', x)$ is defined$) \to \delta(q', x) \in Q'$. The set of transitions is $E' = \{\langle q', x, \lambda(q', x), \delta(q', x) \rangle | q' \in Q, x \in \Sigma\}$. The set of final states is

$$F' = \{\{\langle q_1, c_1 \rangle, \ldots, \langle q_n, c_n \rangle\} \in Q' \, | \, \{q_1, \ldots, q_n\} \cap F \neq \emptyset\}.$$

The final weight function is

$$\rho'(\{\langle q_1, c_1 \rangle, \ldots, \langle q_n, c_n \rangle\}) = \min\{c_i \, | \, 1 \leq i \leq n, q_i \in F\}.$$

3   Method Description

We use a beam search based on three features: n-gram language model probabilities, acoustic probabilities and word insertion penalty. The lattice resulted from the beam search represents a nondeterministic acyclic string-to-weight transducer $\mathcal{A}$. Each transition of the transducer is labeled with a word and weight which represents a combination of the three features.

First, we sketch a method, which we call "the N-best method". It is a variant of the method presented in [2] for finding the N-best hypotheses in the transducer. Afterwards we describe the new method as a derivation of the N-best method. Since a full determinizaion of the lattice $\mathcal{A}$ is unfeasible, both methods do not generate the whole set Q' of states of the deterministic transducer $\mathcal{A}'$ (see Sect. 2) but only a part of it.

Fig. 1 The N-best method—
finding the N-best hypotheses
in the lattice $\mathcal{A}$

$FindNBestHypotheses(\mathcal{A})$
1   $i' \leftarrow \{\langle i, 0 \rangle\}$
2   $n \leftarrow 0$
3   $S \leftarrow \{\langle i', 0, n \rangle\}$
4   $\pi(n) \leftarrow$ NIL
5   **while** $S \neq \emptyset$
6   **do** $\langle p', c, n' \rangle \leftarrow Pop(S)$
7   **if** $p' \in F'$ **then output** $\langle GetHypothesis(n'), c + \Phi(p') \rangle$
8   **if** *the allowed time elapsed* **then exit**
9   **for each** $\langle x, w, q' \rangle$ *such that* $\langle p', x, w, q' \rangle \in E'$
10   **do** $k \leftarrow c + w$
11                  $n \leftarrow n + 1$
12                  $Push(S, \langle q', k, n \rangle)$
13                  $\pi(n) \leftarrow \langle x, n' \rangle$

$GetHypothesis(k)$
1   $v \leftarrow \epsilon$
2   **while** $\pi(k) \neq$ NIL
3   **do** $\langle x, m \rangle \leftarrow \pi(k)$
4           $v \leftarrow x \cdot v$
5           $k \leftarrow m$
6   **return** v

3.1 The N-best Method

A *hypothesis* is a sequence v of words, such that $F_{\mathcal{A}}(v)$ is defined. The N-best method generates from the lattice the N-best hypotheses and then rescores them one by one using an n-gram language model with higher n. In the experiments, presented in Sect. 4, we used 2-gram model to build the lattice and 3-gram model for rescoring. The hypothesis with the best rescored value is selected as a recognition output.

3.1.1 Finding the N-best Hypotheses

On Fig. 1 we present a variant of the algorithm in [2] for finding the N-best hypotheses from the lattice $\mathcal{A}$. The main difference between the vairant presented here and the original in [2] is that we do not fix apriori N to generate the N-best hypotheses, but we generate as much as possible in a given time limit.[2]

The algorithm does not determinize the whole lattice $\mathcal{A}$, i.e. does not generate the whole set Q' (see Sect. 2), but only that part of it that contains the N-best candidates. The algorithm maintains a priority queue S, which is initialized on line 3 in *FindN BestHypotheses*. The priority queue keeps triples $\langle p, c, n' \rangle$, where $p' \in Q'$ is a state reached during the determinization, c is the weight with which p' has been reached and n' is a number used for restoring the sequence of words with which p' has been reached. The order in the priority queue S is defined as follows:

---

[2]For the experiments presented in this paper the time for the determinization procedure is 10 % of the time of the input signal.

$\langle p', c', n' \rangle \leq \langle p'', c'', n'' \rangle$ iff $c' + \Phi(p') \leq c'' + \Phi(p'')$, where $\Phi(\{\langle q_1, c_1 \rangle, \ldots, \langle q_n, c_n \rangle\}) = \min\{c_i + \phi(q_i) | 1 \leq i \leq n\}$ and $\phi(q) = \min\{w(\pi) + \rho(f) | f \in F, \pi \in P(q, f)\}$. Since the lattice represents an acyclic transducer the set of values $\{\phi(q) | q \in Q\}$ can be precomputed in time $O(|Q| + |E|)$. For each number n' in a triple $\langle p, c, n' \rangle$ we keep a predecessor $\pi(n')$ which represents a pair of a word with which p has been reached and a preceding number. The procedure *GetHypothesis* uses π to restore a sequence of words.

The procedure *FindN BestHypotheses* outputs N-best hypotheses with their weights, sorted by weights. It combines the determinzation algorithm for string-to-weight transducers with the A^* search algorithm for finding N-best paths with heuristic function $h(p', c') = c' + \Phi(p')$. In practice the procedure is very efficient, because $\Phi(p')$ is the weight of the best path from p' to the goal. For more details we refer the reader to [2].

3.1.2 Rescoring in the N-best Method

In the N-best method we rescore the N-best hypotheses one by one and select the one with highest rescored weight. The weight c of a hypothesis v has the type

$$c = -\log P_A - \lambda_{LM}\log P_2(v) + \lambda_{WP}|v|,$$

where P_A is acoustic probability, $P_2(v)$ is the probability of v according to a 2-gram language model, $|v|$ is the length of v, λ_{LM} and λ_{WP} are parameters. The rescored weight of v has the type

$$c' = -\log P_A - \lambda'_{LM}\log P_3(v) + \lambda'_{WP}|v|,$$

where $P_3(v)$ the probability of v according to a 3-gram language model, λ'_{LM} and λ'_{WP} are other parameters. The values of the parameters λ_{LM}, λ_{WR}, λ'_{LM} and λ'_{WR} are chosen to optimize the word accuracy on a development set. To obtain the rescored weight c' from c and v we compute $\log P_2(v)$ and $\log P_3(v)$ by v and then we use the following formula:

$$c' = c + \lambda_{LM}\log P_2(v) - \lambda'_{LM}\log P_3(v) + (\lambda'_{WP} - \lambda_{WP})|v|.$$

3.2 The New Method

In the new method instead of generating explicitly N-best hypotheses we generate a deterministic transducer which contains much more hypotheses besides N-best hypotheses. Eventually we find the best rescored hypothesis among all hypotheses the deterministic transducer represents.

3.2.1 Building a Deterministic Transducer for Rescoring

We start to determinize the lattice by building the initial state of the deterministic
transducer. On each iteration we choose one unexpanded state and expand it, i.e. we
generate all its outgoing transitions along with the states they lead to. We continue in
this way iteration after iteration until a timing criterion is satisfied. On each iteration
we select an unexpanded state of minimum weight. Here, the weight of the state is
the minimum weight of successful path through this state. In this way we manage to
determinize the most plausible part of the lattice in the given time constraints. The
result of the procedure is a determinized string-to-weight transducer.

While building it each traversed transition of $\mathcal{A}'$ is traversed only once. In
contrast in the N-best method one and the same transition of $\mathcal{A}'$ may be traversed
many times. This allows us in comparison with the N-best method to determinize a
bigger part of the lattice in the same time limit.

On Fig. 2 we present the procedure that builds the deterministic transducer for
rescoring. The procedure returns a part of the determinized automaton $\mathcal{A}'$ with
states $Q'' \subseteq Q'$ and transitions $E'' \subseteq E'$. We keep all unexpanded states in a priority
queue S. The order in the priority queue S is defined as follows: $p' \leq p''$ iff
$c(p') + \Phi(p') \leq c(p'') + \Phi(p'')$. The number $c(p')$ represents the weight of the
currently best path from the initial state i' to p' in the deterministic transducer $\mathcal{A}$.
The priority $c(p') + \Phi(p')$ of the state p' with the best priority in the queue is best
path from the initial state i' to p'. For each generated state $p' \in Q'$ we keep a value
$expanded(p')$ which indicates whether p' is expanded. Once a state is expanded it
cannot be pushed in the queue again. We use a heap to realize the priority queue S.
On line 16 the priority of q' may be improved to k. In this case the update of the
heap requires time $O(\log|S|)$.

```
BuildDeterministicTransducerForRescoring(A)
 1   i' ← {⟨i, 0⟩}
 2   c(i') ← 0
 3   S ← {i'}
 4   Q'' ← {i'}; E'' ← ∅
 5   expanded(i') ← false
 6   while S ≠ ∅
 7       do p' ← Pop(S)
 8           if the allowed time elapsed then exit
 9           for each ⟨x, w, q'⟩ such that ⟨p', x, w, q'⟩ ∈ E'
10               E'' ← E'' ∪ {⟨p', x, w, q'⟩}
11               do if q' ∉ Q'' then Q'' ← Q'' ∪ {q'}
12                                       expanded(q') ← false
13                   if expanded(q') then continue
14                   k ← c(p') + w
15                   if q' ∉ S then c(q') ← k
16                                 Push(S, q')
17                                 continue
18                   if k < c(q') then c(q') ← k
19           expanded(p') ← true
20   return A'' = ⟨Σ, Q'', E'', i', F'', ρ''⟩ where F'' = F' ∩ Q'' and ρ'' = ρ' ↾ F''
```

Fig. 2 The new method—building a deterministic transducer for rescoring

3.2.2 Rescoring in the New Method

We apply the final rescoring on all hypotheses represented by the deterministic transducer $\mathcal{A}''$ constructed by the procedure on Fig. 2. For this purpose this transducer is intersected with two string-to-weight transducers: a transducer which represents the 2-gram language model used in the beam search and a transducer which represents the rescoring 3-gram language model. In the intersection the weights of the former transducer are subtracted while the weights of the latter are added.

The intersection $I = \langle \Sigma, Q''', E''', i''', F''', \rho''' \rangle$ can be constructed as follows. Let $\mathcal{A}_2 = \langle \Sigma, Q_2, E_2, i_2, F_2, \rho_2 \rangle$ and $\mathcal{A}_3 = \langle \Sigma, Q_3, E_3, i_3, F_3, \rho_3 \rangle$ be the string-to-weight deterministic transducers representing the 2-gram and the 3-gram language models respectfully, i.e. $F_{\mathcal{A}_2}(v) = -\log P_2(v)$ and $F_{\mathcal{A}_3}(v) = -\log P_3(v)$. Let $S = Q'' \times Q_2 \times Q_3$ and

$$T = \{ \langle p'', p_2, p_3 \rangle, x, w, \langle q'', q_2, q_3 \rangle \mid x \in \Sigma, \exists w'', w_2, w_3 \in \mathbb{R}_+ :$$

$$\langle p'', x, w'', q'' \rangle \in E'', \langle p_2, x, w_2, q_2 \rangle \in E_2, \langle p_3, x, w_3, q_3 \rangle \in E_3,$$

$$w = w'' - \lambda_{LM} w_2 + \lambda'_{LM} w_3 + (\lambda'_{WP} - \lambda_{WP}) \}.$$

The[3] initial state is $i''' = \langle i', i_2, i_3 \rangle$. The set of states Q''' is the smallest set such that $i''' \in Q'''$ and $(q''' \in Q''' \wedge \langle q''', x, w, p''' \rangle \in E''') \rightarrow p''' \in Q'''$. The set of final states is $F''' = Q''' \cap (F'' \times F_2 \times F_3)$ and the set of transitions is $E''' = T \cap (Q''' \times \Sigma \times \mathbb{R}_+ \times Q''')$.

The best path in the intersection I represents the best rescored hypothesis. We found experimentally that the time needed for the build the intersection transducer I is negligible with respect to the time needed for the procedure that builds the deterministic transducer for rescoring.

4 Evaluation

We have tested the new method using our LVCSR system for Bulgarian [5]. For training and testing the Bulgarian Phonetic Corpus[4] created in the framework of the AComIn project was used. All tests are performed using a speaker independent (SI) acoustic model. The n-gram language models were constructed using a ~ 250 M words legal corpus for Bulgarian. The test set consists of 9 speakers with 50 long legal utterances each. We have varied the beam width between 1000 and

---

[3]Let us note that transition weights of the lattice are obtained by combining acoustic weights with 2-gram language weights, which are computed through $\mathcal{A}_2$. For this reason $w = w'' - \lambda_{LM} w_2 + \lambda'_{LM} w_3 + (\lambda'_{WP} - \lambda_{WP})$ is always positive.

[4]Bulgarian Phonetic Corpus: http://lml.bas.bg/BulPhonC/.

3000 states by a step of 500. For building the lattice a 2-gram language model was used. The rescoring has been performed using a 3-gram language model. The rescoring process time for both—the n-best method and the proposed new method —has been constrained to 0.1× of the duration of the utterance. The experiments were run on a machine with 2.4 GHz Intel Xeon E5645 processor.

Table 1 presents the resulting word accuracy, word error rate reduction with respect to the n-best method, letter error rate, letter error rate reduction, the average number of hypotheses per utterance considered in the rescoring process and the time ratio for the recognition. The measurements are performed on the recognition without rescoring, with rescoring using the n-best method and with rescoring using the newly proposed method by varying the beam width.

Table 2 presents test results for statistical significance of the word accuracy. We computed the word accuracy for each one of the 450 test sentences (9 speakers, 50 sentences for each speaker) with the N-best method and the new method. We performed paired t-test of the two resulted sets of values. The null hypothesis was that the difference of the two means—the mean accuracy per sentence for the new

Table 1 Comparison of speech recognition with no rescoring, n-best rescoring and the new method

	Word accuracy			WER reduction w.r.t. n-best		
	No rescoring (%)	N-best (%)	New method (%)	No rescoring (%)	N-best (%)	New method (%)
Beam = 1000	88.77	92.61	92.63	−52.04	0.00	0.16
Beam = 1500	89.40	93.53	93.57	−63.87	0.00	0.56
Beam = 2000	89.57	93.64	93.78	−64.02	0.00	2.27
Beam = 2500	89.76	93.64	93.82	−60.98	0.00	2.84
Beam = 3000	89.85	93.61	93.85	−58.87	0.00	3.77
	Letter accuracy			LER reduction w.r.t. n-best		
	No rescoring (%)	N-best (%)	New method (%)	No rescoring (%)	N-best (%)	New method (%)
Beam = 1000	96.01	97.09	97.11	−36.97	0.00	0.77
Beam = 1500	96.35	97.53	97.55	−48.04	0.00	0.77
Beam = 2000	96.50	97.64	97.69	−48.61	0.00	2.05
Beam = 2500	96.55	97.66	97.71	−47.23	0.00	2.36
Beam = 3000	96.57	97.68	97.75	−47.84	0.00	3.13
	Number of hypotheses			Time ratio		
	No rescoring	N-best	New method	No rescoring	N-best	New method
Beam = 1000	1	1527	3.99E + 17	0.31×	0.39×	0.40×
Beam = 1500	1	378	3.04E + 15	0.45×	0.54×	0.54×
Beam = 2000	1	155	3.12E + 12	0.65×	0.74×	0.74×
Beam = 2500	1	78	1.12E + 09	0.92×	1.02×	1.02×
Beam = 3000	1	44	1.52E + 07	1.28×	1.37×	1.37×

Table 2 Paired t-test for the difference of the word accuracy per sentence between the new method and the N-best method

	p-value (%)
Beam = 1000	38.76
Beam = 1500	30.62
Beam = 2000	6.67
Beam = 2500	2.84
Beam = 3000	2.99

method and the mean accuracy per sentence for N-best method—is 0 and the alternative hypothesis was that the difference is strongly positive. Table 2 presents the resulting p-values obtained for the beam widths between 1000 and 3000.

5 Conclusion

The evaluation clearly shows that considering more hypotheses in the rescoring process within a fixed time constraint reduces the word error rate by up to 3.77 % with respect to the n-best rescoring method at a statistical significance p-value below 3 %. The improvement value depends on the particular setup—the beam width, the time constraint for rescoring, the language models used for building the lattice and for rescoring. A question for future investigation is how those factors influence the improvement of the recognition accuracy by the newly proposed method. Another direction might be the improvement of the new method efficiency by reducing the number of generated non-coaccessible states and implementing the priority queue using Fibonacci heaps.

Acknowledgments The research work reported in the paper is supported by the project AComIn "Advanced Computing for Innovation", grant 316087, funded by the FP7 Capacity Programme (Research Potential of Convergence Regions). We would like to thank the reviewers for their useful comments and suggestions.

References

1. Huang, Xuedong, Acero, Alex, Hon, Hsiao-Wuen: Spoken Language Processing: A Guide to Theory, Algorithm, and System Development, 1st edn. Prentice Hall PTR, Upper Saddle River, NJ, USA (2001)
2. Mohri, M., Riley, M.: An efficient algorithm for the N-best-strings problem. In: Proceedings of the International Conference on Spoken Language Processing 2002 (ICSLP '02), Denver, Colorado, Sept 2002
3. Sak, H., Saraclar, M., Güngör, T.: On-the-fly lattice rescoring for real-time automatic speech recognition. INTERSPEECH, ISCA (2010)

4. Hori, T., Hori, C., Minami, Y., Nakamura, A.: Efficient WFST-based one-pass decoding with on-the-fly hypothesis rescoring in extremely large vocabulary continuous speech recognition. IEEE Trans. Audio Speech Lang. Process. **15**, 1352–1365 (2007)
5. Mitankin, P., Mihov, S., Tinchev, T.: Large vocabulary continuous speech recognition for Bulgarian. In: Proceedings of the RANLP 2009, Sept 2009
6. Mohri, M.: Finite-state transducers in language and speech processing. Comput. Linguist. **23**(2) (1997)

Part IV

Optimisation and Intelligent Control

Metaheuristic Method for Transport Modelling and Optimization

Stefka Fidanova

Abstract Public transport is a shared passenger transport service, which is available for use by general public. The operational efficiency of public transport is essential to provide good service. Therefore it needs to be optimized. The main public transport between cities, up to 1000 km, are trains and buses. It is important for transport operators to know how many peoples will use it. In this paper we propose a model of public transport. The problem is defined as multi-objective optimization problem. The two goals are minimum transportation time for all passengers and minimal price. We apply ant colony optimization approach to model the passenger flow. The model shows how many passengers will use a train and how many will use a bus according what is more important for them, the price or the time.

1 Introduction

Nowadays the people use different ways for transportation from one city to another one. It is important to model the use of public transport and thus to optimize it. Normally the railway transport (excluding the super fast trains with velocity more than 200 km/h) is cheaper, but slower. The fast trains and the buses are faster, but more expensive. Therefore preparing a transportation model we need to take into account all this. The innovation in this paper is modeling the passenger flow, describing the transportation problem as optimization. Thus we define the problem for modeling the public transport as multi-objective optimization problem with two objective functions, total price of all tickets and total traveling time of all passengers. Our aim is to minimize the both objective functions. These objectives are controversial, when one of them decreases the other increases. In the case the optimization problem is multi-objective we receive a set of non-dominated solutions. We can analyze them and according to the preferences of the potential

S. Fidanova (✉)
Institute of Information and Communication Technologies,
Bulgarian Academy of Sciences, Acad. G. Bonchev str. bl25A, 1113 Sofia, Bulgaria
e-mail: stefka@parallel.bas.bg

© Springer International Publishing Switzerland 2016
S. Margenov et al. (eds.), *Innovative Approaches and Solutions in Advanced Intelligent Systems*, Studies in Computational Intelligence 648,
DOI 10.1007/978-3-319-32207-0_19

passengers we can see how many of them will use train and how many of them will use a bus (or fast train if it exists).

The railroads are the oldest public transport which is used till now. Now days the bus transport is a concurrent of the rail transport especially in the place with highways. However, the analytical models are very important for further planing and decision making.

There exist different kinds of transportation models [1]. The importance and role of each type of models is discussed in relation with its function. Some of the models are concentrated on scheduling [5]. Other models are focused on simulation to analyze the level of utilization of different types of transportation [11]. There are models which goal is optimal transportation network design [8]. Our model is focused on modeling the passenger flow according their preferences. By the model, can be seen the weak points of every mode of transportations. The model can show the distribution of the passenger flow when the schedule is changed or when the capacity of the vehicles is changed.

Multi-objective optimization is difficult in computational point of view. We propose ant colony optimization algorithm to solve the problem. The ant colony optimization is a methodology for solving optimization problems. The main contributions applying ant colony optimization are representation of the problem with graph and construction of appropriate heuristic information. The model is tested on one artificially made problem with one train and one bus which start from the first station on the same time. After, the model is tested on real problem, the transportation between Sofia and Varna, one of the longest destination in Bulgaria.

The rest of the paper is organized as follows. In Sect. 2 the ant colony optimization algorithm is described. In Sect. 3 the transportation problem is formulated and an ACO algorithm which solve it is proposed. Section 4 shows experimental results their analysis on artificial and real data. In Sect. 5 we draw some conclusions and directions for future work.

2   Ant Colony Optimization

The proposed problem is very hard of computational point of view. Such kind of problems usually are solved with some metaheuristic method. Therefore we apply Ant Colony Optimization (ACO) algorithm, one of the best metaheuristics, to solve it.

The idea for ant algorithm comes from the real ant behavior. Ants put on the ground chemical substance called pheromone, which help them to return to their nest when they look for a food. The ants smell the pheromone and follow the path with a stronger pheromone concentration. Thus they find shorter path between the nest and the source of the food. The ACO algorithm uses a colony of artificial ants that behave as cooperating agents. With the help of the pheromone they try to construct better solutions and to find the optimal ones. The problem is represented by a graph and the solution is represented by a path in the graph or by tree in the graph. For the successes of the algorithm, it is very important how the graph will be constructed. Ants start from random nodes of the graph and construct feasible solutions. When all ants construct their solution the pheromone values are updated. Ants compute a set of feasible moves and select the best one, according to the transition probability rule. The transition probability p_{ij}, to chose the node j when the current node is i, is based on the heuristic information η_{ij} and on the pheromone level τ_{ij} of the move, where $i, j = 1, \ldots, n$. α and β shows the importance of the pheromone and the heuristic information respectively.

$$
p_{ij} = \frac{\tau_{ij}^{\alpha} \eta_{ij}^{\beta}}{\sum_{k \in \{allowed\}} \tau_{ik}^{\alpha} \eta_{ik}^{\beta}} \tag{1}
$$

The heuristic information is problem dependent. It is appropriate combination of problem parameters and is very important for ants management. The ant selects the move with highest probability. The initial pheromone is set to a small positive value τ_0 and then ants update this value after completing the construction stage [2, 4, 6]. The search stops when $p_{ij} = 0$ for all values of i and j, which means that it is impossible to include new node in the current partial solution.

The pheromone trail update rule is given by:

$$
\tau_{ij} \leftarrow \rho \tau_{ij} + \Delta \tau_{ij}, \tag{2}
$$

where $\Delta \tau_{ij}$ is a new added pheromone and it depends of the quality of achieved solution.

The pheromone is decreased with a parameter $\rho \in [0, 1]$. This parameter models evaporation in the nature and decreases the influence of old information in the search process. After that, we add the new pheromone, which is proportional to the quality of the solution (value of the fitness function). There are several variants of ACO algorithm. The main difference is the pheromone updating.

Multi-Objective Optimization (MOP) has his roots in the nineteenth century in the work of Edgeworth and Pareto in economics [9]. The optimal solution for MOP is not a single solution as for mono-objective optimization problems, but a set of solutions defined as Pareto optimal solutions. A solution is Pareto optimal if it is not possible to improve a given objective without deteriorating at least another objective. The main goal of the resolution of a multi-objective problem is to obtain the Pareto optimal set and consequently the Pareto front. One solution dominates another if minimum one of its components is better than the same component of other solutions and other components are not worse. The Pareto front is the set of non dominated solutions. After that, the users decide which solution from the Pareto front to use according additional constraints, related with their specific application. When metaheuristics are applied, the goal becomes to obtain solutions close to the Pareto front.

3   Problem Formulation

There are different kinds of problems arising in transportation. Some of them concern optimal scheduling [7]. Others model the passenger flow on stations and their optimal management [10]. There are problems concerning only one type of vehicles and other problems combining different types of vehicles [3]. The common is that all they are difficult in computational point of view.

Our problem concerns passengers traveling in a given direction, covered with several types of vehicles, train and bus. The problem is how many of the potential passengers will use a train and how many will use a bus. In our case there is destination from station A to station B. There are two kinds of vehicles, trains and buses, which travel between station A and station B. Every vehicle has his set of stations where it stops, between the terminus. Some of the stations can be common for some of the vehicles. Let the set of all stations is $S = \{s_1, \ldots, s_n\}$ and on every station s_i, $i = 1, \ldots, n - 1$, n is the number of stations, at every time slot there are number of passengers which want to travel to station s_j, $j = i+1, \ldots, n$. Every vehicle travel with different speed and the price to travel from station s_i to station s_j can be different.

The input data are set of stations S, starting time of every vehicle from the first station, time for every vehicle to go from station s_i to station s_j, the capacity of every vehicle, the price to travel with some of the vehicles from one station to another one, number of passengers which want to travel from one station to another one at every moment. Our algorithm calculates how many passengers will get on every of the vehicles on station s_i to station s_j at every time slot. There are two objectives, the total price of all tickets and the total travel time. If some vehicle does not stop on some station, we put the travel time and the price to this destination to be 0.

The output data are number of passengers in every vehicle from the station s_i to the station s_j at every time slot and the values of the two objectives.

In our ACO implementation the time is divided to time slots ($N \times 24$ time slots correspond to 60/N min; for example $2 \times 24 = 48$ time slots, correspond to 30 min). We solve the problem for one line. A set of $N \times 24$ nodes corresponds to every station, showing the time slots. Thus our graph of the problem shows if there is any vehicle on station s_i, $i = 1, \ldots, n$, at fixed time moment. We will put the pheromone on the nodes of the graph. Our artificial ants start to create their solutions from the first station. They chose in a random way how many passengers will get on every vehicle. The upper bound of the passengers in one vehicle is the maximum between the capacity of the vehicle and the number of passengers which want to travel at this time slot. The number of all passengers getting vehicle in some time moment is maximal possible. If there is only one vehicle at this moment the maximal possible number of passengers get on this vehicle. We apply probabilistic rule called transition probability to model how many passengers will get any of the vehicles on the next stations. The heuristic information, we construct, is the sum of the reciprocal values of the two objective functions.

The problem is two-objective, therefore there is not one optimal solution and a set of non-dominated solutions. One solution dominates other if it is better to one of the components and not worst to other. Set of solutions where none of them dominates other is a set of non-dominated solutions. In our variant of ACO algorithm first we decrease the all pheromone multiplying it with evaporation coefficient. After we add new pheromone on the elements of non-dominated solutions. The new added pheromone is one over the sum of the values of the two objective functions.

4 Experimental Results

We have prepared a software, which realizes our algorithm. The algorithm parameters are set as follows: the evaporation parameter $\rho = 0.5$; $\alpha = \beta = 1$; initial pheromone value is set to be $\tau_0 = 0.5$. We prepare a small example with one train and one bus starting at the same time from the first station and having 4 stations. The bus is faster and the train is chipper. We can see in different cases how many passengers will use the train and how many of them will use a bus, according their preference to pay less or to arrive faster.

On this example we use 5 ants and 5 iterations. The algorithm achieves 18 non-dominated solutions. The Table 1 shows achieved solutions ordered by increasing order of the price and decreasing order of the time. The forth column

Table 1 Experimental results

No	Price	Time	Train
1	265	451	86
2	270	450	79
3	272	448	86
4	274	446	79
5	277	443	76
6	282	438	66
7	289	431	78
8	291	429	69
9	292	428	64
10	295	425	66
11	301	419	65
12	302	418	60
13	307	413	62
14	309	411	64
15	313	407	67
16	321	399	60
17	324	396	63
18	331	389	61

shows the number of passengers, used train. We observe that when the price increases, the number of passengers used train decreases. There is several exceptions. In solution No. 6, the price is lower than in solution Nos. 7 and 8, but the number of passengers used train is less. When we checked the details we sow that in this case more passengers prefer to use train for long destination. In solution 16 the passengers used train is less than the passengers in solutions 17 and 18, because the passengers used train prefer short destinations.

We apply the algorithm on one real example. The first station is Sofia, the capital of Bulgaria and the last station is Varna, the maritime capital of Bulgaria. The destination is about 450 km. There are 5 trains and 23 buses which travel from Sofia to Varna, but they move with different speed, the prices are different and they stop on different stations between Sofia and Varna. We have not exact data about the number of passengers which travel from one station to another one between Sofia and Varna. Therefore we estimated the number of passengers from one station to another one taking into account the number of residents of every place where some of the vehicles stop.

On this example we use 10 ants and 100 iterations. The algorithm achieves 5 non-dominated solutions. The Table 2, similar to the Table 1, shows achieved solutions ordered by increasing order of the price and decreasing order of the time. The forth column shows the number of passengers, used train. The number of passengers used train, respectively bus, vary only when on the same station at the same time there is more than one type of vehicle. Like in a previous example we observe that when the price decreases, the number of the passengers used train increases and when the time decreases the number of the passengers used train decreases and the number of passengers used bus increases respectively.

On the Fig. 1 is shown the number of passengers used train. Every column series corresponds to one of the solutions, the left series corresponds to the first solution from the Table 2 and the right series corresponds to the solution 5 from the Table 2. Every series consists of three columns. The left column corresponds to the all number of passengers used train. The middle column corresponds to the passengers used train for a long distances, more than 300 km. The right column corresponds to the number of passengers used train for a short distances, less than 300 km. We observe that in the solutions with less value for price and greater value for time the all number of passengers increase on accordance of increase of the number of passengers on long destinations. The number of passengers on short destinations is the same for all solutions. It means that when the destination is short the difference

Table 2 Experimental results Sofia Varna	No	Price	Time	Train
	1	51,843	25,840	1951
	2	51,797	25,842	1952
	3	51,579	25,862	1978
	4	51,571	25,869	1979
	5	51,563	25,870	1980

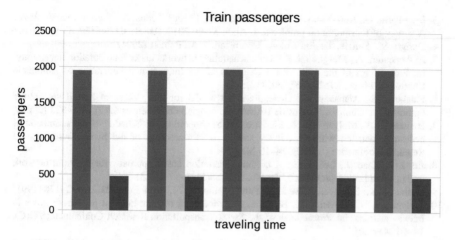

Fig. 1 Passengers in the train Sofia Varna

of prices and traveling time between the trains and the buses is negligible and there are not transfer of passengers between the different types of vehicles.

The algorithm can be applied first on long (national) lines and after fixing them the algorithm can be applied on local lines and to analyze and optimize them too.

5 Conclusion

In this paper we propose a model of the passenger flow when between two destinations there are several kinds of transport. The problem is defined as multi-objective optimization problem. We propose ACO algorithm to solve it. With this model we can analyze existing public transport. We can predict how the passenger flow will change if the timetable of the vehicles will be changed or if new vehicle will be included/excluded. In a future work we will include fuzzy logic. When the difference between time/price of two types of vehicle is too small, the passenger will decide to change the vehicle or not, with some probability.

Acknowledgments This work was partially supported by EC project AcomIn and by National Scientific fund by the grand I02/20.

References

1. Assad, A.A.: Models for rail transportation. Transp. Res. Part A Gen. **143**, 205–220 (1980)
2. Bonabeau, E., Dorigo, M., Theraulaz, G.: Swarm Intelligence: From Natural to Artificial Systems. Oxford University Press (1999)

3. Diaz-Parra, O., Ruiz-Vanoye, J.A., Loranca, B.B., Fuentes-Penna, A., Barrera-Camara, R.A.: A survey of transportation problems. J. Appl. Math. **2014**, Article ID 848129, 17 pp (2014)
4. Dorigo, M., Stutzle, T.: Ant Colony Optimization. MIT Press (2004)
5. El Amaraoui, A., Mesghouni, K.: Train scheduling networks under time duration uncertainty. In: Proceedings of the 19th World Congress of the International Federation of Automatic Control, 2014, pp. 8762–8767 (2014)
6. Fidanova, S., Atanasov, K.: Generalized net model for the process of hybrid ant colony optimization. Comptes Randus de l'Academie Bulgare des Sciences **62**(3), 315–322 (2009)
7. Hanseler, F.S., Molyneaux, N., Bierlaire, M., Stathopoulos, A.: Schedule-based estimation of pedestrian demand within a railway station. In: Proceedings of the Swiss Transportation Research Conference (STRC), 14–16 May 2014
8. Jin, J.G., Zhao, J., Lee, D.H.: A column generation based approach for the train network design optimization problem. J. Transp. Res. **50**(1), 1–17 (2013)
9. Mathur, V.K.: How well do we know Pareto optimality? J. Econ. Educ. **22**(2), 172–178 (1991)
10. Molyneaux, N., Hanseler, F., Bierlaire, M.: Modelling of train-induced pedestrian flows in railway stations. In: Proceedings of the Swiss Transportation Research Conference (STRC), 14–16 May 2014
11. Woroniuk, C., Marinov, M.: Simulation modelling to analyze the current level of utilization of sections along rail rout. J. Transport Lit. **7**(2), 235–252 (2013)

Bi-level Formalization of Urban Area Traffic Lights Control

Todor Stoilov, Krasimira Stoilova and Vassilka Stoilova

Abstract The paper applies new formalization for the control of urban transportation. Bi-level optimization problem is defined, which is a different way to control simultaneously the traffic flows and traffic lights. The bi-level optimization problem uses the well known relations based on the store-and-forward modeling. However, the bi-level approach adds additional control objectives—maximization of traffic flows which cross the urban area. This new formalization of the traffic control allows to be achieved simultaneously the minimization of the waiting vehicles and maximization of the traffic flows. This extension is possible due to the increase of the control space both by the duration of the green lights and appropriate length of the traffic light cycle. The benefit of this new formalization is numerically simulated and applied on real urban area.

Keywords Traffic control · Optimization · Bi-level optimization · Hierarchical systems · Real time control

1 Introduction

The formal approaches for the control of transportation systems are categorized mainly in two general formal descriptions: store and forward modelling and free way traffic control on highways. These two families of models have their domain applications for urban and for high way transportation networks. However, the

T. Stoilov (✉) · K. Stoilova · V. Stoilova
Institute of Information and Communication Technologies - Bulgarian Academy of Sciences, 25A, Acad. G. Bonchev str., 1113 Sofia, Bulgaria
e-mail: todor@hsi.iccs.bas.bg

K. Stoilova
e-mail: k.stoilova@hsi.iccs.bas.bg

V. Stoilova
e-mail: vassilka_stoilova@hotmail.com

© Springer International Publishing Switzerland 2016
S. Margenov et al. (eds.), *Innovative Approaches and Solutions in Advanced Intelligent Systems*, Studies in Computational Intelligence 648, DOI 10.1007/978-3-319-32207-0_20

formal background of these types of models is totally different from mathematical point of view.

For the free traffic model the basis of the control addresses models, applying partial differential equations. The control theory applies first and second order of such equations to make relations between the traffic density, traffic flows and the velocity of the vehicles. These models are mainly used for the definition of ramp metering control policy, where the control influence is the new incoming traffic flow from the corresponding ramp metering point [22, 26]. Despite the complex formalism and the needs for many additional parameters which have to be identified both in real time and in off-line environments, the control algorithms successfully implements feedback control [6], model predictive control [34], optimal, flatness based control [1] etc.

The control of the transportation in urban areas applies relatively simple formalism. It is based on the discontinuity of flows, which can be presented that the input flows are equal in quantities to the output flows of a transport node. The formal description of these relations presents the store and forward model, which is easy to describe in discrete time models. The control low for that case can be defined by solving appropriate optimization problem. In simpler case the solution of the optimization problem can give a feedback control loop, which is easily implemented by the technical systems. The optimization problem is defined mainly assuming the vehicles or waiting queues in the transport network as arguments of the problem. The goal function is a relation of the number of vehicles or queues and can express physical events like waiting time, time for travelling, and others. The constraints of the problem express the store and forward relations for each node of the urban network. The control influences are the green lights on the traffic crossroad points. Additional control influences can be the duration of the traffic lights cycle and the offset between the cycles on different cross sections. By increase of the control space, the control problem becomes very complicated, which constraints its solution in real time and respectively the practical implementation of such a control policy.

This paper introduces an innovation approach for the formal description of the urban traffic control based on a bi-level model. Such an approach gives potential for the increase of the control space of the control problem. Comparisons with the classical optimization problem illustrate the potential of the bi-level optimization.

2   Traffic Control in Urban Area by Bi-level Approach

The idea of the multilevel modelling concerns the solution of interconnected optimization problems. The general description of the multilevel optimization is given in [27]

$$
\begin{aligned}
&\min_{x_k} f_k(x_1, \ldots, x_k) \\
&g_k(x_1, \ldots, x_k) \leq 0,
\end{aligned}
\tag{1a}
$$

where x_{k-1} is the solution of

$$\min_{x_{k-1}} f_{k-1}(x_1,\ldots,x_k)$$
$$g_{k-1}(x_1,\ldots,x_k) \leq 0; \tag{1b}$$

x_1 is the solution of the low level problem

$$\min_{x_1} f_1(x_1,\ldots,x_k)$$
$$g_1(x_1,\ldots,x_k) \leq 0. \tag{1c}$$

Problem (1a) is solved at upper level, where the coordinator controls the variables of the solutions x_k for minimizing the f_k function. Similarly, problem (1c) is at the low-level and corresponds to the lower hierarchical level. The multilevel optimization problem (1a–c) is hard to be solved [3, 13]. Even in the simplest version of two-level optimization it becomes non-convex and/or non-smooth and belongs to the class of global optimization [13]. The evaluation of the global optimum for non-smooth optimization problems can be found by applying penalty functions; by satisfying the Karush-Kuhn-Tucker type conditions which transforms the optimization problem to a set of non-linear inequalities by applying pure non-differentiable optimization technique (bundle optimization algorithm). The multilevel modelling is an opportunity to cope the different aspects of the traffic modelling, control and optimization. The multilevel modelling offers conceptual tool for integration of control influences, sets of state variables, constraints and control goals in a common optimization problem.

The global problem is defined as a set of two interconnected optimization problems, Fig. 1.

The upper level problem assumes the values of $y = y^*$ as known parameters and finds an optimal solution $x^*(y)$ by solving the problem

$$\min_x f_x(x, y^*)$$
$$x \in S_x(y^*).$$

Fig. 1 Bi-level control strategy

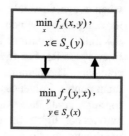

The solution $x^*(y)$ is a function of the parameter y, respectively, the lower level problem assumes $x = x^*$ as known parameter and finds its solution $y^*(x)$ as a function of x. These two interconnected problems give the solution of the global problem

$$
\min_x f_x(x, y)
$$

$$
x \in S_x(y)
$$

$$
y \in \arg \left\{ \begin{array}{l} \min_y f_y(y, x) \\ y \in S_y(x) \end{array} \right\}, \tag{2}
$$

which means x^{opt} is the solution of the optimization problem where y modifies the goal function $f_x(x, y)$ and the admissible area $S_x(y)$. Also, y is a solution of the low level problem, influenced on its turn by x.

The comments below identify the differences between the bi-level optimization problem (2) and the classical optimization problem (one level) in the form

$$
\min_x f_x(x)
$$

$$
x \in S_x. \tag{3}
$$

The bi-level problem (2) extends the space of the arguments of (3) from x^{opt} to (x^{opt}, y^{opt}). The bi-level problem optimizes not only a goal function $f_x(x, y)$, but an additional one $f_y(x, y)$, which means that the both optimization goals can be targeted simultaneously by the bi-level modelling. For the classical case (3) the goal function is only one—$f_x(x)$. The bi-level modelling considers wider set of constraints in (2) in comparison with (3). The multilevel (bi-level) optimization is more general than mathematical programming.

3 Traffic Control in Urban Area

The traffic control in urban areas tries to keep the transportation network capacity on its nominal level. The traffic network capacity can be considerably decreased when congestions take place at network links. The traffic control targets to prevent the congestions, which results in oversaturation and spillback of waiting vehicles at the links. The current traffic control approaches usually focus on either urban or freeway traffic. For the urban areas, the traffic lights are the mainly used control actuators. The parameters, which define the operation of the traffic lights, are the main control influences for the traffic control. These parameters are the green light for appropriate direction, traffic light cycle, comprising the total time for green, red and amber lights of a junction and the offset of time between the cycles of neighbour junctions of the urban network. For the case of freeway traffic, particularly on motorways, it applies ramp metering policies [21].

The most applied performance indices for the traffic flow control in the urban areas are expressed as minimization of the lost/waiting times, the passage/travel times, the number of stops during the driving. The most used criterion is the Total Time Spend (TTS) which comprises the sum of the travel time and waiting time for all vehicles in the network for a predefined time horizon. All these criteria are functions of the numbers of vehicles, which are in the network. Due to the constrained capacity of the junctions to provide smooth character of the traffic flows, queue lengths arise at the links in front of the junctions. Thus, the queue lengths are chosen as state variables for the control problem and they participate in the optimization criterion. As a consequence of these considerations the formal approach, used to describe the traffic dynamics is based on the conservation law of flows. This physical event origins the store-and-forward model, which is in primary consideration in the urban control.

The store-and-forward model has a simple formal description and it is intuitively easy to understand the traffic behaviour. The model is introduced by Gazis [17] and it is actively exploited in previous and recent researches [2, 12, 28, 32]. The model results in a set of linear dynamical discrete type relations, used for the definition and solution of the control optimization problem. The implementation of these optimal solutions influences and manages the transport behaviour, targeting decrease of the traffic congestions. The model can be complicated to deal with priorities for multimodal urban traffic control both for vehicles and buses [4], it is the formal background for real time traffic control systems as SCOOT [19], UTOPIA [20], PRODYN [18], SCATA [7], CRONOS [5], TUC [16].

In general, the state variables x are assumed to be outflows of vehicles or the queue lengths; the control variables are the split g_i or the duration of the green lights for junction i and/or the total sum of the green $GT = \sum g_i$ for different stages of the traffic lights. The time cycle c_i of the traffic light is assumed as a given parameter (not as a problem argument) of the optimization problem. Thus, the available control policies evaluate only the green light duration as a problem argument. Attempts to extend the control space are made in [16]. But the offset and the cycle are implemented as independent control problems. In [4] considerations for the benefits of the usage of the cycle c_i as a control variable are provided, but they have not been used and applied.

This research implements an integration of local optimization problems in hierarchical structure formalizing control problem with control variables both the green light durations and the cycles of the traffic lights. Thus, a bi-level optimization problem is defined, which results in increased control space, wider set of constraints and performer indices.

The attempts to implement bi-level formalism and optimization in traffic control are not new. Integration of dynamic traffic control and assignment of traffic flows was targeted and formalized as a bi-level optimization problem in [8, 30, 33]. In this context the low level problems optimize the assignment of the traffic flows for given matrices of O-D for traffic demands having given fixed signal settings. The optimal flow assignment is used for the upper level problem as given parameters and the optimal signal settings are evaluated from the upper level control problem. Thus, an iterative optimization assignment procedure is used to update alternatively the signal settings for fixed flows and solve the traffic equilibrium problem for fixed signal settings until the solutions of the two problems are considered to be mutually consistent. Such integration between transport flows in urban networks and the optimal signal settings are targeted in [8, 10, 14]. Recently, the bi-level formalism is used for road pricing [15], the usage of traffic control gantries for toll enforcement [25]. Part of the researches addresses the computational requirements, needed for the implementation of bi-level optimization in transport control [9, 10, 23, 31]. For the case of control of transportation system the main control influence, used in the papers above concerns only the duration of the green light for the signal settings. This research is working on direction to extend the control space not only with the green light duration but the traffic cycle's value as well in order to find the optimal problem solution. The expected benefits concern the minimization of the queue lengths and maximization of the traffic flows, which leave the urban network.

4   Formal Model by Bi-level Optimization

A particular case of crossroad section is considered, Fig. 2. Some of the directions are not allowed for the traffic. The notations, applied for this junction are the following:

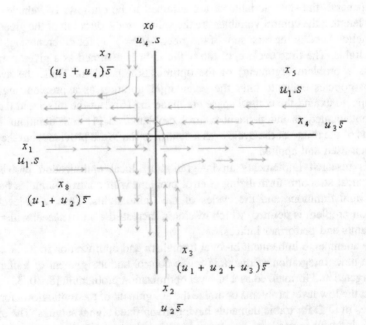

Fig. 2 A particular crossroad section

x_i, $i = 1, .., 8$ denote the queue lengths for each lane of the junction (vehicles); $u_j, j = 1, .., 4$ are the green light durations for each phase of the cycle (s). It has been chosen 4 phases for that junction.

s denotes the saturation flow for the straight directions of the junction (veh/s). For the horizontal and vertical directions it has been chosen as a common known coefficient.

$\bar{s}$ denotes the saturation for the turning flows (veh/s). For the different turning flows it has been assumed as a common known coefficient. The form of the control problem will not be complicated if s and $\bar{s}$ have different values for each lane of the junction. The division of the cycle on 4 phases defines the common constraint:

$$u_1 + u_2 + u_3 + u_4 + 0.1c = c, \tag{4}$$

where the total duration of the amber light between the phases is assumed to be 10 % of the cycle duration c. Close to each queue length x_i of Fig. 2 it is given the value of the outgoing vehicles, during the green phase for the corresponding direction. For the case of queue x_1 the volume of the outgoing vehicles is $u_1 s$. For the queue x_3 the volume of the outgoing vehicles is $(u_2 + u_3 + u_4)s$. It is evident that each direction of the junction has different parts of time of the cycle to pass during the green lights.

Following the notations of Fig. 2 and applying the store and forward model, a low level optimization problem is defined. For this problem the control arguments are the durations of the green phases $u_j, j = 1, .., 4$. The duration of the cycle c is assumed as a constant value, which will be evaluated by the upper optimization problem. The formal description of the low level control problem is the following:

$$
\begin{aligned}
x_1 &= x_{10} + x_{1in} - u_1 s \\
x_2 &= x_{20} + x_{2in} - u_2 \bar{s} \\
x_3 &\leq x_{30} + x_{3in} - (u_2 + u_3 + u_4)\bar{s} \\
x_4 &= x_{40} + x_{4in} - u_3 \bar{s} \\
x_5 &= x_{50} + x_{5in} - u_1 s \\
x_6 &= x_{60} + x_{6in} - u_4 s \\
x_7 &= x_{70} + x_{7in} - (u_3 + u_4)\bar{s} \\
x_8 &= x_{80} + x_{8in} - (u_1 + u_2)\bar{s} \\
u_1 &+ u_2 + u_3 + u_4 = 0.9c,
\end{aligned}
\tag{5}
$$

where $x_{i0}, i = 1, \ldots, 8$ are the initial values of the queue lengths, which are regarded as disturbances for the control process;

$x_{i,in}$ are the traffic demands. For the case of simplicity of the problem and to assess the bi-level control policy they have been assumed 0 or lack of traffic inputs. The control strategy for this problem is to decrease the values of $x_{i0}, i = 1, \ldots, 8$, which means that the queue lengths will vanish and congestion is missing.

The performance index for the control problem is chosen in a quadratic form

$$f(x, u) = x^T Q_x x + R_x x + \Gamma(u^T q_u u + r_u u). \tag{6}$$

The matrices Q_x, q_u can have values giving priority to a chosen direction. In a simple case these matrices are diagonal which is regarded in this research. The corresponding vectors R_x, r_u are also chosen with the same components. The value of the coefficient Γ provides scaling to have the same order of magnitude for the components of the goal function $f(x, u)$. The solution of this problem will give the durations of the green phases $u_j^{opt}, \forall j = 1, \ldots, 4$ and the lengths of the queues $x_i^{opt}, i = 1, \ldots, 8$, which will result after the implementation of the green phases. For the low level problem the duration of the cycle c is a constant value.

The upper level control problem is defined to maximize the outgoing vehicles of the crossroad directions and to decrease the queues x_2, x_4. The outgoing vehicles (Fig. 2) define the goal function of the upper level problem in a quadratic form

$$F(c, u(c)) = \{(u_2(c)\bar{s})^2 + (u_3(c)\bar{s})^2 - Gc^2\}, \tag{7}$$

where the notation $u_2(c)$ means that the green phase duration is a function of the cycle c. The coefficient G preserves the increase to infinity of the duration of the cycle c. From practical considerations, the duration of the cycle is constrained between the upper and lower bounds

$$c_{\min} \leq c \leq c_{\max}. \tag{8}$$

This bi-level formulation derives two interconnected optimization problems. The lower level problem derives the optimal durations of the green phases $u_j^{opt}(c), j = 1, \ldots, 4$ for a given cycle c. The upper level evaluates the duration of the cycle $c^{opt}(u)$ for given u. The low level problems targets minimization of the queue lengths $x_j, j = 1, \ldots, 8$ and the upper level maximizes the outgoing vehicles for the turning directions.

5 Numerical Solution Algorithm

The terms *bi-level* and *multilevel programming* are called *mathematical programs with optimization problems in the constraints* which reflect the situation for solving interconnected optimization problems [11]. The theoretical approach for the solution of such class of problems origins from the application of Karush-Kuhn-Tucker (KKT) optimality conditions. The KKT conditions are fundamental for many algorithms for constrained optimization problems. They give relaxation for the multilevel optimization problem by its reformulation as a set of nonlinear equalities and inequalities [24]. Thus, for the bi-level case the low level problem becomes from optimization problem to a set of inequalities in the feasible area of the upper level problem. Nevertheless this relaxation, the computational power for solving such nonlinear problem is a constraint for real time implementation. The nonlinear relations do not guarantee the evaluation of the global solution and the KKT conditions are not sufficient. From mathematical point of view the solution of such problems need solvers, which evaluate global minimum (or maximum) for the nonlinear problem.

This research does not develop global optimization solver, but it provides a method finding practical local solutions for the bi-level problem. The method is based on the approximation of inexplicit function in well defined analytical form. Using these analytical descriptions and applying a sequence of iterations by means to approximate the inexplicit relations on each evaluation step, the solution of the bi-level problem is found after a set of numerical iterations. The theoretical background of this method is presented in [29] and the solution of the bi-level problem is presented below. The solutions of the low level problem are different values $u_j^{opt}(c)$ by changing the values of the cycle c. Hence, the relations $u_j^{opt}(c)$ are inexplicit functions of c. To derive explicit analytical relations, Taylor series are applied:

$$u_2(c) \approx u_2(c^*) + \frac{du_2}{dc}|_{c^*}(c - c^*) \quad \text{and} \; u_3(c) \approx u_3(c^*) + \frac{du_3}{dc}|_{c^*}(c - c^*). \quad (9)$$

The values of the unknown matrix derivatives $\frac{du}{dc}$ can be calculated from the relation $\frac{du}{dc} = Q^{-1}A^T(AQ^{-1}A^T)^{-1}$ [29] for a linear–quadratic optimization problem in the form $\min(x^T Q x + R^T x)$, subject to constraints $Ax = B$.

By substituting the approximations (9) in the upper goal function (7) and after few arrangements, the upper level goal function has well defined analytical structure

$$F(c, u(c)) = \{\frac{du_2}{dc}\Big|_{c^*}c^2 - 2\alpha_2(c^*)\frac{du_2}{dc}|_{c^*}c + \frac{du_3}{dc}\Big|_{c^*}c^2 - 2\alpha_3(c^*)\frac{du_2}{dc}|_{c^*}c - 2Gc\},$$

$$(10)$$

where $\alpha_i = u_i(c^*) - \frac{du_i}{dc}|_{c^*}c^*, i = 1, 2$ are coefficients. The upper level problem concerns the maximization of (10), subject to the constraints (8). Due to the particular analytical form of $F(c, u(c))$ the solution can be found by putting the first derivative of F towards c to zero or

$$\max_c F(c) \equiv \frac{dF(c)}{dc} = 0, \quad (11)$$

which gives the optimal solution of the upper level problem

$$c^{opt} = \frac{\alpha_2(c^*)\frac{du_2}{dc}|c^* + \alpha_3(c^*)\frac{du_3}{dc}|c^* - G}{\left(\frac{du_2}{dc}|c^*\right)^2 + \left(\frac{du_3}{dc}|c^*\right)^2}. \tag{12}$$

Taking into account constraints (8), this optimal value c^{opt} must be corrected to belong between the upper and lower bounds of (8).

To express the derivatives $\frac{du_i}{dc}, i = 1, 2$ in terms of the low level problem, the last must be described explicitly from the parameters of the low level problem. The low level problem can be reformulated using only the arguments of the control variables $u_j, \forall j = 1, \ldots, 4$. For a given initial queue lengths $x_{i0}, i = 1, \ldots, 8$, the state variables $x_i, i = 1, \ldots, 8$ can be expressed by (5) and substituted in the goal function of the low level problem (6). Thus, the cost criteria $f(u) = u^T Q_u u + R_u u$ can be expressed only by the control variables $u_j, \forall j = 1, \ldots, 4$, and the feasible area is $A_u u = B_u$, where the matrices Q_u, R_u, A_u, B_u have appropriate structure. Substituting (5) in (6) and after rearrangements the explicit descriptions of Q_u, R_u and constraints A_u, B_u are

$$Q_u = \begin{vmatrix} 2s^2 + \bar{s} & \bar{s}^2 & 0 & 0 \\ \bar{s}^2 & 3\bar{s}^2 & \bar{s}^2 & 0 \\ 0 & \bar{s}^2 & 3\bar{s}^2 & 2\bar{s}^2 \\ 0 & \bar{s}^2 & 2\bar{s}^2 & 2\bar{s}^2 + s^2 \end{vmatrix}; \quad R_u = \begin{vmatrix} -x_{10}s - 2x_{50}s - 2x_{80}\bar{s} \\ -x_{20}\bar{s} - 2x_{30}\bar{s} - 2x_{80}\bar{s} \\ -x_{90}\bar{s} - 2x_{60}s - 2x_{70}\bar{s} \end{vmatrix} \tag{13}$$

$$A_u = |\,1.11 \quad 1.11 \quad 1.11 \quad 1.11\,|, \qquad B_u = c.$$

For the low level problem the solutions $u_j(c), j = 1, \ldots, 4$ are functions of the cycle c. Using [29], the relation $u = u(c)$ has the form

$$u(c) = u(c^*) + Q_u^{-1} A_u^T (A_u Q_u^{-1} A_u^T)^{-1} (c - c^*), \tag{14}$$

where c^* is an initial feasible point for the cycle. From the matrix Eq. (14) the needed numerical relations for $u_2(c)$, $u_3(c)$ can be easily defined.

Using relations (12)–(14) the algorithm for the solution of the bi-level problem is:

1. Choose an initial feasible cycle c^*.
2. The low level problem (13) is solved with the initial values of the queue lengths $x_{i0}, i = 1, \ldots, 8$ and c^*. The solutions u^* of the low level problem are found.
3. Evaluation of the coefficients $\frac{du_2}{dc}|_{u^*}, \frac{du_3}{dc}|_{u^*}$ at point u^*.
4. Evaluation of new c^* according to (12).
5. If for a given scalar ε the convergence test $|c^{*,s+1} - c^{*,s}| \leq \varepsilon$ is satisfied, stop; otherwise start a new iteration $s = s+1$ with the new c^{*s} and go back to step2.

This algorithm will be efficient if both the low and upper level problems always solve and generate their appropriate optimization solutions $u_j^{opt}(c)$ $c^{opt}(u)$. From theoretical point of view this cannot be guaranteed, which provides difficulties for the practical applications of the bi-level optimization. For the particular case of this research, the upper level problems, defined and tested for three cases can be defined in explicit analytical forms. Respectively, the solutions of the upper level problems have been derived in analytical relations (12, 16, 18). Additionally, for the low level problem the particular constraint (4) is not restrictive and we can find solutions for the low level problem for each $c > 0$. From practical considerations, the defined upper and low level problems provide appropriate solutions. The convergence of these solutions to a common one is not studied theoretically in this paper. However, the results from the simulations, given below, have not met a lack of convergence.

The numerical simulations contain solutions of optimization problem only for the low level. As an analytical solution (12) has been derived for the upper level problem, this considerably reduces the need for additional computational power for solving the bi-level problem. MATLAB tool has been used for the numerical simulations.

6 Simulation Results

This section presents numerical simulation results for the implementation of the bi-level modelling of the control process. The results have been compared with a classical optimal control problem, where the queue lengths are minimized by changing the green light duration, but the cycle of the traffic light is kept as a constant value. The potential of the bi-level formalism has been illustrated additionally by changing the upper level problem. Three kinds of upper level problems have been simulated:

Case 1: The currently presented goal for maximization of the outgoing flows from queues x_2, x_3. The analytical form of the upper level goal function is (7). Comparisons between the bi-level and single level control policies are done by two criteria: total waiting vehicles, defined by the low level problem; the potential number of vehicles, which are able to make cross turns on the junctions decreasing queues x_2, x_4

Case 2: The goal of the upper level problem is changed to minimize the waiting time for the vehicles, which make turns from queues x_2, x_4. The upper level goal function for that case has analytical form:

$$F(c, u(c)) = \{(c - u_2(c))^2 + (c - u_3(c))^2 - \beta c\}, \qquad (15)$$

where the coefficient β is used to reduce the increase of c to infinity. Following the approach for the approximation of the inexplicit functions $u_j^{opt}(c)$ in Taylor series and substituting them in (15), explicit analytical relation for $F(c)$ is found. Due to the particular simple case of the upper level problem, applying (11) for the case of minimization of $F(c)$ the analytical solution of c^{opt} is

$$c^{opt} = \begin{cases} c^* = c_{\min}, & \text{if } c^* < c_{\min}; \\ c^* = \dfrac{\alpha_2(c^*)\frac{du_2}{dc}|c^* + \alpha_3(c^*)\frac{du_3}{dc}|c^* - G}{(\frac{du_2}{dc}|c^*)^2 + (\frac{du_3}{dc}|c^*)^2}, & \text{if } c_{\min} < c^* < c_{\max} \\ c^* = c_{\max}, & \text{if } c^* > c_{\max}; \end{cases} \quad (16)$$

This value is applied for step 4 of the numerical algorithm for solving the bi-level problem. Comparisons between the single and bi-level control policies are done again by two criteria: total number of waiting vehicles; total waiting time for vehicles in queues x_2, x_4.

Case 3: The third form of the upper level goal function is defined to minimize the queues x_2, x_4 by changing c. The upper level goal function is

$$F(c, x(c)) = \{(x_2(c))^2 + (x_4(c))^2\}. \quad (17)$$

Using the approximations of the inexplicit functions $u_j^{opt}(c)$ from (15) and substituting them in (5), explicit analytical relations about $x_2(c)$, $x_4(c)$ are found, which are used for (17). Thus, analytical form of $F(c)$ is found, which is used in (11) for its minimization towards c. The corresponding solution of c^{opt} is

$$c^{opt} = \begin{cases} c^* = c_{\min}, & \text{if } c^* < c_{\min}; \\ c^* = \dfrac{\delta_1(c^*)\bar{s}\frac{du_2}{dc}|c^* + \delta_3(c^*)\bar{s}\frac{du_3}{dc}|c^*}{(\bar{s}\frac{du_2}{dc}|c^*)^2 + (\bar{s}\frac{du_3}{dc}|c^*)^2}, & \text{if } c_{\min} < c^* < c_{\max} \\ c^* = c_{\max}, & \text{if } c^* > c_{\max}; \end{cases} \quad (18)$$

where δ_1, δ_2, are coefficients, evaluated at each iteration step

$$\delta_1 = x_{20} - \bar{s}u_2(c^*) + \bar{s}\frac{du_2}{dc}|c^*.c^*, \quad \delta_2 = x_{40} - \bar{s}u_3(c^*) + \bar{s}\frac{du_3}{dc}|c^*.c.$$

At each simulation step it has been evaluated both the cycle and the green light values for the different phases. For the next phase the initial queue lengths $x_{i0}, i = 1, \ldots, 8$ are assumed as states $x_i, i = 1, \ldots, 8$, obtained from the solution of the previous control problem. For all cases the parameters, chosen for the simulation are: $s = 1$ (veh/s); $\bar{s} = 0.25$ (veh/s); $c = 60$ s for the single level optimization and initial value for the bi-level one; $c_{\min} = 20$ s, $c_{\max} = 90$ s; $u_{\min} = 10$ s, $u_{\max} = 60$ s; $x_0^T = |70 \quad 60 \quad 70 \quad 60 \quad 70 \quad 60 \quad 70 \quad 60|$.

Results for Case 1:

The bi-level problem gives benefits both for the total number of waiting vehicles and for the increase of the capacity of the junction for the turning flows x_2, x_4. The decrease of the total number of waiting vehicles is faster in comparison with the

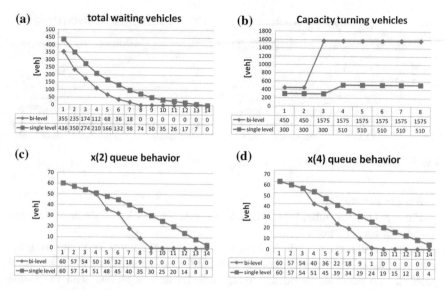

Fig. 3 Comparisons for Case 1

single optimization policy, Fig. 3a. Figure 3b illustrates the increase in the capacity of the flows that make turns in comparison with the single level control. For illustration purposes on Fig. 3c, d are given the behaviours of the flow queues x_2, x_4, which are addressed by the upper-level problem. The simulation results prove faster decrease of these queues in comparison with a single control policy.

Results of Case 2:

For that case the upper level minimizes the waiting times of the queues x_2, x_4. Figure 4a illustrates again faster behaviour in decrease of the total waiting vehicles of the bi-level control. Figure 4b gives information how the waiting time sharply decreases for the queues x_2, x_4. It is interesting that the bi-level control does not give priority to queues x_2, x_4 for the first two cycles. But having integrated the delays, the next cycles result to big decrease, which is bigger in comparison with the single control policy.

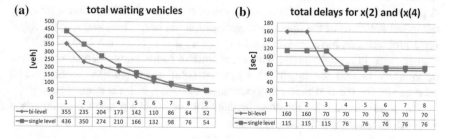

Fig. 4 Comparisons for Case 2

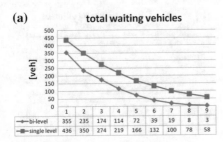
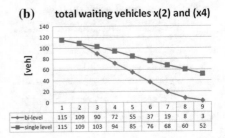

Fig. 5 Comparisons for Case 3

Results for case 3:

For that case the upper level problem supports the low level one in minimizing the queues for directions x_2, x_4.

Figure 5a illustrates again the benefit of the bi-level control for minimizing the total number of waiting vehicles. Having additional support from the upper level problem, the decrease of queues x_2, x_4 is faster in comparison with the single control policy, Fig. 5b.

7 Conclusions

The formal model, which the bi-level approach uses, allows more control variables to be defined as solutions of the control problem. For the research above this results in extended control space including both the duration of the green lights and the traffic cycle. The increased space of the control allows being satisfied extended set of constraints and additional control goals, which cannot be tackled if a single optimization problem is applied. For the particular case, due to extension of the control space the bi-level control optimization can optimize additionally the waiting time, the predefined queue lengths of vehicles, the maximization of the output flows from a part of the traffic network. The research illustrates only a simple cross road section, but it can be extended to an urban network. This will result in increase of the dimensions of the optimization problem. Respectively, the computational power for real time implementation of the bi-level optimization will be also increased. A potential for the real time implementation of the bi-level optimization is given by analytical solutions of the upper level problem. It has been illustrated that the upper level problem can have several forms (cases 1, 2, 3). Thus, the control unit can change them in an appropriate manner, according to the current situation. But the opportunity to find the upper level optimal solution analytically is the general prerequisite for the real time implementation of such a bi-level control strategy.

The simulation results illustrate benefits for the control process in comparison with the classical optimization approach.

Acknowledgments This work is partly supported by project AComIn-"Advanced Computing for Innovation", grant 316087, funded by the FP7 Capacity Programme (Research Potential of Convergence Regions).

References

1. Abouaïssa, H., Fliess, M.: Cédric Join Fast parametric estimation for macroscopic traffic flow model. In: 17th IFAC World Congress (2008). https://hal.inria.fr/inria-00259032
2. Aboudolas, K., Papageorgiou, M., Kosmatopoulas, E.: Store and forward base methods for the signal control problem in large-scale congested urban road networks. Transp. Res. Part C: Emerg. Technol. **17**, 163–174 (2009)
3. Bard, J., Plummer, J., Sourie, J.: Determining tax credits for converting non-food crops to biofuels: an application of bi-level programming. In: Multilevel Optimization: Algorithms and Applications, pp. 23–50 (1998)
4. Bhouri, N., Mayorano, F., Lotito, P., Haj Salem, H., Lebacque, J.: Public transport for multimodal urban traffic control. J. Cybern. Inf. Technol. **15**(5), 17–36 (2015). ISSN 1311-9702
5. Boillot, F., Midenet, S., Pierrelee, J.: Real life CRONOS evaluation. In: Proceedings of the 10th International Conference on Road Traffic Information and Control, N 472, pp. 182–186. IEEE, London (2000)
6. Borga, D., Scerria, K.: Efficient traffic modelling and dynamic control of an urban region. In: 4th International Symposium of Transport Simulation-ISTS'14, 1–4 June 2014, Corsica, France. Transp. Res. Proc. **6**, 224–238 (2015)
7. Chen, W., Jarjees, G., Drane, C.: A new approach for bus priority at signalized intersections. In: Proceedings of 19th ARRB Transport Research LTD Conference, Sydney, Australia (1998)
8. Chen, O.: Integration of dynamic traffic control and assignment, 198 pp. Massachusetts Institute of Technology (1998)
9. Chiou, S.-W.: A cutting plane projection method for bi-level area traffic control optimization with uncertain travel demand. Appl. Math. Comput. **266**, 390–403 (2015)
10. Clegg, J., Smith, M., Xiang, Y., Yarrow, R.: Bi-level programming applied to optimizing urban transportation. Transp. Res. Part B, **35**, 41–70 (2001)
11. Colson, B., Marcotte, P., Savard, G.: An overview of bilevel optimization. Ann. Oper. Res. **153**, 235–256 (2007). doi:10.1007/s10479-007-0176-2
12. De Oliveira, L., Camponogara, E.: Multiagent model predictive control of signaling split in urban traffic networks. Transp. Res. part C: Emerg. Technol. **18**(1), 120–139 (2010)
13. Dempe, S.: Annotated bibliography on bi-level programming and mathematical programs with equilibrium constraints. J. Optim. **52**(3), 333–359 (2003)
14. Dempe, S., Zemkoho, A.B.: A Bilevel Approach for Traffic Management in Capacitated Networks. ISSN 1433–9307 (2008)
15. Dempe, S.: The bi-level road pricing problem. Int. J. Comput. Optim. **2**(2), 71–92 (2015)
16. Diakaki, C., Papageorgiou, M., Aboudolas, K.: A multivariable regulator approach to traffic-responsive network –wide signal control. Control Eng. Pract. **10**(2), 183–195 (2002)
17. Gazis, D., Potts, R.: The oversaturated intersection. In: Proceedings of the Second International Symposium on Traffic Theory, London, UK, pp. 221–237 (1963)
18. Henry, J., Farges, J.: Priority and PRODYN. In: Proceedings of 1st World Congress on Applications of Transport Telematics and Intelligent Vehicle Highway Systems, Paris, France, pp. 3086–3093 (1994)
19. Hunt, P., Robertson, D., Bretherton, R., Role, M.: The SCOOT on-line traffic signal optimization technique. J. Traffic Eng. Control **23**, 190–199 (1982)

20. Mauro, V., Tranto, C.: UTOPIA. In: Proceedings of 6-th IFAC/IFORS/IFIP Symposium on Control, Computers, Communications on Transportation, Paris, France, pp. 245–252 (1989)
21. Papageorgiou, M., Kotsialos, A.: Freeway ramp metering. An overview. In: IEEE Intelligent Transportation Systems. Conference Proceedings, pp. 228–239 (2000)
22. Papageorgiou, M., Diakaki, C., Dinopoulou, V., Kotsialos, A., Wang, Y.: Review of road traffic control strategies. Proc. IEEE **91**(12), 2043–2067 (2003)
23. Patriksson, M., Rockafellar, R.: Mathematical model and descent algorithm for bi-level traffic management. Transp. Sci. **36**(3), 271–291 (2002)
24. Rauova, I.: Evaluation of Different Bilevel Optimization Algorithms with Applications to Control. Diploma thesis FCHPT-5414–28119, p. 69. Slovak University of Technology in Bratislava (2010)
25. Schwartz, S.: Bi-level Programming to Optimize the Use of Traffic Control Gantries for Toll Enforcement. Freie Universitat, Berlin (2013)
26. Spiliopoulou, A., Papamichail, I., Papageorgiou, M., Tyrinopoulosb, I., Chrysoulakis, J.: Macroscopic traffic flow model calibration using different optimization algorithms. Transp. Res. Proc. **6**, 144–157 (2015)
27. Stackelberg, H.: The Theory of the Market Economy. Oxford University Press (1952)
28. Stoilov, T., Stoilova, K., Papageorgiou, M., Papamichail, I.: Bi-level optimization in a transport network. J. Cybern. Inf. Technol. **15**(5), 37–49 (2015). Online ISSN: 1314-4081. doi:10.1515/cait-2015-0023
29. Stoilova, K., Stoilov, T.: Goal and predictive coordination in two level hierarchical systems. Int. J. Gen. Syst. **37**(2), 181–213 (2008)
30. Taale and van Zuylen: The effects of anticipatory traffic control for several small networks, Europe. J. Transp. Infrastruct. Res. **3**(1), 61–76 (2003)
31. Talbi, E.G. (ed.): Metaheuristics for Bi-level Optimization, 286 pp. Springer. ISSN 1860-9503 (electronic), ISBN 978-3-642-37838-6 (eBook) (2013)
32. Tamura, H.: Decentralized optimization for distributed lag models of discrete systems. Automatica **11**, 593–602 (1975)
33. Yang and Yagar: Traffic assignment and signal control in saturated road. Transp. Res. Part A **29**(2), 125–139 (1995)
34. Zhou, Z., De Schutter, B., Lin, S., Xi, Y.: Multi-agent model-based predictive control for large-scale urban traffic networks using a serial scheme. IET Control Theory Appl. **9**(3), 475–484 (2015)

Building of Numerically Effective Kalman Estimator Algorithm for Urban Transportation Network

Aleksey Balabanov

Abstract An effective numerical algorithm of building steady-state Kalman estimator for the urban transportation store-and-forward network model is represented. The proposed approach use special structure of the model and is based on resolvent method to construct a stabilizing solution of discrete algebraic Riccati equation.

Keywords Steady-state Kalman observer · Discrete algebraic Riccati equation · Resolvent method · Urban transportation network

1 Introduction

An urban transportation network (UTN) is a network of linked junctions which are usually controlled by the traffic lights [1]. Traffic lights control is provided by intelligent transportation system (ITS). To improve the UTN traffic management and monitoring, the ITS requires current accurate traffic information [2, 3]. Traffic data may be obtained from various types of road sensors and devices such as inductive loop detectors, microwave detectors, cameras, etc. The data may be also estimated by some methods which use GPS devices or drivers mobile phones. Unfortunately the accuracy of mentioned traffic data sources is not enough and received data must be filtered and estimated by ITS.

There are many approaches presented by different authors to solve this issue, see for example [2, 4, 5] and the following references within. Good estimation results showed approaches which used some variant of the Kalman Filter [2, 6]. The Kalman filter or Kalman estimator can be adapted to track various categories of traffic information of UTN given multisource noisy sensor data.

This paper considers estimation of vehicles amount in queues which are appeared in front of controlled junctions in saturated UTN. The paper proposes to

A. Balabanov (✉)
Institute of Information and Communication Technologies, 1113 Sofia, Bulgaria
e-mail: alexey.balabanov83@gmail.com

© Springer International Publishing Switzerland 2016   319
S. Margenov et al. (eds.), *Innovative Approaches and Solutions in Advanced
Intelligent Systems*, Studies in Computational Intelligence 648,
DOI 10.1007/978-3-319-32207-0_21

use a steady-state Kalman estimator. The main result of the paper is an approach to build steady-state Kalman estimator which is applicable for on-line computations. The proposed approach is based on special structure of UTN model and use resolvent method to find stabilizing solution of continuous type algebraic Riccati equation as solution of discrete algebraic Riccati equation. The paper presents evaluation of the numerical algorithm to show its efficiency. In a final part of the paper a numerical example was presented to illustrate main stages of computation process.

2   Problem Formulation

The well-known Gazis and Potts store-and-forward model [1] of the UTN is considered. This model describes in a simple way the flow process in an UTN and represents UTN as discrete linear state-space plant. When in the plant are presented disturbances and sensors noise, the plant state should be estimated. As it is known [7], state estimation of the plant might be provided by Kalman estimator. The become sections briefly overview steady-state Kalman estimation problem and UTN state-space model building. The problem formulation is given in the end of Sect. 2.2.

2.1   Overview of Steady-State Kalman Estimator Building

For a given discrete plant

$$\left.\begin{array}{l}\mathbf{x}(k+1) = \mathbf{A}_d\mathbf{x}(k) + \mathbf{B}_d\mathbf{u}(k) + \mathbf{F}_d\mathbf{w}(k), \quad \mathbf{x}(0) = \mathbf{x}_0; \\ \mathbf{y}(k) = \mathbf{C}_d\mathbf{x}(k) + \mathbf{v}(k)\end{array}\right\} \qquad (1)$$

and the processes $\mathbf{w}(k)$, $\mathbf{v}(k)$ covariance data $E\{\mathbf{w}(k)\mathbf{w}^T(k)\} = \mathbf{Q}_d \geq 0$, $E\{\mathbf{v}(k)\mathbf{v}^T(k)\} = \mathbf{R}_d > 0$ the optimal steady-state Kalman estimator [7] constructs a state estimate $\hat{\mathbf{x}}(k) \in R^n$ that minimizes the mean square error $E\{e(k)^T e(k)\}$ in the limit $k \to \infty$, where vector $e(k) = \hat{\mathbf{x}}(k) - \mathbf{x}(k) \in R^n$—reconstruction error; $\mathbf{x}(k) \in R^n$—state; $\mathbf{u}(k) \in R^m$—control; $\mathbf{x}_0$—initial state; $\mathbf{y}(k) \in R^r$, $\mathbf{v}(k) \in R^r$—measurement and measurement noise respectively; $\mathbf{w}(k) \in R^l$—disturbance. When the processes $\mathbf{w}(k)$, $\mathbf{v}(k)$ are uncorrelated, the Kalman estimator is represented as state equation

$$\hat{\mathbf{x}}(k+1) = \mathbf{A}_d\hat{\mathbf{x}}(k) + \mathbf{B}_d\mathbf{u}(k) + \mathbf{L}_d(\mathbf{y}(k) - \mathbf{C}_d\hat{\mathbf{x}}(k)), \quad \hat{\mathbf{x}}(0) = \hat{\mathbf{x}}_0, \qquad (2)$$

where estimator gain matrix is given by $\mathbf{L}_d = \mathbf{A}_d \mathbf{X} \mathbf{C}_d^T (\mathbf{C}_d \mathbf{X} \mathbf{C}_d^T + \mathbf{R}_d)^{-1}$ and $\mathbf{X}$ is derived by solving a discrete algebraic Riccati equation (DARE)

$$\mathbf{X} = \mathbf{A}_d \mathbf{X} \mathbf{A}_d^T + \bar{\mathbf{Q}}_d - \mathbf{A}_d \mathbf{X} \mathbf{C}_d^T (\mathbf{C}_d \mathbf{X} \mathbf{C}_d^T + \mathbf{R}_d)^{-1} \mathbf{C}_d \mathbf{X} \mathbf{A}_d^T, \quad \bar{\mathbf{Q}}_d = \mathbf{F}_d \mathbf{Q}_d \mathbf{F}_d^T. \quad (3)$$

The Kalman estimator approach has the following limitations: (1) the pair $(\mathbf{C}_d, \mathbf{A}_d)$ should be detectable; (2) the pair $(\mathbf{A}_d, \mathbf{N})$ has to have no uncontrollable unstable mode, where $\mathbf{N}$ corresponds to factorization $\bar{\mathbf{Q}}_d = \mathbf{N}\mathbf{N}^T$.

2.2 Overview of UTN State-Space Model Building

Let the considered UTN has n^{src} external vehicles sources f_l, $l \in \{1, 2, \ldots, n^{src}\}$ and n^{snk} outgoing of UTN vehicles sinks h_r, $r \in \{1, 2, \ldots, n^{snk}\}$. Let also, the UTN consists of p traffic light controlled junctions and all junctions are connected by n links. The traffic lights change their phases (stages) periodically to provide each queue right of way. For sake of simplicity, we will suppose that each link include only one queue with x_i, $i \in \{1, 2, \ldots, n\}$ vehicles in it and also suppose that each traffic light cycle time is a constant value and equal to c seconds. Let for j-th junction cycle is split by m_j, $j \in \{1, 2, \ldots, p\}$ phases. Each r-th phase of j-th junction provides for queues set $\Omega_{r,j}$ right of way during an effective green time $g_{r,j}^e$. The phase also includes a lost time $l_{r,j}$ then traffic flow can't be used.

A junction control may be implemented by changing effective green times duration within the cycle (it is so called splits control [1]). There are only $m_j - 1$ effective green time variables to change. The last effective green time dependent from other one. Generally, UTN has $m = \sum_{j=1}^{p} m_j - p$ independent variables $g_{1,1}^e, g_{2,1}^e, \ldots, g_{m_1-1,1}^e, g_{1,2}^e, \ldots, g_{m_p-1,p}^e$. Let g_i is a green time for i-th queue. According to sets $\Omega_{r,j}$, to each g_i corresponds appropriate effective green time $g_{r,j}^e$ (or sum of them). For relative green times variables $\hat{g}_i = g_i/c$, $\hat{g}_{i,j}^e = g_{i,j}^e/c, i, j = 1, 2, \ldots$ and vectors $\hat{\mathbf{g}} = [\hat{g}_1, \hat{g}_2, \ldots, \hat{g}_n]^T$, $\hat{\mathbf{g}}^e = [\hat{g}_{1,1}^e, \hat{g}_{2,1}^e, \ldots, \hat{g}_{m_1-1,1}^e, \hat{g}_{1,2}^e, \ldots, \hat{g}_{m_p-1,p}^e]^T$ there is a linear relation

$$\hat{\mathbf{g}} = \mathbf{G}_{TN} \hat{\mathbf{g}}^e + \mathbf{L}_{TN}, \quad \mathbf{G}_{TN} \in R^{n \times m}, \quad \mathbf{L}_{TN} \in R^n. \quad (4)$$

The change of i-th queue length (vehicles amount) in k-th cycle can be represented by the equation

$$x_i(k+1) = x_i(k) + q_i^{in}(k) - q_i^{out}(k) + d_i^{in}(k) - d_i^{out}(k), i \in \{1, 2, \ldots, n\}, \quad (5)$$

where $x_i(k)$, $x_i(k+1)$ are the i-th queue lengths at the beginning of k-th and $(k+1)$-th cycle respectively; q_i^{in}, q_i^{out} queue outside input and output respectively; d_i^{in}, d_i^{out} are input and output within i-th link respectively. Let variance $d_i^{dif} = d_i^{in} - d_i^{out}$. Then, let $s_{i,j}$ denotes vehicles exchange from j-th to i-th queue; $f_{i,j}$ means vehicles came from j-th source to i-th queue; $h_{i,j}$ means vehicles outcome from j-th queue to i-th sink. Obviously, value q_i^{in} consists of external sources and from other queues income, $q_i^{in} = \sum_{l=1}^{n^{src}} f_{i,l} + \sum_{j=1, j \neq i}^{n} s_{i,j}$. The same holds for value $q_i^{out} = \sum_{r=1}^{n^{snk}} h_{r,i} + \sum_{e=1, e \neq i}^{n} s_{e,i}$.

A passed through vehicles amount z during time τ, (s), may be expressed by vehicles flow v, (veh/s), as $z = v\tau$. We suppose that outgoing flow can be split toward n directions as $z = \sum_{i=1}^{n} z_i = \sum_{i=1}^{n} \lambda_i v\tau$, where λ_i is a proportion toward i-th direction. It is assumed that $0 \leq \lambda_i \leq 1$, $\sum_{i=1}^{n} \lambda_i = 1$ and values v, λ_i are random numbers with known statistic characteristics. Let $v = \bar{v} + \Delta v$, $\lambda_i = \bar{\lambda}_i + \Delta\lambda_i$, where $\bar{v} = E\{v\}$, $\bar{\lambda}_i = E\{\lambda_i\}$ are average of values v, λ_i respectively and Δv, $\Delta\lambda_i$ are the appropriate centered random numbers. The denotation $E\{\}$ means mathematical expectation.

Then, similarly to equation $z = \sum_{i=1}^{n} \lambda_i v\tau = \sum_{i=1}^{n} (\bar{\lambda}_i + \Delta\lambda_i)(\bar{v} + \Delta v)\tau$, the above values can be represented as $f_{i,l} = (\bar{\lambda}_{i,l}^f + \Delta\lambda_{i,l}^f)(\bar{v}_l^f + \Delta v_l^f) c = \bar{\lambda}_{i,l}^f \bar{v}_l^f c + \psi_{i,l}^f c$, $s_{i,j} = (\bar{\lambda}_{i,j}^s + \Delta\lambda_{i,j}^s)(\bar{v}_j^s + \Delta v_j^s) g_j = \bar{\lambda}_{i,j}^s \bar{v}_j^s g_j + \psi_{i,j}^s g_j$, $h_{r,i} = (\bar{\lambda}_{r,i}^h + \Delta\lambda_{r,i}^h)(\bar{v}_i^h + \Delta v_i^h) g_i = \bar{\lambda}_{r,i}^h \bar{v}_i^h g_i + \psi_{r,i}^h g_i$, $s_{e,i} = (\bar{\lambda}_{e,i}^s + \Delta\lambda_{e,i}^s)(\bar{v}_i^s + \Delta v_i^s) g_i = \bar{\lambda}_{e,i}^s \bar{v}_i^s g_i + \psi_{e,i}^s g_i$, $d_i^{dif} = (\bar{v}_i^d + \Delta v_i^d) c$, where $\bar{\lambda}_{i,l}^f$, $\bar{v}_l^f$ are average values of splits and flows from source f_l toward i-th queue, $\Delta\lambda_{i,l}^f$, Δv_l^f are deviations of these values, and $\psi_{i,l}^f = \Delta\lambda_{i,l}^f \bar{v}_l^f + \bar{\lambda}_{i,l}^f \Delta v_l^f + \Delta\lambda_{i,l}^f \Delta v_l^f$. The similar meanings have the rest of values.

By collecting the above, we can rewrite Eq. (5) in matrix form

$$\mathbf{x}(k+1) = \mathbf{x}(k) + c\mathbf{S}_{TN}\hat{\mathbf{g}}(k) + c\mathbf{T}_{TN} + c\mathbf{D}_{TN} + c\mathbf{w}(k), \qquad (6)$$

where $\mathbf{x}(k) = [x_1(k), x_2(k), \ldots, x_n(k)]^T$, $\mathbf{S}_{TN} \in R^{n \times n}$ with elements $\mathbf{S}_{TN}(i,j) = \begin{cases} \bar{\lambda}_{i,j}^s \bar{v}_j^s, & i \neq j; \\ -\sigma_i, & i = j, \end{cases}$ $\mathbf{T}_{TN} = [\phi_1 \quad \phi_2 \quad \cdots \quad \phi_n]^T$, $\mathbf{D}_{TN} = [\bar{v}_1^d \quad \bar{v}_2^d \quad \cdots \quad \bar{v}_n^d]^T$, $\mathbf{w}(k) = [w_1(k) \ w_2(k) \ \ldots \ w_n(k)]^T$, $\phi_i = \sum_{l=1}^{n^{src}} \bar{\lambda}_{i,l}^f \bar{v}_{i,l}^f$, $\sigma_i = (\sum_{r=1}^{n^{snk}} \bar{\lambda}_{r,i}^h \bar{v}_i^h + \sum_{e=1, e \neq i}^{n} \bar{\lambda}_{e,i}^s \bar{v}_i^s)$, $w_i(k) = \sum_{l=1}^{n^{src}} \psi_{i,l}^f + \sum_{j=1, j \neq i}^{n} \psi_{i,j}^s \hat{g}_j - \sum_{r=1}^{n^{snk}} \psi_{r,i}^h \hat{g}_i - \sum_{e=1, e \neq i}^{n} \psi_{e,i}^s \hat{g}_i + \Delta v_i^d(k)$, $i \in \{1, 2, \ldots, n\}$.

Substitution (4) to Eq. (6) yields

$$\mathbf{x}(k+1) = \mathbf{x}(k) + c\mathbf{S}_{TN}\mathbf{G}_{TN}\hat{\mathbf{g}}^e(k) + c\mathbf{S}_{TN}\mathbf{L}_{TN} + c\mathbf{T}_{TN} + c\mathbf{D}_{TN} + c\mathbf{w}(k).$$

According to [1], there exists a nominal situation when the input of each queue equals to its output, $\mathbf{x}(k+1) = \mathbf{x}(k)$. This condition holds when

$$c\mathbf{S}_{TN}\mathbf{G}_{TN}\hat{\mathbf{g}}^N + c\mathbf{N}_{TN} = 0, \tag{7}$$

where $\hat{\mathbf{g}}^N$ is a nominal green time vector and $\mathbf{N}_{TN} = \mathbf{S}_{TN}\mathbf{L}_{TN} + \mathbf{T}_{TN} + \mathbf{D}_{TN}$ comprises network demand vector. Obviously, this situation exists when the nominal green time vector $\hat{\mathbf{g}}^N$ is a unique solution of linear Eq. (7), i.e. when $rank(\mathbf{S}_{TN}\mathbf{G}_{TN}) = rank([\mathbf{S}_{TN}\mathbf{G}_{TN}\,\mathbf{N}_{TN}]) \leq m$. By presenting $\hat{\mathbf{g}}^e(k) = \hat{\mathbf{g}}^N + \mathbf{u}(k)$ the whole UTN model can be represented as a discrete-time linear state-space model (1) which specified by matrices

$$\mathbf{A}_d = \mathbf{I}_n, \ \mathbf{B}_d = c\mathbf{S}_{TN}\mathbf{G}_{TN} \in R^{n \times m}, \ \mathbf{F}_d = c\mathbf{I}_n, \ \mathbf{C}_d \in R^{n \times r}, \tag{8}$$

where $\mathbf{I}_n$ is a $n \times n$ identity matrix, $\mathbf{u}(k) = \hat{\mathbf{g}}^e(k) - \hat{\mathbf{g}}^N$, $\mathbf{u} \in R^m$.

For further Kalman estimator synthesis issue, we must discuss limitations mentioned in the end of Sect. 2.1. For the first condition matrix $\mathbf{V}^T = \left[\mathbf{C}_d^T, \mathbf{A}_d^T\mathbf{C}_d^T, \ldots, \left(\mathbf{A}_d^T\right)^{n-1}\mathbf{C}_d^T\right] \in R^{n \times (n \times r)}$ has to have full $rank(\mathbf{V}) = n$. Since $\mathbf{A}_d$ is an identity matrix, this condition holds when matrix $\mathbf{C}_d \in R^{r \times n}$ size $r \geq n$ and $rank(\mathbf{C}_d) = n$. The second condition requires that the pair $(\mathbf{A}_d, \mathbf{N})$, has to have no uncontrollable unstable mode, i.e. the pair $(\mathbf{A}_d, \mathbf{N})$ should be stabilizable, $\mathbf{N}$ corresponds to $\bar{\mathbf{Q}}_d = c^2\mathbf{Q}_d = \mathbf{N}\mathbf{N}^T$. Since all matrix $\mathbf{A}_d$ modes equal to one we will suppose a strict inequality for disturbance covariance matrix $\mathbf{Q}_d > 0$. The last condition is natural since each queue length equation includes value $w_i(k)$. It is worth to mention that if there is a way from i-th to j-th queue, then processes w_i and w_j may correlate to each other due to fact that value $\psi_{i,j}^s$ is included to both values w_i, w_j with different sign.

When there is necessary information, for given store-and-forward model (1) and (8) and disturbances covariance data $\mathbf{Q}_d\,\mathbf{R}_d$ the state variables (vehicles amount which stay in the queues) can be estimated by steady-state Kalman estimator. For a certain period of time, matrices $\mathbf{Q}_d$, $\mathbf{R}_d$ can be left unchanged. But since of weather conditions or traffic demand changing, the disturbances characteristics (matrices $\mathbf{Q}_d$, $\mathbf{R}_d$ elements) may become significantly changed. In this case steady-state Kalman estimator should be rebuilt for a new data. Since a real UTN may be significantly large, a problem may appear and the ITS may require a fast algorithm to estimate the UTN conditions.

The paper proposes the Kalman estimator (2) building algorithm for UTN model (1) and (8) which numerically reduce problem.

3 Building of Discrete Kalman Estimator for UTN by Resolvent Method

3.1 Numerical Development of Resolvent Method

The resolvent method [8] propose methodology for finding stabilizing solution $\mathbf{X}$ of continuous algebraic Riccati equation (CARE)

$$\mathbf{A}^T\mathbf{X} + \mathbf{X}\mathbf{A} + \mathbf{P} - \mathbf{X}\mathbf{Q}\mathbf{X} = \mathbf{0}_{n,n}. \tag{9}$$

It is known [9] that there is a relation between discrete and continuous type algebraic Riccati equations solutions. This relation allows using numerical algorithm to solve DARE (3) by finding CARE (9) solution with known matrices defined as

$$\mathbf{A} = \mathbf{I}_n - 2\mathbf{\Delta}^{-T}, \ \mathbf{Q} = 2(\mathbf{I}_n + \mathbf{A}_d^T)^{-1}\bar{\mathbf{R}}_d^{-1}\mathbf{\Delta}^{-1}, \ \mathbf{P} = 2\mathbf{\Delta}^{-1}\bar{\mathbf{Q}}_d(\mathbf{I}_n + \mathbf{A}_d^T)^{-1}, \tag{10}$$

where $\mathbf{\Delta} = \mathbf{I}_n + \mathbf{A}_d + \bar{\mathbf{Q}}_d(\mathbf{I}_n + \mathbf{A}_d^T)^{-1}\bar{\mathbf{R}}_d$, $\bar{\mathbf{Q}}_d = \mathbf{F}_d\mathbf{Q}_d\mathbf{F}_d^T$, $\bar{\mathbf{R}}_d = \mathbf{C}_d^T\mathbf{R}_d^{-1}\mathbf{C}_d$.

A developed numerical resolvent method consists of the following procedures. For given Hamilton matrix $\mathbf{H} = \begin{bmatrix} \mathbf{A} & -\mathbf{Q} \\ -\mathbf{P} & -\mathbf{A}^T \end{bmatrix}$ a characteristic polynomial

$$\Delta(s) = \det(s\mathbf{I}_{2n} - \mathbf{H}) = \delta(x)|_{x=s^2} = x^n + \delta_1 x^{n-1} + \cdots + \delta_{n-1}x + \delta_n \tag{11}$$

has to be found. Then, quadratic functionals

$$\gamma_r = \frac{1}{2\pi j} \int_C \frac{x^r}{\delta(x)} ds, \ x = s^2, \ r = 0, 1, \ldots, n-1 \tag{12}$$

have to be computed and their linear combinations

$$\eta_m = \gamma_0 \delta_m + \gamma_1 \delta_{m-1} + \cdots + \gamma_m \delta_0, \ \delta_0 = 1 \ m = 0, 1, \ldots, n-1. \tag{13}$$

Above C is an anticlockwise closed contour in right half complex plane (s) with shape of semicircle with diameter on imaginary axes. Then by matrices

$$\mathbf{U} = \frac{1}{2}\mathbf{I}_{2n} + \mathbf{H}\mathbf{S}, \ \mathbf{S} = \sum_{k=0}^{n-1} \eta_{n-1-k}\mathbf{G}^k, \ \mathbf{G} = \mathbf{H}^2, \tag{14}$$

the CARE (9) stabilizing solution can be found from linear equation

$$\mathbf{U}_2\mathbf{X} + \mathbf{U}_1 = \mathbf{0}_{2n,n}, \ \mathbf{U}_1, \ \mathbf{U}_2 \in R^{2n \times n}, \ \mathbf{U} = [\mathbf{U}_1 \ \mathbf{U}_2]. \tag{15}$$

For the matrices (14) is convenient to use $n \times n$ size block representations

$$\mathbf{G} = \mathbf{H}^2 = \begin{bmatrix} \mathbf{A}^2 + \mathbf{QP} & \mathbf{QA}^T - \mathbf{AQ} \\ \mathbf{A}^T\mathbf{P} - \mathbf{PA} & \mathbf{PQ} + \mathbf{A}^{2T} \end{bmatrix} = \begin{bmatrix} \mathbf{G}_{11} & \mathbf{G}_{12} \\ \mathbf{G}_{21} & \mathbf{G}_{22} \end{bmatrix}, \tag{16}$$

$$\mathbf{G}^k = \begin{bmatrix} \mathbf{G}_k^{11} & \mathbf{G}_k^{12} \\ \mathbf{G}_k^{21} & \mathbf{G}_k^{22} \end{bmatrix}, \ k = 1, 2, \ldots, \ \mathbf{S} = \begin{bmatrix} \mathbf{S}_{11} & \mathbf{S}_{12} \\ \mathbf{S}_{21} & \mathbf{S}_{22} \end{bmatrix}, \ \mathbf{U} = \begin{bmatrix} \mathbf{U}_{11} & \mathbf{U}_{12} \\ \mathbf{U}_{21} & \mathbf{U}_{22} \end{bmatrix}.$$

The numbers (12), can be computed either as residuals sum inside counter C or as certain integrals

$$\gamma_r = \frac{1}{\pi} \int_0^1 \{h_r(z) - g_r(z)\} \, d\tau, \ z = -\tau^2, \ r = 0, 1, \ldots, n-1, \tag{17}$$

which follow from (12) after contour C transformation to an imaginary axis. Above in (17) subintegral functions $g_r(x) = \frac{x^r}{\delta(x)}$, $h_r(\xi) = \frac{\xi^{n-r-1}}{\mu(\xi)}$, $r = 0, 1, \ldots, n-1$ and $\mu(\xi) = \frac{1}{x^n}\delta(x)$, $\xi = \frac{1}{x}$.

3.2 Resolvent Method Applying for UTN Model Data

This section develops reduction of resolvent method by applying equalities (10) with given matrices (8). Matrices (8) substitution into (10) yields

$$\mathbf{A} = \mathbf{I}_n - 2\mathbf{\Delta}^{-T}, \ \mathbf{Q} = \bar{\mathbf{R}}_d\mathbf{\Delta}^{-1}, \ \mathbf{P} = \mathbf{\Delta}^{-1}\bar{\mathbf{Q}}_d, \ \mathbf{\Delta} = 2\mathbf{I}_n + \frac{1}{2}\bar{\mathbf{Q}}_d\bar{\mathbf{R}}_d, \tag{18}$$

where $\bar{\mathbf{Q}}_d = c^2\mathbf{Q}_d$, $\bar{\mathbf{R}}_d = \mathbf{C}_d^T\mathbf{R}_d^{-1}\mathbf{C}_d$. Matrices (18) transposes are

$$\mathbf{A}^T = \mathbf{I}_n - 2\mathbf{\Delta}^{-1}, \ \mathbf{Q} = \mathbf{Q}^T = \mathbf{\Delta}^{-T}\bar{\mathbf{R}}_d; \ \mathbf{P} = \mathbf{P}^T = \bar{\mathbf{Q}}_d\mathbf{\Delta}^{-T}, \ \mathbf{\Delta}^T = 2\mathbf{I}_n + \frac{1}{2}\bar{\mathbf{R}}_d\bar{\mathbf{Q}}_d. \tag{19}$$

Let us consider matrix

$$\mathbf{AQ} = (\mathbf{I}_n - 2\mathbf{\Delta}^{-T})\bar{\mathbf{R}}_d\mathbf{\Delta}^{-1} = (\bar{\mathbf{R}}_d\mathbf{\Delta}^{-1} - 2\mathbf{\Delta}^{-T}\bar{\mathbf{R}}_d\mathbf{\Delta}^{-1}). \tag{20}$$

Matrix $\mathbf{AQ}$ is symmetric since right part of equality (20) is a difference of two symmetric matrices. The same is easy to show for matrix $\mathbf{PA} = (\mathbf{PA})^T$. Consequently, from (16) and (14) and matrices $\mathbf{AQ}$, $\mathbf{PA}$ symmetry follows $\mathbf{G}_{12} = \mathbf{G}_{21} = \mathbf{G}_k^{12} = \mathbf{G}_k^{21} = \mathbf{S}_{12} = \mathbf{S}_{21} = \mathbf{0}_{n,n}$.

Then, according to equalities (18) and (19) and by sequence of transformations

$$\mathbf{A}^2 + \mathbf{QP} = \left(\mathbf{I}_n - 2\Delta^{-T}\right)^2 + \bar{\mathbf{R}}_d^{-1}\Delta^{-1}\Delta^{-1}\bar{\mathbf{Q}}_d = \mathbf{I}_n - 4\Delta^{-T} + 4\Delta^{-2T} + \Delta^{-2T}\bar{\mathbf{R}}_d^{-1}\bar{\mathbf{Q}}_d$$

$$= \mathbf{I}_n - 4\Delta^{-T} + 2\Delta^{-2T}\left(2\mathbf{I}_n + \frac{1}{2}\bar{\mathbf{R}}_d^{-1}\bar{\mathbf{Q}}_d\right) = \mathbf{I}_n - 2\Delta^{-T} = \mathbf{A}$$

matrix $\mathbf{G}_{11}$ is reduced to $\mathbf{G}_{11} = \mathbf{A}$.

Hence, since $\mathbf{G}_{11} = \mathbf{G}_{22}^T$, $\det(x\mathbf{I}_n - \mathbf{G}) = \delta(x)\delta(x)$, $\delta(\mathbf{G}) = \Delta(\mathbf{H}) = \mathbf{0}$ [8], there is a representation for polynomial (11),

$$\delta(x) = \det(x\mathbf{I}_n - \mathbf{A}) \tag{21}$$

and $\mathbf{S}_{11} = \sum_{k=0}^{n-1} \eta_{n-1-k}\mathbf{A}^k$.

An effective way to find last sum is to apply first for matrix

$$\mathbf{A} = \mathbf{T}\tilde{\mathbf{A}}\mathbf{T}^{-1} \tag{22}$$

linear transformation that makes matrix $\tilde{\mathbf{A}}$ sparse. Then, matrix $\mathbf{S}_{11}$ can be represented as

$$\mathbf{S}_{11} = \mathbf{T}\tilde{\mathbf{S}}_{11}\mathbf{T}^{-1} \tag{23}$$

$$\text{where } \tilde{\mathbf{S}}_{11} = \sum_{k=0}^{n-1} \eta_{n-1-k}\tilde{\mathbf{A}}^k. \tag{24}$$

Let transformation (22) is a transformation to a normal Frobenius form [10]. Then, matrix $\mathbf{T}$ is also a system matrix of linear equation

$$\mathbf{T}\delta = \mathbf{t}_n, \tag{25}$$

of Krylov's method [10]. Krylov's method finds characteristic polynomial of matrix $\mathbf{A}$. In (25) $\delta = [\delta_0, \delta_1, \ldots, \delta_{n-1}]^T$ is vector of unknown characteristic polynomial (21) coefficients and matrix $\mathbf{T} = [\mathbf{t}_0, \mathbf{t}_1, \ldots, \mathbf{t}_{n-1}]$ is built by vector series

$$\mathbf{t}_k = \mathbf{A}\mathbf{t}_{k-1}, \; k = 1, 2, \ldots, n, \; \mathbf{t}_0 = \mathbf{t}, \tag{26}$$

where initial vector $\mathbf{t}$ should be chosen to produce invertible matrix $\mathbf{T}$. According to [10] such vector always exists.

Finally, matrix (14) can be represented in a simplified way as

$$\mathbf{U} = \begin{bmatrix} 0.5\mathbf{I}_n + \mathbf{AS}_{11} & -\mathbf{QS}_{11}^T \\ -\mathbf{PS}_{11} & 0.5\mathbf{I}_n - \mathbf{A}^T\mathbf{S}_{11}^T \end{bmatrix}$$

and the stabilizing solution of CARE can be found as a least squares solution of linear Eq. (15).

Alternatively, the CARE stabilizing solution may be found by using representation $\mathbf{X} = \mathbf{Y} + \mathbf{Q}^{-1}\mathbf{A}$. It may exist since $\mathbf{Q}$ is invertible matrix $(\mathbf{Q}^{-1} = \Delta\bar{\mathbf{R}}_d^{-1}, \bar{\mathbf{R}}_d > 0)$ and $\mathbf{Q}^{-1}\mathbf{A}$ is a symmetric positive defined matrix which equal to

$$\mathbf{Q}^{-1}\mathbf{A} = \Delta\bar{\mathbf{R}}_d^{-1}\left(\mathbf{I}_n - 2(\Delta^{-1})^T\right) = \bar{\mathbf{R}}_d^{-1}\Delta^T\left(\mathbf{I}_n - 2(\Delta^{-1})^T\right)$$

$$= \left(\bar{\mathbf{R}}_d^{-1}\Delta^T - 2\bar{\mathbf{R}}_d^{-1}\right) = \bar{\mathbf{R}}_d^{-1}\left(\Delta^T - 2\mathbf{I}_n\right) = \frac{1}{2}\bar{\mathbf{Q}}_d.$$

The representation leads to reduced CARE

$$\bar{\mathbf{P}} - \mathbf{YQY} = \mathbf{0}_{n,n}, \tag{27}$$

where $\bar{\mathbf{P}} = \mathbf{A}^T\mathbf{Q}^{-1}\mathbf{A} + \mathbf{P}$. Resolvent method applying for CARE (27) gives appropriate matrices $\bar{\mathbf{G}}_{11} = \mathbf{Q}\bar{\mathbf{P}}$, $\bar{\mathbf{S}}_{11}$, $\bar{\mathbf{G}}_{12} = \bar{\mathbf{G}}_{21} = \bar{\mathbf{G}}_k^{12} = \bar{\mathbf{G}}_k^{21} = \bar{\mathbf{S}}_{12} = \bar{\mathbf{S}}_{21} = \mathbf{0}_{n,n}$. Since $\mathbf{AQ} = \mathbf{QA}^T$ matrix $\bar{\mathbf{G}}_{11} = \left(\mathbf{QA}^T\mathbf{Q}^{-1}\mathbf{A} + \mathbf{QP}\right) = \left(\mathbf{A}^2 + \mathbf{QP}\right) = \mathbf{A}$. It yields $\mathbf{G}_{11} = \bar{\mathbf{G}}_{11}$ and $\mathbf{S}_{11} = \bar{\mathbf{S}}_{11}$. Then, CARE (27) solution can be expressed as $\mathbf{Y} = 2\bar{\mathbf{P}}\mathbf{S}_{11}$. With respect to equality $\bar{\mathbf{P}} = \mathbf{A}^T\mathbf{Q}^{-1}\mathbf{A} + \mathbf{P} = \frac{1}{2}\bar{\mathbf{Q}}_d$ the CARE (9) solution can be presented as

$$\mathbf{X} = \bar{\mathbf{Q}}_d(0.5\mathbf{I}_n + \mathbf{S}_{11}). \tag{28}$$

Finally, Kalman estimator gain matrix can be computed as $\mathbf{L}_d = \mathbf{XC}_d^T(\mathbf{C}_d\mathbf{XC}_d^T + \mathbf{R}_d)^{-1}$. By summarizing the above thoughts we can suggest the following

Algorithm 1:

1. For given matrices $\mathbf{C}_d$, $\mathbf{Q}_d$, $\mathbf{R}_d$ find matrices $\bar{\mathbf{Q}}_d$, $\bar{\mathbf{R}}_d$, Δ, $\mathbf{A}$ form (18).
2. Build vector series (26) and find matrix $\mathbf{T}^{-1}$.
3. Find polynomial (21) coefficients from Eq. (25) on base $\mathbf{T}^{-1}$.
4. Find numbers (12) as certain integrals (17) by Simpson quadrature or alternatively as sum of residuals.
5. Find numbers (13).
6. Find sum (24) and find matrix (23).
7. Find stabilizing solution of CARE (9) from representation (28).
8. Find Kalman estimator gain matrix $\mathbf{L}_d$.

3.3   Flops Evaluation

This section evaluates numerical performance of the proposed algorithm by calculating flops. Let a flop is one of the following operation $+, -, \times, /, >$, etc., under floating point numbers. We will take into account just $O(n^3)$ complexity procedures of the proposed algorithm. The third and fifth steps are $O(n^2)$ complexity processes.

From [10, 11] we know the following flops evaluations, see Table 1. Let $\mathbf{C}_d$ is a real $n \times n$ matrix. Then, let for the forth step of Algorithm 1 the values (12) are found as sum of residuals. Here a polynomial roots search is a dominant computational procedure. It is possible to construct compact numerical procedure ξ_7 for sum (24) computation which requires about $2n^3$ flops.

Then, the Algorithm 1 requires: for first step $2\xi_1 + \xi_2 + \xi_4 + \xi_5 = 10n^3$ flops, for second $n\xi_3 + \xi_4 = 4n^3$ flops, for forth $\xi_6 = 13n^3$ flops, the sixth $2\xi_1 + \xi_7 = 6n^3$, for seventh $\xi_2 = \frac{2}{3}n^3$ flops and the last eighth step requires $2\xi_1 + \xi_2 + \xi_5 = 5\frac{1}{3}n^3$ flops. Totally, the proposed algorithm requires approximately $39n^3$ flops. The CARE (and appropriate DARE) solution computation requires about $24n^3$ flops.

The performance of the proposed approach may be compared by others algorithms to solve CARE/DARE. To find CARE/DARE solution Schur approach is usually used. According to [12], Schur approach costs about $240n^3$ flops. The last evaluation is given for Schur approach which process data of general structure. The Schur approach for data (8) is unknown. The DARE solution can be also found as steady-state solution of time-varying DARE. This approach is easy to adapt for data (8) which will cost about $(6k + 4)n^3$ flops, where k is a number of steps which we need to reach steady-state solution.

Table 1 Flops evaluations

Operation	Denotation	Flops
Two square $n \times n$ matrices multiplication	ξ_1	$2n^3$
Two square $n \times n$ matrices multiplication when the result is a symmetric matrix	ξ_2	$\frac{2}{3}n^3$
$n \times n$ matrix and n size vector multiplication	ξ_3	$2n^2$
$n \times n$ matrix inversion (in case LU decomposition use)	ξ_4	$2n^3$
$n \times n$ symmetric matrix inversion (in case LU decomposition use)	ξ_5	$\frac{2}{3}n^3$
n-th degree polynomial roots searching (in case using technique of finding companion matrix eigenvaluen	ξ_6	$13n^3$

3.4 Some Remarks About the Approach

Remark 1 The reduction of resolvent method procedures is due to the fact that the system (1) matrix $\mathbf{A}_d$ is an identity matrix. Clearly that these kind of reductions can be also reached in the dual problem of linear-quadratic optimal control.

Remark 2 It is worth to mention that the Kalman estimator building has sense when queue length measurements are not perfect. In the UTN model where there are noise free components, the appropriate states should be excluded [7]. The reduced UTN model will keep the initial properties and the proposed approach will be also realizable.

Remark 3 A linear transformation (22) to normal Frobenius form may be source of possible significant numerical error. It can be shown that matrix $\mathbf{A}$ specter belong to unit cycle situated in the center of the complex plane. Since that, the vectors $\mathbf{t}_k$ of series (26) become fast close to zero vector and an appropriate matrix $\mathbf{T}$ becomes close to singular. In this case a linear transformation to Jordan normal form can be preferable.

Remark 4 It is known [13] that if both of symmetric matrices $\mathbf{A}$, $\mathbf{B}$ are positive definite, the same is true for product $\mathbf{AB}$. Then it is easy to see that polynomial

$$\delta(x) = \det(x\mathbf{I}_n - \mathbf{A}) = \det\left((x-1)\mathbf{I}_n + 2\left(2\mathbf{I}_n + \frac{1}{2}\bar{\mathbf{R}}_d\bar{\mathbf{Q}}_d \right)^{-1} \right)$$

roots are also positive and, since matrix $\mathbf{A}$ specter belong to unit cycle, the polynomial (21) roots belong to range $[0..1)$. This property can seriously reduce procedure which finds polynomial roots.

Remark 5 According to flops evaluation, up to 50 % of time may be taken by values (12) computation procedure. That computation process may be easily parallelized by parallelizing certain integrals (17) computation (by splitting integration interval) or by parallelizing roots finding process (by splitting search interval).

Remark 6 The problem can be reduced by origin state vector permutation. The permutation matrix $\mathbf{C}_r$ of new state vector $\tilde{\mathbf{x}} = \mathbf{C}_r\mathbf{x}$ is chosen to present covariance matrices $\tilde{\mathbf{Q}}_d = \mathbf{C}_r^T\bar{\mathbf{Q}}_d\mathbf{C}_r$, $\tilde{\mathbf{R}}_d = \mathbf{C}_r^T\bar{\mathbf{R}}_d\mathbf{C}_r$, in a block-diagonal form

$$\tilde{\mathbf{Q}}_d = \begin{bmatrix} \tilde{\mathbf{Q}}_d^1 & 0 & \cdots & 0 \\ 0 & \tilde{\mathbf{Q}}_d^2 & \cdots & 0 \\ \vdots & \vdots & \ddots & \vdots \\ 0 & 0 & \cdots & \tilde{\mathbf{Q}}_d^N \end{bmatrix}, \quad \tilde{\mathbf{R}}_d = \begin{bmatrix} \tilde{\mathbf{R}}_d^1 & 0 & \cdots & 0 \\ 0 & \tilde{\mathbf{R}}_d^2 & \cdots & 0 \\ \vdots & \vdots & \ddots & \vdots \\ 0 & 0 & \cdots & \tilde{\mathbf{R}}_d^N \end{bmatrix}$$

with equal $n_i \times n_i$ size square matrices $\tilde{\mathbf{Q}}_d^i$, $\tilde{\mathbf{R}}_d^i$, $i = \{1, 2, \ldots, N\}$, where $\sum_{i=1}^{N} n_i = n$ and N is an amount of block-diagonal matrices.

Then, the original problem may be split to N lower size problems with appropriate solutions matrices $\tilde{\mathbf{X}}_d^i \in R^{n_i \times n_i}$, $i = \{1, 2, \ldots, N\}$. The existence of block-diagonal form evidences about independent parts of network which can be estimated separately. When the i-th subproblem size $n_i = 1$ then the solution is defined in the most simple way as

$$x_i = \frac{q_i + \sqrt{q_i^2 + 4q_i r_i}}{2}, \tag{29}$$

where x_i, q_i, r_i are the only coefficients of matrices $\tilde{\mathbf{X}}_d^i$, $\tilde{\mathbf{Q}}_d^i$, $\tilde{\mathbf{R}}_d^i$ respectively.

4 Numerical Example

To illustrate the presented approach, we consider a simple UTN situated along "Yosif Gurko" street, Sofia, Bulgaria and crossed by "Vasil Levski" and "Evlogi and Hristo Georgiev" boulevards. The traffic conditions were collected by natural observing and by using AIMSUN microscopic traffic flow simulator. The UTN observing was evaluated on evening-peak at 6:00 P.M., 8:00 P.M.. For this period long queues are persisted and the area may require optimization. The simulation model was little simplified for example case. The AIMSUN simulation were done under fixed-time signal control settings with cycle time c = 60 s. Schematically the discussed area may be presented as shown in Fig. 1.

The considered UTN has three traffic light controlled junctions ($p = 3$), nine queues ($n = 9$) with x_i, $i = 1, 2, \ldots, 9$ vehicles in it, thirty four paths (marked by symbols 1–9, a–y in cycles). Every junction has two phases. We suppose also that every queue length in the UTN is measured by one sensor. Then a discrete-time linear state-space model (1) is given by values $\mathbf{x} = [x_1 \ x_2 \ \ldots \ x_9]^T$, $\mathbf{w} = [w_1 \ w_2 \ \ldots \ w_9]^T$, $\mathbf{u} = [u_1 \ u_2 \ u_3]^T$, $\mathbf{y} = [y_1 \ y_2 \ldots y_9]^T$, $\mathbf{v} = [v_1 \ v_2 \ldots v_9]^T$,

$$\mathbf{A}_d = \mathbf{C}_d = \mathbf{I}_9, \quad \mathbf{F}_d = 60\mathbf{I}_9, \tag{30}$$

$$\mathbf{B} = \begin{bmatrix} -18.84 & -16.17 & 45.43 & 52.7 & 9.27 & 0 & 0 & 0 & 0 \\ 0 & 6.49 & 0 & 0 & -20.12 & -9.13 & 15.12 & 1.22 & 0 \\ 0 & 0 & 0 & 0 & 0 & 6.39 & 0 & 3.06 & -33.28 \end{bmatrix}^T,$$

where the network flow demand vector $\mathbf{N} = [0.109 \quad 0.037 \quad -0.26 \quad -0.307 \ 0.113 \ 0.001 \ -0.105 \ -0.033 \ 0.25]^T$ veh/s, and the nominal green times vector $g^N = [0.35 \quad 0.5 \quad 0.5]^T$ s.

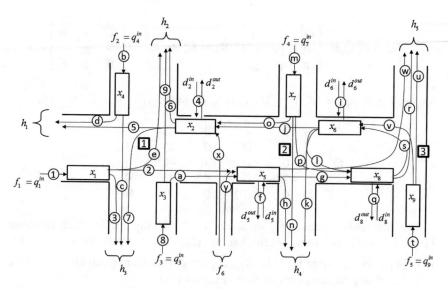

Fig. 1 Three junction urban network

Since the sixth and the eighth network links are small, we suppose that appropriate queue lengths are measured noise free means, i.e. components v_6, v_8 have zero standard deviation. According to UTN geometry we accepted the following covariance matrices

$$\mathbf{Q}_d = \frac{1}{60^2} \begin{bmatrix} 9 & 0 & 0 & 0 & -2.4 & 0 & 0 & 0 & 0 \\ 0 & 16 & 0 & 0 & 0 & -1.2 & -3.2 & 0 & 0 \\ 0 & 0 & 36 & 0 & -4.8 & 0 & 0 & 0 & 0 \\ 0 & 0 & 0 & 36 & 0 & 0 & 0 & 0 & 0 \\ -2.4 & 0 & -4.8 & 0 & 16 & 0 & 0 & -1.2 & 0 \\ 0 & -1.2 & 0 & 0 & 0 & 2.25 & 0 & -0.45 & -1.2 \\ 0 & -3.2 & 0 & 0 & 0 & 0 & 16 & -1.2 & 0 \\ 0 & 0 & 0 & 0 & -1.2 & -0.45 & -1.2 & 2.25 & 0 \\ 0 & 0 & 0 & 0 & 0 & -1.2 & 0 & 0 & 16 \end{bmatrix},$$

$$\tag{31}$$

$$\mathbf{R}_d = diag([4 \quad 9 \quad 16 \quad 16 \quad 9 \quad 0 \quad 4 \quad 0 \quad 4]). \tag{32}$$

It is seen that matrix $\mathbf{R}_d$ is not invertible. To make the problem nonsingular we have to exclude sixth and eighth state vector components from the UTN model. Then, for new covariance matrices it is able to apply permutation

$$\tilde{\mathbf{Q}}_d = \mathbf{C}_r^T \bar{\mathbf{Q}}_d \mathbf{C}_r \begin{bmatrix} \tilde{\mathbf{Q}}_d^1 & 0 & 0 \\ 0 & q_2 & 0 \\ 0 & 0 & q_3 \end{bmatrix}, \quad \tilde{\mathbf{R}}_d = \mathbf{C}_r^T \bar{\mathbf{R}}_d^{-1} \mathbf{C}_r = \begin{bmatrix} \tilde{\mathbf{R}}_d^1 & 0 & 0 \\ 0 & r_2 & 0 \\ 0 & 0 & r_3 \end{bmatrix},$$

where $q_2 = 36$, $q_3 = 16$, $r_2 = 16$, $r_3 = 4$, $\tilde{\mathbf{R}}_d^1 = diag([4 \quad 9 \quad 16 \quad 9 \quad 4])$,

$$\tilde{\mathbf{Q}}_d^1 = \begin{bmatrix} 9 & 0 & 0 & -2.4 & 0 \\ 0 & 16 & 0 & 0 & -3.2 \\ 0 & 0 & 36 & -4.8 & 0 \\ -2.4 & 0 & -4.8 & 16 & 0 \\ 0 & -3.2 & 0 & 0 & 16 \end{bmatrix}.$$

The permutation matrix $\mathbf{C}_r$ equal to 7×7 identity matrix with reordered $\{1-3, 5, 6, 4, 7\}$ columns. According to (29) scalar solutions $x_4 = 48$ and $x_7 = 19.31$. For matrices $\mathbf{C}_d = \mathbf{I}_5$, $\tilde{\mathbf{R}}_d^1$, $\tilde{\mathbf{Q}}_d^1$, the proposed approach can be applied. The first step requires computation of matrix (18),

$$\mathbf{A} = \begin{bmatrix} 0.35 & 0 & -0.0057 & -0.067 & 0 \\ 0 & 0.3 & 0 & 0 & -0.03 \\ -0.0014 & 0 & 0.35 & -0.033 & 0 \\ -0.029 & 0 & -0.059 & 0.3 & 0 \\ 0 & -0.069 & 0 & 0 & 0.49 \end{bmatrix}$$

Then, for initial vector $\mathbf{t}_0 = 1024[1 \quad 1 \quad 1 \quad 1 \quad 1]^T$ by series (26) were found matrix $\mathbf{T}$ and its inverse

$$\mathbf{T}^{-1} = \begin{bmatrix} -0.53 & -0.32 & 0.58 & 0.3 & -0.029 \\ 6.31 & 3.62 & -6.72 & -3.52 & 0.3 \\ -27.57 & -14.87 & 28.44 & 15.12 & -1.11 \\ 52.28 & 26.62 & -52.23 & -28.36 & 1.69 \\ -36.27 & -17.55 & 35.14 & 19.55 & -0.87 \end{bmatrix}.$$

The polynomial (21) coefficients were given from Eq. (25), $\delta(x) = x^5 - 1.81x^4 + 1.3x^3 - 0.45x^2 + 0.079x - 0.0055$. The numbers (12) were found as (17), $\gamma_{1-5} = \{14.62 \quad -0.72 \quad 0.15 \quad -0.09 \quad 0.22\}$ and numbers (13) $\eta_{1-5} = \{14.62 \quad -27.28 \quad 20.49 \quad -8.03 \quad 2.09\}$. The matrix sum (23) is

$$\mathbf{S}_{11} = \begin{bmatrix} 0.84 & 0 & 0.018 & 0.091 & 0 \\ 0 & 0.91 & 0 & 0 & 0.032 \\ 0.0046 & 0 & 0.84 & 0.045 & 0 \\ 0.04 & 0 & 0.081 & 0.92 & 0 \\ 0 & 0.072 & 0 & 0 & 0.71 \end{bmatrix}$$

and the stabilizing solution of CARE is given from representation (28),

$$
\mathbf{X} = \begin{bmatrix}
11.98 & 0 & -0.029 & -2.59 & 0 \\
0 & 22.37 & 0 & 0 & -3.36 \\
-0.029 & 0 & 47.94 & -5.18 & 0 \\
-2.59 & 0 & -5.18 & 22.34 & 0 \\
0 & -3.36 & 0 & 0 & 19.3
\end{bmatrix}.
$$

Finally, the whole Kalman estimator gain matrix can be written as

$$
\mathbf{L}_d = \begin{bmatrix}
0.74 & 0 & -0.0018 & 0 & -0.021 & 0 & 0 \\
0 & 0.70 & 0 & 0 & 0 & -0.042 & 0 \\
-0.0073 & 0 & 0.74 & 0 & -0.042 & 0 & 0 \\
0 & 0 & 0 & 0.75 & 0 & 0 & 0 \\
-0.047 & 0 & -0.023 & 0 & 0.7 & 0 & 0 \\
0 & -0.018 & 0 & 0 & 0 & 0.82 & 0 \\
0 & 0 & 0 & 0 & 0 & 0 & 0.82
\end{bmatrix}.
$$

5 Conclusion

The problem of steady-state Kalman estimator building for store-and-forward UTN model was discussed in the paper. As it was shown, steady-state Kalman estimator improves observed information of network (information about amount of vehicles which stay in the queues).

The main contribution of the paper is the algorithm given in the end of Sect. 3.2. This algorithm numerically reduced the formulated in the end of Sect. 2.2 problem. The proposed Kalman estimator (2) building algorithm for UTN model (1) respected special structure of known matrices (8) and was focused to effective numerical solution of DARE (3). The paper developed resolvent method [8] to solve the problem and reduced computational effort. More precisely speaking, DARE solving is presented in the algorithm just formally, since DARE solution is given by representation (28).

The algorithm is supposed to be applied in ITS for on-line computation.

Acknowledgments The research work presented in this paper is partially supported by the FP7 grant **AComIn** No. 316087, funded by the European Commission in Capacity Programme in 2012–2016.

References

1. Papageorgiou, M., Diakaki, C., Dinopoulou, V., Kotsialos, A., Wang, Y.: Review of road traffic control strategies. Proc. IEEE **91**(12), 2043–2067 (2003)
2. Abouaïssa, H., Majid, H.: Macroscopic traffic flow control via state estimation. Cybern. Inf. Technol. **15**(5), 5–16 (2015)
3. Stoilov, T., Stoilova, K., Papageorgiou, M., Papamichail, I.: Bi-level optimization in a transport network. Cybern. Inf. Technol. **15**(5), 37–49 (2015)
4. Rewadkar, D.N., Dixit, T.: Review of different methods used for large-scale urban road networks traffic state estimation. Int. J. Emerg. Technol. Adv. Eng. **3**(10), 369–373 (2013)
5. Kong, Q.-J., Zhao, Q., Wei, C., Liu, Y.: Efficient traffic state estimation for large-scale urban road networks. IEEE Trans. Intell. Transp. Syst. **14**(1), 398–407 (2013)
6. Pueboobpaphan, R., Nakatsuji, T.: Real-time traffic state estimation on urban road network: the application of unscented Kalman filter. In: Applications of Advanced Technology in Transportation, pp. 542–547 (2006)
7. Simon, D.: Optimal State Estimation: Kalman, H Infinity, and Nonlinear Approaches, 552 pp. John Wiley & Sons, Inc., Hoboken (2006)
8. Barabanov, A.T.: A stabilizing solution to the algebraic Riccati equation. The resolvent method. J. Comput. Syst. Sci. Int. **47**(3), 362–373 (2008)
9. Kondo, P., Furuta, K.: On the bilinear transformation of Riccati equations. IEEE Trans. Autom. Control **31**(1), 50–54 (1986)
10. Numerical Methods and Algorithms. http://old.vscht.cz/mat/NM-Ang/NM-Ang.pdf
11. Golub, G.H., Van Loan, C.F.: Matrix computations, 694 pp, 3rd edn. Johns Hopkins Studies in the Mathematical Sciences, Johns Hopkins University Press, Baltimore (1996)
12. Lanzon, A., Yantao, F., Anderson, B.D.O., Rotkowitz, M.: Computing the positive stabilizing solution to algebraic Riccati equations with an indefinite quadratic term via a recursive method. IEEE Trans. Autom. Control **53**, 2280–2291 (2008)
13. Korn, G.A., Korn, T.M.: Mathematical Handbook for Scientists and Engineers: Definitions, Theorems, and Formulas for Reference and Review, 1130 pp. Dover Publications (2000)

Printed in the United States
By Bookmasters